Methods in Molecular Biology™

Series Editor
John M. Walker
School of Life Sciences
University of Hertfordshire
Hatfield, Hertfordshire, AL10 9AB, UK

For further volumes:
http://www.springer.com/series/7651

Proteoglycans

Methods and Protocols

Edited by

Françoise Rédini

INSERM UMR957-EA3822, Nantes, France

Humana Press

Editor
Françoise Rédini, PhD
INSERM UMR957-EA3822
Nantes, France
francoise.redini@univ-nantes.fr

ISSN 1064-3745 e-ISSN 1940-6029
ISBN 978-1-4939-5879-5 ISBN 978-1-61779-498-8 (eBook)
DOI 10.1007/978-1-61779-498-8
Springer New York Dordrecht Heidelberg London

Printed on acid-free paper

Humana Press is part of Springer Science+Business Media (www.springer.com)

Dedication

To Marc Padrines,
This book is the testimony of a friendly collaboration of more than ten years on proteoglycans and bone. Thank you for the lively and fruitful discussions that we could share during this period, and for your unwavering cheerfulness.

Preface

Proteoglycans: Complex Diversity in Structure and Functions

Scientific interest and curiosity for proteoglycan research has dramatically increased over the past decades, these molecules being no more considered as a scaffold for the cells of a given tissue, but more as a reservoir for growth factors and cytokines modulating their activation status and turnover. For example, heparin and other related heparan sulfate molecules are increasingly being recognized as important modulators of many signaling pathways.

The evolution and accuracy of methodologies specific to their complex structure allow a better knowledge of their structure together with a more precise definition of their functions and involvement in physiology and pathologies. The huge increase of original research articles on proteoglycans reflects the increasing interest for these molecules.

There is no unifying structure for proteoglycans (PGs), such as collagen triple helix for example, and they display a great diversity of protein forms. However, their basic structure is defined as a protein portion and long unbranched polysaccharides (named glycosaminoglycans or GAGs). PGs were initially grouped together because of the high negative charges of their GAG chains; it makes them easily separable from other molecules by ion-exchange chromatography. However, PGs are not that similar. The core protein size ranges from 10 to >500 kDa, and the number of GAG chains attached varies from 1 to >100. In addition, several PGs carry GAG chains of more than one type (hybrid PGs: aggrecan, syndecans…) and/or have additional *N*- or *O*-linked sugar modifications. Not all PGs are "full-time" PGs. There are also a growing number of matrix molecules which may or may not be linked with GAG chains, depending on the developmental stage or due to regulatory factors. They are called "part-time" PGs, such as MHC class II invariant chain, thrombomodulin, CD44, macrophage colony-stimulating factor, amyloid precursor protein, collagen type IX, XII, XIV, and XVIII, and the transferring receptor, with alternatively spliced variants having GAG-initiation sites. Some PGs such as versican or CD44 also occur as alternatively spliced forms with varying sugar modifications. Versican can also be considered as part-time PGs because a variant of versican without GAG attachment sites has been discovered.

The protein forms have complex modular structures with protein motifs that are of similar sequence to those found in other protein families: several PGs thus contain distinct protein and carbohydrate domain structures that confer specific functional properties. The protein domains are often the products of separate exons. Recent studies have identified approximately 30 different PG protein cores; these cores are not only scaffolds for GAGs but they also contain domains that have particular biological activities. Many PGs are thus multifunctional molecules that engage in several different specific interactions at the same time.

In addition, numerous variations also occur in GAG chain structure; GAGs are large extended structures with highly charged sulfate and carbohydrate groups, and they dominate the physical properties of the protein to which they are attached. PGs in the extracellular matrix thus function physically as creators of a water-filled compartment. Their high fixed negative charge attracts counter ions, and the osmotic imbalance caused by a local high concentration of ions draws water from the surrounding areas. PGs thus keep the matrix hydrated and create a water compartment because they exclude other macromolecules while retaining permeability to low molecular weight solutes. This property increases the concentration of the macromolecules and therefore may increase reaction rates and promote all interactions that are concentration-dependent. Thus, PGs have important physical effects on events in the concentrated milieu around the cells and in the extracellular matrix.

The GAG side chains covalently linked to the core protein may be chondroitin sulfate (CS), or its epimerized homolog dermatan sulfate (DS), or keratan sulfate (KS), heparan sulfate (HS), or heparin (HP). Except KS, GAG synthesis is initiated by sequential addition of four monosaccharides: xylose (xyl), galactose (Gal), galactose and glucuronic acid (GlcUA). From this linker tetrasaccharide, the sugar chains are extended by addition of two alternating monosaccharides: an aminosugar and GlcUA. In HP and HS, the aminosugar is *N*-acetyl-glucosamine (GlcNAc) and in CS/DS, it is *N*-acetyl-galactosamine (GalNAc). The extent of epimerization of GlcUA to iduronic acid (IdUA) and the sulfation pattern of the disaccharide units distinguish HP from HS, and CS from DS. In KS, the GAGs are initiated as *N*-linked or *O*-linked oligosaccharides and extended by addition of GlcNAc and Gal. There is also regional variability to the epimerization and sulfation in each GAG chain. Studies of these patterns have defined the motifs required for specific interactions with growth factors, cytokines, matrix components, enzymes, and other proteins.

Divided into three categories, the volume first covers issues of basic concepts and up to date analysis methods for (I)proteoglycan and (II)glycosaminoglycan respectively at the protein and saccharide levels. Then the multifunctional aspect of proteoglycans is highlighted through three relevant examples of proteoglycans with highly different structures: serglycin, aggrecan, and heparin sulfate proteoglycans. The final chapter describes proteoglycan involvement in the pathogenesis of various disorders (kidney, corneal epithelial wound healing,…) and their potential therapeutic value in osteo-articular diseases.

Nantes, France ***Françoise Rédini, PhD***

Contents

Contributors

PHILIPPE AUZELOUX • *INSERM, UMR 990, Clermont-Ferrand, France; Imagerie moléculaire et thérapie vectorisée, Clermont Université, Université d'Auvergne, Clermont-Ferrand, France*

MARC BAUD'HUIN • *INSERM, UMR 957, Nantes, France; Physiopathologie de la Résorption Osseuse et Thérapie des Tumeurs Osseuses Primitives, Université de Nantes, Nantes Atlantique Universités, Nantes, France*

SZABOLCS BENI • *Department of Chemistry, University of California—Riverside, Riverside, CA, USA; Department of Pharmaceutical Chemistry, Semmelweis University, Budapest, Hungary*

GWENOLA BOUGRAS • *INSERM, UMR 957, Nantes, France; Physiopathologie de la Résorption Osseuse et Thérapie des Tumeurs Osseuses Primitives, Université de Nantes, Nantes Atlantique Universités, Nantes, France*

STÉPHANE BRÉZILLON • *Laboratoire de Biochimie Médicale et de Biologie Moléculaire, CNRS UMR 6237—MEDyC, Université de Reims-Champagne-Ardenne, Reims, France*

RÉGIS BRION • *INSERM, UMR 957, Nantes, France; Physiopathologie de la Résorption Osseuse et Thérapie des Tumeurs Osseuses Primitives, Université de Nantes, Nantes Atlantique Universités, Nantes, France*

JON M. CARTHY • *Department of Pathology and Laboratory Medicine, The James Hogg iCAPTURE Centre for Cardiovascular and Pulmonary Research, Institute for Heart + Lung Health, University of British Columbia, Vancouver, BC, Canada*

CÉLINE CHARRIER • *INSERM, UMR 957, Nantes, France; Physiopathologie de la Résorption Osseuse et Thérapie des Tumeurs Osseuses Primitives, Université de Nantes, Nantes Atlantique Universités, Nantes, France*

JEAN-MICHEL CHEZAL • *INSERM, UMR 990, Clermont-Ferrand, France; Imagerie moléculaire et thérapie vectorisée, Clermont Université, Université d'Auvergne, Clermont-Ferrand, France*

MAURICIO CORTES • *Departments of Pediatrics, The University of Chicago, Chicago, IL, USA*

MIRIAM DOMOWICZ • *Departments of Pediatrics, The University of Chicago, Chicago, IL, USA*

MARÍA T. DOURS-ZIMMERMANN • *Institute of Surgical Pathology, University Hospital Zurich, Zurich, Switzerland*

DAVID G. FERNIG • *Institute of Integrative Biology, University of Liverpool, Liverpool, UK*

CORINA FLANGEA • *Department of Chemical and Biological Sciences, "Aurel Vlaicu" University of Arad, Arad, Romania*

SYLVIE FOURNEL-GIGLEUX • *UMR 7561 CNRS-Université Henri Poincaré Nancy I, Vandoeuvre-lès-Nancy, France*

STEVEN GEORGES • *INSERM, U957, Nantes, France; Laboratoire de Physiopathologie de la Résorption Osseuse et Thérapie des Tumeurs Osseuses Primitives, Université de Nantes, Nantes Atlantique Universités, Nantes, France*
MARCO GUERRINI • *Ronzoni Institute for Chemical and Biochemical Research, Milan, Italy*
FRED K. HAGEN • *Department of Biochemistry and Biophysics, Proteomics Center, University of Rochester Medical Center, Rochester, NY, USA*
SCOTT J. HARVEY • *INSERM Avenir U983, Hôpital Necker-Enfants Malades, Paris, France*
DOMINIQUE HEYMANN • *INSERM, U957, Laboratoire de Physiopathologie de la Résorption Osseuse et Thérapie des Tumeurs Osseuses Primitives, Université de Nantes, Nantes Atlantique Universités, Nantes, France; Centre Hospitalier Universitaire de Nantes, Nantes, France*
MARION JASNIN • *Department of Molecular Structural Biology, Max Planck Institute of Biochemistry, Martinsried, Germany*
CHRISTOPHER J. JONES • *Department of Chemistry, University of California—Riverside, Riverside, CA, USA*
WINSTON WHEI-YANG KAO • *Department of Ophthalmology, College of Medicine, Edith J. Crawley Vision Research Center, University of Cincinnati, Cincinnati, OH, USA*
DEREK J. LANGESLAY • *Department of Chemistry, University of California—Riverside, Riverside, CA, USA*
CYNTHIA K. LARIVE • *Department of Chemistry, University of California—Riverside, Riverside, CA, USA*
FRÉDÉRIC LEZOT • *INSERM, UMR 957, Nantes, France; Physiopathologie de la Résorption Osseuse et Thérapie des Tumeurs Osseuses Primitives, Université de Nantes, Nantes Atlantique Universités, Nantes, France*
JIN-PING LI • *Department of Medical Biochemistry and Microbiology, Uppsala University, Uppsala, Sweden*
CHIA-YANG LIU • *Department of Ophthalmology, College of Medicine, Edith J. Crawley Vision Research Center, University of Cincinnati, Cincinnati, OH, USA*
JEAN-CLAUDE MADELMONT • *INSERM, UMR 990, Clermont-Ferrand, France; Imagerie moléculaire et thérapie vectorisée, Clermont Université, Université d'Auvergne, Clermont-Ferrand, France*
NATHALIE MAINRECK • *Laboratoire de Biochimie Médicale et de Biologie Moléculaire, CNRS UMR 6237—MEDyC, Université de Reims-Champagne-Ardenne, Reims, France*
MICHEL MANFAIT • *Equipe MEDIAN, CNRS UMR 6237—MEDyC, Université de Reims-Champagne-Ardenne, Reims, France*
FRANÇOIS-XAVIER MAQUART • *CHU de Reims, CNRS UMR 6237—MEDyC, Université de Reims-Champagne-Ardenne, Reims, France*
BRUCE M. MCMANUS • *Department of Pathology and Laboratory Medicine, The James Hogg iCAPTURE Centre for Cardiovascular and Pulmonary Research, Institute for Heart + Lung Health, University of British Columbia, Vancouver, BC, Canada*

James R. Mensch • *Departments of Pediatrics, The University of Chicago, Chicago, IL, USA*

Elisabeth Miot-Noirault • *INSERM, UMR 990, Clermont-Ferrand, France; Imagerie moléculaire et thérapie vectorisée, Clermont Université, Université d'Auvergne, Clermont-Ferrand, France*

Shuji Mizumoto • *Laboratory of Proteoglycan Signaling and Therapeutics, Frontier Research Center for Post-Genomic Science and Technology, Hokkaido University, Sapporo, Japan*

John S. Mort • *Research Unit, Shriners Hospital for Children, Montreal, QC, Canada*

Eon Jeong Nam • *Division of Respiratory Diseases, Children's Hospital, Harvard Medical School, Boston, MA, USA*

Shoko Nishihara • *Department of Bioinformatics, Laboratory of Cell Biology, Soka University, Tokyo, Japan*

Alessandro Ori • *Structural and Computational Biology Unit, EMBL, Heidelberg, Germany*

Mohamed Ouzzine • *UMR 7561 CNRS-Université Henri Poincaré Nancy I, Vandoeuvre-lès-Nancy, France*

Marc Padrines • *INSERM, U957, Nantes, France; Laboratoire de Physiopathologie de la Résorption Osseuse et Thérapie des Tumeurs Osseuses Primitives, Université de Nantes, Nantes Atlantique Universités, Nantes, France*

Pyong Woo Park • *Division of Respiratory Diseases, Children's Hospital, Harvard Medical School, Boston, MA, USA*

Gunnar Pejler • *Department of Anatomy, Physiology and Biochemistry, Swedish University of Agricultural Sciences, Uppsala, Sweden*

Caroline Peyrode • *INSERM, UMR 990, Clermont-Ferrand, France; Imagerie moléculaire et thérapie vectorisée, Clermont Université, Université d'Auvergne, Clermont-Ferrand, France*

Maziar Rahmani • *Department of Pathology and Laboratory Medicine, The James Hogg iCAPTURE Centre for Cardiovascular and Pulmonary Research, Institute for Heart + Lung Health, University of British Columbia, Vancouver, BC, Canada*

Uwe Rauch • *Department of Vascular Wall Biology, Institute of Experimental Medical Sciences, Lunds University, Lund, Sweden*

Andreas Rizzi • *Institute of Analytical Chemistry and Food Chemistry, University of Vienna, Vienna, Austria*

Elin Rönnberg • *Department of Anatomy, Physiology and Biochemistry, Swedish University of Agricultural Sciences, Uppsala, Sweden*

Peter J. Roughley • *Research Unit, Shriners Hospital for Children, Montreal, QC, Canada*

Timothy R. Rudd • *Institute of Integrative Biology, University of Liverpool, Liverpool, UK; Ronzoni Institute for Chemical and Biochemical Research, Milan, Italy*

Norihiko Sasaki • *Department of Bioinformatics, Laboratory of Cell Biology, Soka University, Tokyo, Japan*

NANCY B. SCHWARTZ • *Departments of Pediatrics, and Biochemistry and Molecular Biology, The University of Chicago, Chicago, IL, USA*
DANIELA G. SEIDLER • *Institute for Physiological Chemistry and Pathobiochemistry, University Hospital of Münster, Münster, Germany*
ALINA SERB • *Department of Biochemistry, "Victor Babes" University of Medicine and Pharmacy, Timisoara, Romania*
EUGEN SISU • *Department of Biochemistry, "Victor Babes" University of Medicine and Pharmacy, Timisoara, Romania*
DOROTHE SPILLMANN • *Department of Medical Biochemistry and Microbiology, Uppsala University, Uppsala, Sweden*
GANESH D. SOCKALINGUM • *Equipe MEDIAN, CNRS UMR 6237—MEDyC, Université de Reims-Champagne-Ardenne, Reims, France*
KAZUYUKI SUGAHARA • *Laboratory of Proteoglycan Signaling and Therapeutics, Frontier Research Center for Post-Genomic Science and Technology, Hokkaido University, Sapporo, Japan*
KATARZYNA A. UNIEWICZ • *Institute of Integrative Biology, University of Liverpool, Liverpool, UK; PromoCell GmbH Sickingenstr, Heidelberg, Germany*
NARAYANAN VENKATESAN • *UMR 7561 CNRS-Université Henri Poincaré Nancy I, Vandoeuvre-lès-Nancy, France*
AURÉLIEN VIDAL • *INSERM, UMR 990, Clermont-Ferrand, France; Clermont Université, Université d'Auvergne, Imagerie moléculaire et thérapie vectorisée, BP 10448, 63000, Clermont-Ferrand, France*
YANUSZ WEGROWSKI • *Laboratoire de Biochimie Médicale et de Biologie Moléculaire, CNRS UMR 6237—MEDyC, Université de Reims-Champagne-Ardenne, Reims, France*
MARK C. WILKINSON • *Institute of Integrative Biology, University of Liverpool, Liverpool, UK*
EDWIN A. YATES • *Institute of Integrative Biology, University of Liverpool, Liverpool, UK*
LEON ZAGREAN • *Neuroscience Laboratory, "Carol Davila" University of Medicine and Pharmacy, Bucharest, Romania*
ALINA D. ZAMFIR • *Department of Chemical and Biological Sciences, "Aurel Vlaicu" University of Arad, Arad, Romania; Mass Spectrometry Laboratory, National Institute for Research and Development in Electrochemistry and Condensed Matter, Timisoara, Romania*
DIETER R. ZIMMERMANN • *Institute of Surgical Pathology, University Hospital Zurich, Zurich, Switzerland*

Part I

Proteoglycans

Chapter 1

Proteoglycans: Gene Cloning

Mauricio Cortes, James R. Mensch, Miriam Domowicz, and Nancy B. Schwartz

Abstract

Aggrecan is a large proteoglycan that plays roles in numerous tissues during vertebrate development and adult life. The 6,327-nt chick aggrecan coding sequence had been determined from overlapping clones, but a full-length cDNA, needed for use in transgenic expression studies, had not been constructed. The strategy employed to do so was to generate two overlapping cDNA subfragments that shared a unique restriction site in the overlap and then join them at that site. These subfragments were obtained and cloned into the TOPO-TA vector pCR2.1. Digestion of the two constructs with the shared-site enzyme, XbaI, produced vector/5′-cDNA and 3′-cDNA fragments with XbaI-ends; these were ligated to produce the final full-length cDNA.

Key words: Proteoglycan, Aggrecan, Cloning, Strategy, Full-length, cDNA

1. Introduction

Aggrecan is a large molecule found in the extracellular matrix of many vertebrate tissues, notably cartilage and brain (1). It plays a variety of roles during development and in the adult; some are still being elucidated. The proteoglycan aggrecan consists of a core protein to which numerous carbohydrate chains are attached; these include chondroitin sulfate, keratan sulfate, and various oligosaccharides (see Fig. 1) (2). Our laboratory has studied many aspects of the physiological, biochemical, and structural properties of aggrecan, with emphasis on its roles during development (3–11). Chicken was the initial organism employed, affording ease of producing staged embryonic material and of experimental access to live embryos; a lethal recessive mutation, *nanomelia*, provided a natural aggrecan knock-out model (12, 13). Much work has also been done in mice, which also have natural knock-out mutant models

Françoise Rédini (ed.), *Proteoglycans: Methods and Protocols*, Methods in Molecular Biology, vol. 836,
DOI 10.1007/978-1-61779-498-8_1, © Springer Science+Business Media, LLC 2012

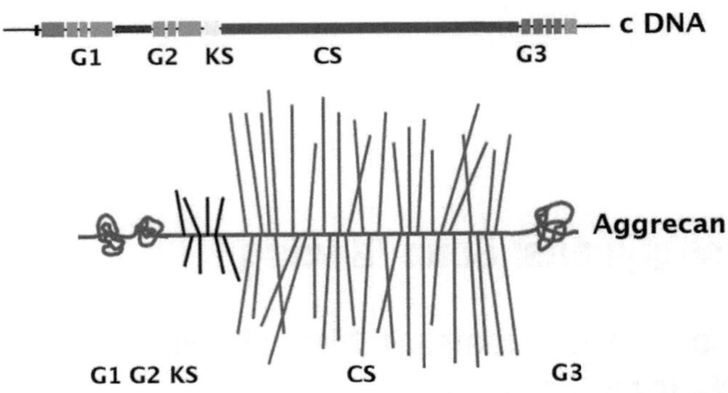

Fig. 1. Schematic representations of the aggrecan cDNA domain structure and aggrecan structure. The drawings indicate the approximate locations of the globular domains (G1, G2 and G3) and the keratan sulfate (KS) and chondroitin sulfate (CS) attachment regions. An arbitrary number of CS and KS chains is represented by *vertical lines*.

available: the *cartilage matrix deficiency* (*cmd*) alleles (14, 15). These models and the ability to genetically manipulate mice through controlled breeding and by creation of transgenic animals have enabled the study of aggrecan function in a mammalian system. We wanted to construct transgenic mice bearing the chick aggrecan coding sequence, as an antibody, S103L, is available that would permit differentiation of transgenically expressed chick aggrecan from the endogenous mouse protein (16–18). We had determined that the aggrecan coding sequences expressed in chick brain and cartilage are the products of a single gene from the sequence of overlapping cDNA fragments (see Fig. 1) (19), but in order to generate transgenic mice carrying the chick aggrecan coding region we first needed to obtain a full-length chick aggrecan cDNA, which had not previously been done due to its large size, 6,327 bp.

The full-length cloning was accomplished by cloning two overlapping subfragments and then ligating them at a shared restriction enzyme cleavage site in the overlap. First, oligo(dT) was used to prime reverse transcription of chick brain total RNA to produce first-strand cDNA, and then primers chosen from the known chick aggrecan mRNA sequence (GenBank Accession # U78555.1) were used to amplify two aggrecan-specific subfragments with overlapping sequences which shared a unique XbaI restriction site. The subfragments were purified by electrophoresis in and extraction from agarose gels, separately treated with Taq polymerase to add protruding 3′-dA single-nucleotide tails, then incorporated into plasmids using the TOPO-TA cloning system with the vector pCR2.1 from Invitrogen. Briefly, the plasmid pCR2.1 is supplied precleaved by *Vaccinia* virus topoisomerase I, leaving protruding 3′-dT ends with topoisomerases covalently attached via 3′-phosphodiester linkages to the Tyr-274 residues of the enzyme. Importantly, the pCR2.1 plasmid has a unique XbaI restriction site near the TA cloning site; the desired orientation for the aggrecan subfragments in the TA site is for their 5′→3′ sense sequence to be 5′ of this XbaI site. When the vector was mixed with one of the aggrecan subfragments having single-nucleotide 3′-dA tails, the complementary ends annealed and were ligated by the topoisomerases, which were released in the process, generating a circular plasmid. The respective ligated products, designated pCRAgg5′ and pCRAgg3′, were recovered by using them separately to transform competent *E. coli* TOP10F′ cells (Invitrogen). Transformants were selected by plasmid-conferred ampicillin resistance, and insert-bearing plasmid clones were identified via blue/white colony screening. Insert-bearing clones were grown in liquid LB medium, and plasmid DNA was extracted and purified using the QIAprep Spin Miniprep Kit (QIAGEN). Several putative clones of each subfragment were screened for correct insert size by EcoRI digestion and the 5′Agg inserts for correct orientation by XbaI digestion. Positive candidates were sequenced to confirm insert integrity. Confirmed pCRAgg5′ and pCRAgg3′ plasmids were then separately digested with XbaI, the former yielding a near-complete plasmid bearing the 5′ portion of the aggrecan coding sequence and having XbaI ends, and the latter releasing the 3′ portion of the cDNA plus a small part of the plasmid multiple cloning site, also with XbaI ends. These two DNA fragments were gel-purified as before. The pCRAgg5′-XbaI fragment was treated with a 5′-phosphatase to avoid self-ligation and repurified using the QIAquick PCR Purification Kit (QIAGEN). The two fragments were mixed in a 3:1 ratio of Agg3′ to pCRAgg5′ and ligated with T4 DNA ligase. Ligation products were used to transform TOP10F′ cells; insert-positive transformant clones were identified as before then screened for correct overall insert size and 3′-portion orientation by BamHI digestion. The final full-length cDNA construct chosen was sequenced to confirm its integrity.

The chick aggrecan full-length cDNA was subsequently cloned into various mammalian expression plasmids in one-step procedures using conventional restriction enzyme-based cloning or recombination-based technology for expression constructs larger than 10 kb, e.g., tissue-specific transgenic plasmids.

As example, we constructed a plasmid with the chondrocyte-specific Collagen IIa (Col2A) promoter (20) and the resulting vector was named pBSCol2a. To test the specificity of this plasmid, the fluorescent protein EGFP cDNA was cloned in to create the plasmid pBSCol2aEGFP. Specificity of the resulting plasmid was tested by transfecting pBSCol2aEGFP into chicken primary chondrocytes, which revealed distinct expression of EGFP only in chondrocytes (small round cells) and absence in fibroblasts (flat cells) (see Fig. 2a). For generating COL2A aggrecan transgenic mice, Col2aAggrecan DNA was linearized, purified, and injected into C57BL6 eggs, generating founder mice as determined by Southern blot analysis, which then underwent germline transmission. An antibody (S103L)

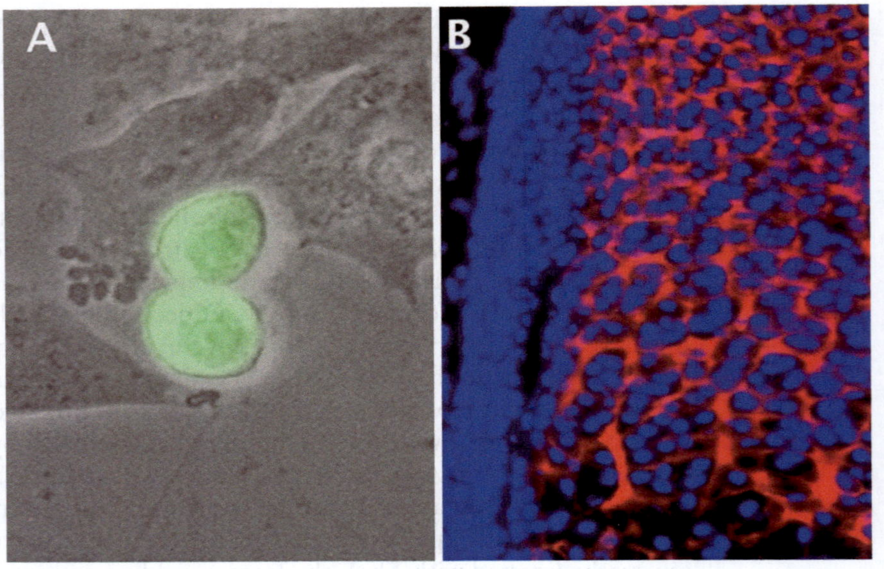

Fig. 2. (**a**) pBSCol2a chondrocyte specificity. The EGFP cDNA sequence was cloned into the pBSCol2a plasmid. E14 chick primary chondrocytes were nucleoporated with a pBSCol2aEGFP plasmid and the expression of EGFP was analyzed by fluorescence. EGFP expression was observed only in the small round cells, which is the characteristic morphology of chondrocytes, and absent in the flat cells which resemble fibroblasts. (**b**) Determination of the chick aggrecan transgene expression in Col2Agc transgenic mice. Hind-limb sections of postnatal day 3 mice were treated with chondrotinase prior to immunofluorescence with the S103L chick aggrecan antibody. Immunostaining revealed chondrocyte-specific staining of the transgene in the extracellular matrix.

against chick aggrecan revealed expression of the transgene in the extracellular matrix of the developing growth plates of the transgenic mice (see Fig. 2b).

In sum, we have successfully cloned the full-length aggrecan coding sequence and have generated transgenic animals to study aggrecan function *in vivo*. This methodology could be adapted to cloning other cDNA sequences, including those of other large proteoglycans.

2. Materials

General:

1.5-mL microcentrifuge tubes and 1–200- and 100–1,000-μL pipettor tips were from USA Scientific (all are RNase/DNase/pyrogen-free as supplied, autoclave if sterility is desired).

Sterile 17×100-mm polypropylene snap-cap tubes (presumed to be RNase-free) were from Sarstedt or Fisher.

10-mL sterile (presumed RNase-free), individually wrapped pipets were from Falcon (Fisher).

2.1. RNA Extraction from Embryonic Chick Brain

– TRIzol RNA Extraction Reagent was from Invitrogen.

– DEPC-treated water was prepared by adding 1 μL diethylpyrocarbonate (Sigma) per mL to deionized-distilled water (ddH_2O), shaking the solution vigorously, allowing the solution to stand overnight at room temperature, and then autoclaving it. This treatment inactivates RNases.

– Isopropanol (2-propanol) and chloroform were from Fisher.

– 1.2 M NaCl/0.8 M sodium citrate (Fisher) was prepared using DEPC-treated water.

– Formamide was from Sigma.

2.2. Reverse Transcription for First-Strand cDNA Synthesis

– The SuperScript II First-Strand Synthesis System from Invitrogen includes SuperScript II reverse transcriptase, 10× RT reaction buffer, 0.5 μg/μL oligo(dT)$_{12-18}$ primer, 10 mM dNTP mix, 40 U/μL RNaseOUT, 25 mM $MgCl_2$, and 0.1 M dithiothreitol (DTT) for performing reverse transcription from mRNA templates; also includes 2 U/μL RNase H for removal of template RNA from first-strand cDNA product.

– DEPC-treated H_2O.

2.3. PCR of Aggrecan cDNA Subfragments

– Oligonucleotide primers for the subfragment PCRs (two pairs) were ordered from Integrated DNA Technologies.

– PfuUltra High-Fidelity polymerase for PCR of aggrecan cDNA subfragments (Stratagene), supplied with 10× reaction buffer.

– 10 mM dNTP solution from SuperScript kit.

2.4. Agarose Gel Purification of PCR Products

– Agarose was from Invitrogen. Preparation of a 0.75% agarose gel is detailed in Subheading 3.4.

– Ethidium bromide was from Sigma. Stock solution is 10 mg/mL in ddH_2O. (WARNING: This compound is toxic/mutagenic; handle with care!)

– EDTA (ethylenediaminetetracetic acid, disodium salt), UltraPure was from Invitrogen.

– TAE gel buffer: 50× stock solution is 242 g Tris base (UltraPure, Invitrogen), 57.1 mL glacial acetic acid (Fisher), and 100 mL of 0.5 M EDTA made to 1 L with ddH_2O.

– QIAquick Gel Extraction Kit (QIAGEN) for recovery of gel-purified DNA fragments. Kit contains: Buffers QG, PE, and EB; QIAquick spin columns and 2-mL collection tubes; and 6× gel loading buffer.

2.5. Treatment of Gel-Purified Fragments with Taq Polymerase

Taq polymerase and 10 mM dATP were from Invitrogen.

2.6. TOPO-TA Insertion of PCR Products into pCR2.1 Plasmids

TOPO-TA Cloning Kit (Invitrogen), containing linearized pCR2.1 vector with topoisomerase covalently attached, 1.2 M NaCl/0.06 M $MgCl_2$ (Salt solution).

2.7. Transformation of Competent Cells and Subsequent Plating

- TOP10F′ One Shot competent cells for recovering plasmid constructs by transformation. S.O.C. medium for posttransformation cell recovery is included with cells.
- Sterile plastic Petri dishes (Falcon 1029) were from Fisher.
- Tryptone, yeast extract, and NaCl to prepare LB medium; agar for preparing LB plates (Fisher).
- LB medium is 10 g Tryptone, 5 g yeast extract, and 10 g NaCl per liter in ddH_2O, pH adjusted to 7.5–8 with 5 N NaOH (Fisher), then autoclaved 25 min at ≥121°C.
- For LB agar (1.5% w/v), 15 g agar per liter of LB is added prior to autoclaving. Place the flask of LB agar in an autoclave-proof tray (stainless-steel or polypropylene) to catch any overflow.
- Antibiotic for selection of transformed bacterial clones: ampicillin sodium (Sigma). A 100 mg/mL stock solution (1,000×) is prepared in ddH_2O and sterilized by filtration through a Millipore GV 0.22-μm syringe filter.
- Ampicillin stock solution is added (1 mL/L of agar) after agar has cooled to ~45°C (or to liquid LB at room temperature).
- The molten agar is swirled gently to mix (avoid foaming) then poured into sterile plastic Petri dishes (Fill dishes about half-way, 20–25 mL each for 100-mm diameter plates (Falcon 1029)).
- IPTG (isopropyl-β-D-thiogalactopyranoside, Research Products International) solution was made 100 mM in ddH_2O (23.8 mg/mL) and filter-sterilized.
- X-gal (5-bromo-4-chloro-3-indolyl-β-D-galactopyranoside) solution was made 40 mg/mL in dimethylformamide (DMF, Sigma), solution need not be sterilized.
- IPTG/X-gal solutions and plating aliquots of cells are spread on agar plates using flame-bent glass rods or Pasteur pipets, or disposable plastic loops bent against the inside of the plate lid. Glass spreaders are sterilized for each plate by dipping in ethanol and passing through a flame. Touch the spreader to the agar surface to cool.

2.8. Picking Transformants and Growth of Small Liquid Cultures

- Fisherbrand sterile plastic loops (Fisher)
- Tube cultures: Pipet 5 mL of LB + 100 μg/mL ampicillin into sterile 17 × 100-mm snap-cap tubes.

2.9. Small-Scale Plasmid DNA Preparations

Plasmid Spin Miniprep Kit (QIAGEN) contains: Buffers P1, RNase solution to add to P1, P2, N3, PE, and EB; QIAprep spin columns.

2.10. Screening Recovered Plasmids for Aggrecan cDNA Subfragment Constructs

- EcoRI and XbaI restriction enzymes, associated 10× reaction buffers, and 10 mg/mL bovine serum albumin (BSA) were from New England Biolabs.
- An agarose gel was prepared as before (see Subheading 2.4).

2.11. Large-Scale Preparations of pCRAgg5′ and pCRAgg3′ DNA

- LB medium: 5 mL in tubes, 250 mL in 1-L flasks (autoclaved) with 100 μg/mL ampicillin added.
- The QIAfilter Plasmid Maxi Prep Kit (QIAGEN) was used; it contains: Buffers P1 (and RNase A to be added to P1), P2, P3, QBT, QC, and QF; QIAfilter cartridges; and QIAGEN-tips (500 size).
- Isopropanol was from Fisher.
- TE buffer: 10 mM Tris–HCl/1 mM EDTA (pH 8) Tris solution is made from Tris base in ddH_2O and pH is adjusted with HCl. EDTA is added from 0.5 M stock solution

2.12. Digestion of Plasmid Constructs with XbaI

XbaI, 10× NEBuffer 4, and 10 mg/mL BSA were from New England Biolabs.

2.13. Purification of XbaI-Ended DNAs

- 0.75% agarose gel with EtBr (see Subheadings 2.4 and 3.4).
- QIAquick Gel Extraction Kit (QIAGEN) was again used for recovery of gel-purified DNA fragments.

2.14. Calf Intestinal Phosphatase (CIP) Treatment of pCRAgg5′-XbaI Ends

Calf intestinal phosphatase and NEBuffer 3 were from New England Biolabs.

2.15. Removal of CIP from 7.3-kb pCRAgg5′ Fragment

QIAquick Nucleotide Removal Kit (QIAGEN) was used to purify the CIP-treated DNA. The kit contains: Buffers PN, PE, and EB, and QIAquick spin columns.

2.16. Ligation to Produce Full-Length cDNA

T4 DNA ligase and 10× T4 ligase reaction buffer were from New England Biolabs.

2.17. Transformation into Competent Cells and Small-Scale Plasmid DNA Preparations

The materials used were as described in Subheadings 2.7–2.9.

2.18. Screening for Correct Full-Length cDNA Constructs

BamHI, NEBuffer 3, and 10 mg/mL BSA were from New England Biolabs.
An agarose gel was prepared as before (see Subheading 2.4).

2.19. Large-Scale Preparation of pCkAggFull DNA

The QIAfilter Plasmid Maxi Prep Kit (QIAGEN) was again used (see Subheading 2.11).

3. Methods

3.1. Preparation of Total RNA from Embryonic Chick Brain

1. Fertilized chicken eggs were obtained from a commercial vendor and placed in a Jamesway incubator cabinet with automatic turning, at 38°C. On day 14 of incubation (E14), an egg was opened, the embryo extracted, and the brain dissected from the skull.
2. The brain was placed into a sterile 17 × 100-mm polypropylene snap-cap tube containing 4 mL of TRIzol. (The TRIzol amount varies according to the weight of the specimen: 1 mL per 50–100 mg of tissue).
3. The tissue was disrupted with a Polytron homogenizer, and then the homogenate was allowed to stand for 10 min at room temperature.
4. The homogenate was then centrifuged at 12,500 × *g* (11,000 rpm in a SA-600 Sorvall rotor) for 10 min at 4°C. The resulting top layer of fat was removed, after which the clear supernatant was collected into a fresh 17 × 100-mm tube, avoiding the pelleted tissue debris.
5. In a fume hood, 0.4 mL of chloroform was added; the tube was capped and shaken vigorously for 15 s, then maintained at room temperature for 2.5 min.
6. The tube was centrifuged at 12,500 × *g* (11,000 rpm) in an SA-600 Sorvall rotor at 4°C for 10 min to separate the aqueous and organic phases.
7. The upper (aqueous) phase containing the extracted RNA was transferred to a new sterile 17 × 100-mm polypropylene snap-cap tube, 1 mL isopropanol and 1 mL of 1.2 M NaCl/0.8 M sodium citrate were added, the contents mixed, and the tube let stand for 10 min at room temperature (see Note 1).

8. The tube was centrifuged at 12,500 × *g* (11,000 rpm) in an SA-600 Sorvall rotor at 4°C for 10 min to pellet the precipitated RNA.
9. The supernatant was removed and the pellet was washed by adding 1 mL of 75% ethanol, then centrifuged at 7,500 × *g* (8,500 rpm) in an SA-600 Sorvall rotor at 4°C for 5 min to repellet the RNA.
10. The supernatant was removed and the pellet was air-dried briefly, then 100 μL of formamide was added and the tube placed at 4°C overnight to allow the pellet to dissolve. The formamide/RNA solution was stored at −20°C until use (see Note 2).

3.2. Reverse Transcription to Produce First-Strand cDNA

1. To a 1.5-mL RNase-free microcentrifuge tube was added 1 μL of 1 mg/mL E14 brain RNA (diluted from formamide solution with DEPC-treated H_2O), 1 μL of 10 mM dNTPs, 1 μL of 0.5 μg/mL oligo(dT), and 5 μL of DEPC-treated H_2O. This mixture was incubated at 65°C for 5 min, and then placed on ice for 2+ min.
2. In a separate tube, 2 μL of 10× RT buffer, 4 μL of 25 mM $MgCl_2$, 2 μL of 0.1 mM DTT, and 1 μL of 40 U/μL RNaseOUT were combined, then this mixture was added to the RNA/dNTP/oligo(dT) solution in the first tube, and the final mixture was incubated at 42°C for 2 min.
3. One μL of 50 U/μL SuperScript II reverse transcriptase was then added, the contents mixed, and the tube incubated at 42°C for 50 min.
4. The reaction was terminated by incubating the tube at 70°C for 15 min, then placing it on ice.
5. One μL of 2 U/μL RNaseH was added, and the tube was incubated at 37°C for 20 min.
6. The cDNA preparation was stored at −20°C until further use.

3.3. High-Fidelity PCR of Aggrecan cDNA Subfragments

Two PCR reactions were assembled, one to produce the 3.4-kb 5′ portion (Agg5′) of the aggrecan cDNA sequence and the other to generate the 2.9-kb 3′ portion (Agg3′). The primer pairs used were:

Agg5′:
Forward: 5′-CACTGAACTGTTAAAGGTAAACTATG-3′

Reverse: 5′-CTGATCTCTAGAATTTCTTTTGCTTCT-3′

Agg3′:
Forward: 5′-AAATTCTAGAGATCAGTGGACTGCCTTCAGG-3′

Reverse: 5′-CTAATGGGTGGGTCTGTGCACGACACC-3′

1. For each PCR, 40 μL of H_2O, 1 μL (25 pmol) of a forward primer, 1 μL (25 pmol) of a reverse primer, 1 μL of 10 mM

dNTPs, 5 μL of 10× Pfu reaction buffer, 1 μL of PfuUltra polymerase and 1 μl of the first-strand cDNA product (see subheading 3.2) were combined in a 0.3 mL PCR tube on ice and mixed.

2. The tubes were placed in a thermal cycler and subjected to a PCR program including: an initial denaturation period of 4 min at 95°C; 25 cycles of 95°C for 30 s, 51°C for 30 s, and 72°C for 4 min; and an additional 2 min at 72° followed by cooling to 4°C (see Note 3).

3.4. Agarose Gel Purification of PCR Products

Purification of DNA fragments from agarose gels was done with the QIAquick Gel Extraction Kit from QIAGEN, using the supplied buffers and spin columns, according to the kit instructions which are recapitulated below. Please consult www.qiagen.com for further details.

1. A 0.75% agarose gel was prepared: 0.75 g of agarose was melted in 100 mL of TAE in a microwave oven; the solution was cooled to about 40°C, then 3 μL of 10 mg/mL ethidium bromide was added prior to pouring into an 11 × 14-cm gel tray. A comb was used that produced wells with a capacity of approximately 50 μL.
2. A 3-μL aliquot of 6× loading buffer was added to 15 μL of each subfragment PCR product and the samples were mixed well.
3. The samples were loaded onto the agarose gel submerged in TAE buffer and electrophoresed at 60–70 mA until the blue-green marker dye (xylene cyanol) had migrated about half the length of the gel.
4. The gel was placed on a UV transilluminator and photographed. Fragment sizes were determined by comparison with the bands in 2 μL of 1 kb Plus DNA Ladder solution loaded alongside the cDNA samples. The 3.4- and 2.9-kb bands were excised from their respective lanes; the slices were trimmed of excess agarose and placed into separate, previously weighed 1.5-mL microcentrifuge tubes. The tubes with slices were weighed and the weight of the gel slices calculated.
5. Buffer QG from the QIAquick Gel Extraction Kit was added: 3 μL per mg of gel slice. (If a slice weighs more than 400 mg, it is divided into two or more tubes, and a corresponding number of spin columns is used).
6. The tubes were capped and incubated for 10 min at 50°C. The tube contents were examined to check that the gel slices had completely dissolved and that the color of the solution remained a yellow similar to the initial Buffer QG (see Notes 4 and 5).
7. Two QIAquick spin columns were inserted into 2-mL collection tubes from the kit, and the two dissolved gel slices were each

applied to a column. The column/tube assemblies were centrifuged at 13,000 rpm in an Eppendorf microcentrifuge for 1 min, and the flow-through effluents were discarded. The columns were reinserted into the same collection tubes. (If the volume of a sample is greater than 800 μL, the sample should be applied in two or more aliquots, centrifuging the columns and discarding the flow-through effluent as described. A total of 400 mg of gel slice can be processed per spin column).

8. To remove trace agarose, 500 μL of Buffer QG was applied to each spin column, the column/tube assemblies were centrifuged and the effluents discarded.
9. A 750-μL aliquot of Buffer PE from the kit was applied to each spin column. The column/tube assemblies were centrifuged and the wash effluents were discarded.
10. The columns were centrifuged again for 1 min to remove residual ethanol. The collection tubes and residual liquid were discarded.
11. The spin columns were mounted onto clean 1.5-mL microcentrifuge tubes; 30 μL of kit Buffer EB was applied to the center of each column membrane and let stand for 1 min; then the column/tube assemblies were centrifuged for 1 min to elute the DNA fragments.

3.5. Treatment of Gel-Purified Fragments with Taq Polymerase

The PfuUltra polymerase does not leave 3′-dA overhangs on the fragment ends; these are needed for the TOPO-TA ligation described in Subheading 3.6. Therefore, the fragments were treated briefly with Taq polymerase in the presence of dATP to add the dA overhangs.

1. In separate, clean microcentrifuge tubes, 10 μL each of the gel-purified Agg5′ and Agg3′ fragments were combined with: 6 μL H_2O, 2-μL of 10× PCR buffer, 1 μL of 2 mM dATP, and 1 μL Taq polymerase (0.5 U/μL).
2. The two 20-μL reactions were incubated at 72°C for 15 min.

3.6. TOPO-TA Insertion of PCR Products into pCR2.1 Plasmids

The Taq-treated aggrecan subfragments were used directly in TOPO-TA cloning ligation reactions.

1. In separate, clean microcentrifuge tubes, 4.5 μL of either Agg5′ or Agg3′ fragments with 3′-dA overhangs were combined with 1 μL of 1.2 M NaCl/0.06 M $MgCl_2$ and 0.5 μL of linear pCR2.1 TOPO plasmid.
2. The reactions were mixed, and then let stand at room temperature for 40 min (see Note 6).

3.7. Transformation of Competent Cells and Subsequent Plating

The protocols used for transformation and plating of TOP10F′ competent cells were those of the supplier, Invitrogen. Please consult www.invitrogen.com for additional information.

1. Two sterile 1.5-mL microcentrifuge tubes were placed on ice. Four LB plates containing 100 μg/mL ampicillin were each coated with 40 μL of 40 mg/mL X-gal and 40 μL of 100 mM IPTG, distributing the liquid evenly on the plate surface with a spreader; the plates were then placed at 37°C to prewarm them. A vial of S.O.C. medium was warmed to room temperature. A water bath was equilibrated at 42°C.
2. Two 1.5-mL tubes containing 50 μL of chemically competent *E. coli* TOP10F′ cells were thawed on ice and the contents resuspended by stirring gently with the pipette tip.
3. A 1.5-μL aliquot of each TOPO-TA ligation reaction was added to a tube containing competent cells, the contents were mixed by gently stirring, and the mixtures were let stand on ice for 20 min.
4. The tubes were placed in the 42°C water bath for 30 s (exactly), then immediately placed on ice.
5. A 200 μL aliquot of room-temperature S.O.C medium was added to each transformation, after which the tubes were incubated at 37°C with shaking at 200 rpm for 1 h.
6. Two aliquots of each transformation, 50 and 100 μL, were pipetted onto the prewarmed LB/ampicillin/IPTG/X-gal plates and spread evenly.
7. The plates were incubated overnight, inverted, at 37°C.

3.8. Picking Transformants and Growth of Small Liquid Cultures

Colonies of *E. coli* transformed by a pCR2.1-based plasmid are selected for by the ampicillin in the plate medium. Colonies harboring pCR2.1 plasmids which have an insert in their TOPO cloning site are white on IPTG/X-gal agar, due to disruption of the *lacZ*α coding region on the plasmid; plasmids with no insert produce blue colonies.

1. A sterile disposable plastic bacteriological loop was touched to a white colony that was well separated from others on one of the transformation plates.
2. The loop tip was rubbed on the inner wall of a 17 × 100-mm sterile polypropylene snap-cap tube that contained 5 mL of LB-100 μg/mL ampicillin medium, dipping the loop into the liquid and drawing it up the wall of the tube to help suspend the bacterial sample. The tube was labeled with the plasmid/insert and pick-number information.
3. Several colonies from each transformation were picked and used to inoculate tube cultures in this manner.
4. The inoculated tubes, with their caps on but not tightly closed, were incubated overnight at 37°C with shaking at 230 rpm.

3.9. Small-Scale Plasmid DNA Preparations

Plasmid DNA was isolated using the QIAprep Spin Miniprep Kit solutions, materials, and protocol (QIAGEN).

1. The 5-mL overnight cultures were pelleted at ~1,000×*g* for 8 min in a bench-top clinical centrifuge.
2. The supernatants were collected for autoclaving or bleach treatment prior to disposal; the cell pellets were each resuspended in 250 μL of kit Buffer P1 (per kit instructions, ensure that kit RNase has been added to the P1 solution prior to use), then transferred to clean 1.5-mL microcentrifuge tubes.
3. 250 μL of Buffer P2 was added to each tube, and the tubes were gently inverted 6 times to mix the contents.
4. 350 μL of Buffer N3 was added to each tube, immediately gently inverting the tubes 6 times to mix.
5. The lysis mixtures were centrifuged at 13,000 rpm (about 17,900×*g*) in an Eppendorf microcentrifuge for 10 min.
6. QIAprep Spin Columns, one per prep, were fitted onto supplied 2-mL collection tubes.
7. The supernatants from step 5 were pipetted into the spin columns, and the column/collection tube assemblies were centrifuged at 13,000 rpm for 1 min. The resulting flow-throughs were discarded.
8. 750 μL of Buffer PE was added to each column, the column/tubes were centrifuged at 13,000 rpm for 1 min, and the flow-throughs were discarded.
9. The column/tubes were centrifuged again at 13,000 rpm for 1 min to remove residual wash buffer.
10. The spin columns were fitted onto clean 1.5-mL microcentrifuge tubes
11. 30 μL of Buffer EB was added to each column, let stand for 1 min, and the column/tube assemblies were centrifuged at 13,000 rpm for 1 min to collect the eluted plasmid DNA.

3.10. Screening Recovered Plasmids for Aggrecan cDNA Subfragment Constructs

Plasmid clones with inserts of the correct sizes, 3.4 kb for Agg5′ and 2.9 kb for Agg3′, were identified by digesting plasmid miniprep DNA with EcoRI restriction endonuclease, which cleaves at sites that closely flank the TOPO cloning site (by 11 and 7 bp). Agg5′ inserts in the desired orientation, with the internal XbaI site proximal to the XbaI site in the vector, were identified by XbaI digestion; the correct orientation yields a 7.3-kb band while the incorrect orientation produces 3.9- and 3.4-kb bands. The digestion products were sized on agarose gels.

1. For each clone tested, 16.5 μL of ddH_2O, 2 μL of 10× EcoRI buffer, 1 μL of plasmid miniprep DNA, and 0.5 μL of (20 U/μL) EcoRI enzyme were pipetted into a 1.5-mL microcentrifuge tube. The contents were mixed gently, and the tube was placed in a 37°C water bath for 1 h.
2. For the Agg 5′ clones tested, 16.3 μL of ddH_2O, 2 μL of 10× NEBuffer 2, 0.2 μL of 10 mg/mL BSA, 1 μL of plasmid miniprep DNA, and 0.5 μL of XbaI were pipetted into a 1.5-mL microcentrifuge tube. The contents were mixed gently, and the tube was placed in a 37°C water bath for 1 h.
3. The digestions were analyzed by agarose gel electrophoresis as described in Subheading 3.4, steps 1–3, except that the combs used produced smaller (~15 μL) but more numerous wells. To prepare the samples for loading, 10 μL of each digested plasmid was mixed with 2 μL of 6× loading buffer. Fragment sizes were determined by comparison with the bands in a 2 μL 1 kb Plus DNA Ladder aliquot loaded alongside the digest samples.
4. Candidate plasmid clones were sent for automated DNA sequencing at the University of Chicago Cancer Research Center; primers used included the vector-specific insert-flanking M13 reverse and T7 promoter primers, and several internal aggrecan-sequence primers chosen to provide extensive overlaps within each subfragment.
5. A plasmid clone of each subfragment having the correct sequence, and correct orientation in the case of Agg 5′insert, was identified; these were named pCRAgg5′ and pCRAgg3′, respectively.

3.11. Large-Scale Preparations of pCRAgg5′ and pCRAgg3′ DNA

Greater amounts of the two subfragment plasmid constructs were prepared using the Plasmid Maxi Kit from QIAGEN, which included solutions, QIAGEN-tips, and the protocol that was followed. Please see www.qiagen.com for further information.

1. The confirmed plasmid clones were used to transform TOP10F′ cells as was done in Subheading 3.7. The transformation mixtures were plated as before and incubated overnight at 37°C.
2. Early in the day, 5-mL tube cultures of LB containing 100 μg/mL ampicillin were inoculated with a single white colony for each construct and incubated for 8 h at 37°C with shaking at 230 rpm.
3. A 0.5 aliquot of each tube culture was used to inoculate 250 mL of LB-100 μg/mL ampicillin in a 1-L flask. These large cultures were incubated overnight at 37°C with shaking at 230 rpm.
4. The next morning, the cultures were each poured into a 250-mL screw-capped polypropylene centrifuge bottle. The bottles’

weights with caps were equalized by removing liquid from the heavier one, and the bottles were centrifuged in a Sorvall SLA-1500 rotor at 7,000 rpm for 8 min to pellet the cells.

5. The supernatants were collected for decontamination and disposal, and then each pellet was resuspended in 10 mL of Buffer P1 to which RNaseA had been added to 100 μg/mL.
6. The resuspended cells were transferred into 50-mL conical screw-cap tubes, 10 mL of Buffer P2 was added to each; the tubes were capped and inverted 6 times, then allowed to stand at room temperature for 5 min.
7. Two QIAfilter cartridges were readied by screwing the caps onto their outlets and placing the cartridges into 50-mL conical tubes in a rack.
8. 10 mL of 4°C Buffer P3 was added to each mixture; the tubes were capped and immediately inverted 6 times.
9. The mixtures were immediately poured into the QIAfilter cartridges and let stand at room temperature for 10 min.
10. Two QIAGEN-tips were equilibrated by addition of 10 mL of Buffer QBT that was allowed to drain completely by gravity flow.
11. One at a time, the QIAfilter cartridge outlet caps were removed, supplied plungers were inserted into the cartridges, and the lysates were expelled from the cartridges into the equilibrated QIAGEN-tip columns.
12. Two sequential 30 mL washes with Buffer QC under gravity flow were performed on each column.
13. The plasmid DNAs were each eluted from the columns with passage of 15 mL aliquots of Buffer QF collected in 50-mL conical polypropylene screw-cap tubes.
14. To precipitate the plasmid DNA, 10.5 mL of isopropanol was added, the tube contents were mixed, and the tubes were centrifuged at 10,000 rpm in an SA-600 Sorvall rotor for 50 min at 4°C.
15. The supernatants were removed, and the pellets were washed by adding 5 mL of room-temperature 70% ethanol and centrifuging at 10,000 rpm for 15 min at 4°C.
16. The supernatants were removed, the pellets were air-dried for 10 min, and then each was dissolved in 300 μL of Buffer TE. (Typical maxiprep yield is 1–3 μg/mL).

3.12. Digestion of Plasmid Constructs with XbaI

The pCRAgg5′ and pCRAgg3′ plasmids were digested separately to prepare for assembly of the full-length aggrecan cDNA. The pCRAgg5′ plasmid was cleaved at the XbaI site near the 3′-end (with respect to the aggrecan coding strand) of its cDNA insert and at an XbaI site located 58 bp further 3′ in the vector, yielding a large (~7.3 kb) and a very small fragment. The pCRAgg3′ plasmid was cut at the XbaI site very near its 5′-end and at the vector XbaI site, producing a 3.9 kb all-vector piece and a 2.9-kb 3′Agg cDNA piece with a short 3′ vector segment.

1. In each of two 1.5-mL microcentrifuge tubes, 38.5 μL of H_2O, 5 μL of 10× NEBuffer 2, 0.5 μL of 10 mg/mL BSA, and 1 μL of XbaI were combined with 5 μL (~10 μg) of either pCRAgg5′ or pCRAgg3′ maxiprep DNA.
2. The digestions were incubated for 1 h at 37°C.

3.13. Purification of XbaI-Ended DNAs

The 7.3-kb pCRAgg5′ and 2.9-kb Agg3′ Xba-ended DNA fragments were purified by electrophoresis in an agarose gel followed by processing with the QIAquick Gel Extraction Kit as described in Subheading 3.4, with the final elution done in 30 μL of Buffer EB for each.

3.14. Calf Intestinal Phosphatase (CIP) Treatment of pCRAgg5′-XbaI Ends

To prevent excessive transformation background resulting from self-ligation of the linear pCRAgg5′ plasmids without Agg3′ inserts, the 30 μL of purified 7.3-kb, XbaI-ended pCRAgg5′ construct was treated with CIP to remove its 5′-phosphate groups.

1. To the 30 μL of 7.3-kb pCRAgg5′ fragment in a 1.5-mL microcentrifuge tube was added 60 μL of H_2O, 10 μL of 10× NEBuffer 3, and 0.4 μL of CIP.
2. The reaction was incubated for 1 h at 37°C.

3.15. Removal of CIP from 7.3 kb pCRAgg5′ Fragment

CIP must be removed prior to the cDNA ligation reaction to protect the 5′-phosphates on the 2.9 kb 3′Agg fragment. The removal was done by the use of the QIAquick Nucleotide Removal Kit as the source of solutions, materials, and the purification protocol given below. Please consult the QIAGEN QIAquick Spin Handbook for additional information.

1. Ten volumes (1 mL) of Buffer PN were added to the 100-μL CIP reaction from Subheading 3.14 and the solution was mixed well.
2. The solution was pipetted into a QIAquick spin column inserted in a 2-mL collection tube, and the column/tube assembly was centrifuged at 6,000 rpm for 1 min in an Eppendorf microcentrifuge.
3. The flow-through was discarded, and 750 μL of Buffer PE was added to the column. The column/tube assembly was centrifuged as before.
4. The flow-through was discarded, and the column/tube was centrifuged at 13,000 rpm for an additional 1 min to remove trace ethanol solution.

5. The column was inserted into a clean 1.5-mL microcentrifuge tube.

6. The 7.3-kb pCRAgg5′ fragment was eluted by applying 30 μL of Buffer EB to the center of the QIAquick column membrane, allowing the column to stand for 1 min, and then centrifuging the column/tube assembly at 13,000 rpm for 1 min.

3.16. Ligation to Produce Full-Length cDNA

1. On ice in a 1.5-mL microcentrifuge tube, 5 μL of H_2O, 2 μL of 10× T4 ligase buffer, 3 μL of 7.3 kb pCRAgg5′-XbaI-CIP-treated DNA (~40 ng), 9 μL of 2.9 kb 3′Agg-XbaI maxiprep DNA (~120 ng), and 1 μL of T4 DNA ligase were combined and mixed gently.

2. The ligation reaction was incubated overnight at 16°C in a refrigerated water bath.

3.17. Transformation into Competent Cells and Small-Scale Plasmid DNA Preparation

Using 1 μL of the completed ligation reaction, chemically competent TOP10F′ cells were transformed and plated as described in Subheading 3.7. Resulting colonies were picked and 5-mL cultures grown as done in Subheading 3.8, and plasmid DNA was purified as in Subheading 3.9.

3.18. Screening for Correct Full-Length cDNA Constructs

The 2.9-kb 3′Agg-XbaI fragment could ligate with the 7.3-kb pCRAgg5′-XbaI fragment in either the sense or antisense orientation. That only two BamHI sites are present in the construct, one asymmetrically placed site in the Agg3′ sequence and the other in the vector near the TOPO-TA cloning site, provided a simple way to determine the Agg3′orientation in a given plasmid clone. A BamHI digestion of a correctly arranged full-length cDNA clone yields 6.3- and 3.9-kb fragments, while the incorrect orientation produces 5.9- and 4.3-kb pieces.

1. Plasmid clones tested were digested in reactions containing 16.3 μL of H_2O, 2 μL of 10× NEBuffer 3, 0.2 μL 10 mg/mL BSA, 0.5 μL of BamHI, and 1 μL of DNA incubated for 1 h at 37°C.

2. The digestion products were analyzed on an agarose gel as described in Subheading 3.10.

3. A plasmid clone with the correct Agg3′ orientation was sequenced by the University of Chicago Cancer Research Center's automated sequencing facility, using flanking and internal primers as was done for the subfragment clones (see Subheading 3.10).

4. The confirmed clone was named pCkAggFull.

3.19. Large-Scale Preparation of pCkAggFull

This was done by the QIAfilter Plasmid Maxiprep Kit method described in Subheading 3.11.

The final plasmid product was eluted in 300 μL of TE buffer.

4. Notes

1. The salt solution is an optional addition that removes contaminating proteoglycans and polysaccharides, molecules that are abundant in brain. See TRIzol Reagent instructions (Invitrogen).
2. The quality of the RNA preparation will determine whether a long mRNA such as that for aggrecan can be successfully reverse-transcribed; dissolving the RNA in formamide was done to protect its quality.
3. Use of a high-fidelity polymerase is crucial to successful amplification of the aggrecan coding sequence, in particular due to the extensive repeat region in the middle third of the sequence.
4. Continue incubation if a slice is not dissolved. The pH must be ≤7.5 for DNA to bind to the QIAquick membrane in the next step; if the solution is orange or violet (too alkaline), add 10 μL of 3 M sodium acetate, pH 5.0, at which point the solution should turn yellow (QIAGEN manual).
5. If the DNA fragment to be extracted is <500 bp or >4 kb in size, add 1 μL isopropanol per mg of gel slice to improve the yield (QIAGEN manual).
6. The extended incubation time of the TOPO-TA insertion/ligation reaction was critical in obtaining large insert fragments; vectors designed for large inserts are now available.

Acknowledgments

The authors would like to thank Alex Baria and Danijela Maric for technical assistance.

This work was supported by R01 HD17732 (N.B.S.), and T32.G M 008720 and G M 007381 (M.C.).

References

1. Schwartz, N. B., Domowicz, M., Krueger, R. K., Li, H., and Mangoura, D. (1996) Brain Aggrecan, *Perspectives in Dev. Neurobiol.* **3**, 291–306.
2. Schwartz, N. B., Pirok, E. W., III, Mensch, J. R., Jr., and Domowicz, M. S. (1999) Domain organization, genomic structure, evolution, and regulation of expression of the aggrecan gene family, *Prog Nucleic Acid Res Mol Biol* **62**, 177–225.
3. Cortes, M., Baria, A. T., and Schwartz, N. B. (2009) Sulfation of chondroitin sulfate proteoglycans is necessary for proper Indian hedgehog signaling in the developing growth plate, *Development* **136**, 1697–1706.
4. Domowicz, M. S., Cortes, M., Henry, J. G., and Schwartz, N. B. (2009) Aggrecan modulation of growth plate morphogenesis, *Dev Biol* **329**, 242–257.
5. Domowicz, M. S., Sanders, T. A., Ragsdale, C. W., and Schwartz, N. B. (2008) Aggrecan is expressed by embryonic brain glia and regulates astrocyte development, *Developmental Biology* **315**, 114–124.

6. Domowicz, M. S., Mueller, M. M., Novak, T. E., Schwartz, L. E., and Schwartz, N. B. (2003) Developmental expression of the HNK-1 carbohydrate epitope on aggrecan during chondrogenesis, *Dev Dyn* **226**, 42–50.
7. Domowicz, M., Mangoura, D., and Schwartz, N. B. (2000) Cell specific-chondroitin sulfate proteoglycan expression during CNS morphogenesis in the chick embryo, *Int J Dev Neurosci* **18**, 629–641.
8. Domowicz, M. S., Pirok, E. W., III, Novak, T. E., and Schwartz, N. B. (2000) Role of the C-terminal G3 domain in sorting and secretion of aggrecan core protein and ubiquitin-mediated degradation of accumulated mutant precursors, *J Biol Chem* **275**, 35098–35105.
9. Domowicz, M. S., Krueger, R. C., Li, H., Mangoura, D., Vertel, B. M., and Schwartz, N. B. (1996) The nanomelic mutation in the aggrecan gene is expressed in chick chondrocytes and neurons, *Int. J. Dev. Neurosci.* **14**, 191–201.
10. Domowicz, M. S., Li, H., Hennig, A. K., Vertel, B., and Schwartz, N. B. (1995) The biochemically and immunologically distinct CSPG of notochord is a product of the aggrecan gene, *Dev. Biol.* **171**, 655–664.
11. Vertel, B. M., Grier, B. L., Li, H., and Schwartz, N. B. (1994) The chondrodystrophy, nanomelia: biosynthesis and processing of the defective aggrecan precursor, *Biochem J.* **301**, 211–216.
12. Li, H., Schwartz, N. B., and Vertel, B. M. (1993) cDNA cloning of chick cartilage chondroitin sulfate (aggrecan) core protein and identification of a stop codon in the aggrecan gene associated with the chondrodystrophy, nanomelia, *J Biol Chem* **268**, 23504–23511.
13. Vertel, B. M., Walters, L. M., Grier, B., Maine, N., and Goetinck, R. F. (1993) Nanomelic chondrocytes synthesize, but fail to translocate, a truncated aggrecan precursor, *J. Cell Sci.* **104**, 939–948.
14. Kimata, K., Barrach, H.-J., Brown, K. S., and Pennypacker, J. P. (1981) Absence of proteoglycan core protein in cartilage from cmd/cmd (cartilage matrix deficiency) mice., *J. Biol. Chem.* **256**, 6961–6968.
15. Krueger, R. C., Kurima, K., and Schwartz, N. B. (1999) Completion of the mouse aggrecan structure and identification of the defect in the cmd-Bc as a near complete deletion of the murine aggrecan gene., *Mamm. Genome* **10**, 1119–1125.
16. Dorfman, A., Hall, T., Ho, P. L., and Fitch, F. (1980) Clonal antibodies for core protein of chondroitin sulfate proteoglycan, *Proc. Natl. Acad. Sci.* 77, 3971–3973.
17. Dorfman, A. D., Vertel, B. M., and Schwartz, N. B. (1980) Immunological methods in the study of chondroitin sulfate proteoglycans., *Dev. Biol.* **14**, 169–198.
18. Krueger, R. C., Fields, T. A., Mensch, J. R., and Schwartz, N. B. (1990) Chick cartilage chondroitin sulfate proteoglycan core protein II. Nucleotide sequence of cDNA clone and localization of the S103L epitope, *J Biol Chem* **265**, 12088–12097.
19. Li, H., Domowicz, M. S., Hennig, A., and Schwartz, N. B. (1996) S103L reactive chondroitin sulfate proteoglycan (aggrecan) mRNA expressed in developing chick brain and cartilage is encoded by a single gene, *Mol. Brain Res.* **36**, 309–321.
20. Zhou, G., Garofalo, S., Mukhopadhyay, K., Lefebvre, V., Smith, C. N., Eberspaecher, H., and de Crombrugghe, B. (1995) A 182 bp fragment of the mouse pro alpha 1(II) collagen gene is sufficient to direct chondrocyte expression in transgenic mice, *J Cell Sci* **108**, 3677–3684.

Chapter 2

Proteoglycan: Site Mapping and Site-Directed Mutagenesis

Fred K. Hagen

Abstract

Identification of proteoglycan chain modification sites cannot yet be reliably predicted from primary amino acid sequence data. A number of studies have shown that serine is the predominant amino acid that is modified and it is frequently flanked by a C-terminal glycine and proximal N-terminal acidic amino acids; however, not all simple Ser-Gly motifs constitute a modification site. Here we present a rapid method for cloning small, defined segments of putative proteoglycan attachment sites and expressing them as a mini-reporter protein in an insect tissue culture system that is expandable to high throughput analysis. Reporter proteins with attached proteoglycans can be readily discerned from their unmodified form, by a simple gel-shift assay and Western blot detection for an epitope tag engineered into the reporter. Unmodified proteins are generated as a reference standard by treating cells with dsRNA to knock down the endogenous polypeptide xylose transferase, which is responsible for initiating proteoglycan site attachment. Examination of proteoglycan attachment by different metazoan organisms can be studied in the same cell line by cotransfecting a polypeptide xylose transferase expression plasmid and reporter construct from human, mouse, frog, or worm, for example. Reporter proteins engineer with point mutations can be rapidly generated with this system to pinpoint the exact residue that is glycosylated, to verify the mapping data.

Key words: Proteoglycans, Mutagenesis, RNAi, Glycosyltransferase, Site mapping

1. Introduction

Synthesis of the proteoglycans begins with the attachment of xylose to a serine side chain, followed by the stepwise addition of monosaccharides to build the tetrasaccharide linker: GlcA-β1,3-Gal-β1,3-Gal-β1,4-Xyl-1β-*O*-serine-(protein). This tetrasaccharide is the substrate for attachment and polymerization of chondroitin and heparan sulfate chains. The precise amino acid that is first modified with a xylose monosaccharide is defined by the substrate specificity of the polypeptide xylose transferase. Numerous studies have shown prevalence for proteoglycan attachment at a serine-glycine motif (1, 2);

Françoise Rédini (ed.), *Proteoglycans: Methods and Protocols*, Methods in Molecular Biology, vol. 836,
DOI 10.1007/978-1-61779-498-8_2, © Springer Science+Business Media, LLC 2012

however, many studies have shown that not all "SG" motifs are glycosylated (3, 4) and that some modification sites diverge from the SG "rule" (5). Experimental mapping data reported in the literature are difficult to work though, as many sites are qualitatively mapped in a way that does not rank weak proteoglycan modification sites separately from dominant modification sites. Previously, we reported a rapid method for experimentally mapping these sites, by cloning small 12–15 amino acid segments of putative proteoglycan attachment sites and expressing them as part of a minireporter protein in a tissue culture system (3). Predominant proteoglycan attachment sites modify a serine but not a threonine, while mucin glycosylation modifies both serine and threonine, with a slight preference for the latter (6). Glycine residues are present in both proteoglycans and mucins, but are nearly always present C-terminal and next to the serine attachment position of a proteoglycan. While proline activates a mucin glycosylation (7), it inhibits proteoglycan attachment (3) (as in N-glycosylation) (8). One acidic amino acid is required at a proximal position (-4, -3, or -2, N-terminal to the serine) or a pair of acidic residues is required at distal N-terminal positions -6 and -5 from the proteoglycan modification site, whereas negative residues tend to inhibit mucin-type glycosylation. While these general rules can help to reduce the list of potential modified positions in a protein, the interplay of charge density and position and role of proline, glycine, and other amino acids is complicated, as in the case of mucin glycosylation sites (9), emphasizing the importance of experimental mapping data, including a measure of the proteoglycan site's occupancy rate, to obtain a definitive answer.

2. Materials

2.1. Molecular Biologicals, Expression Plasmids, and Oligonucleotides

1. BsaI restriction enzyme (New England Biolabs) for cloning all inserts. This enzyme allows for production of unique 4 bp customized sticky ends from the sequence of the pLC-S2-SP-BsaI plasmid (see Fig. 1c).
2. Polynucleotide kinase (New England Biolabs) and 10 mM rATP (Roche) for phosphorylating synthetic oligonucleotides, prior to cloning.
3. TriReagent and polyacrylamide carrier (Molecular Research Center) for isolation of RNA.
4. Super Script III (Invitrogen) for first-strand cDNA synthesis, to produce a template for PCR amplification of glycosyltransferases.
5. A high fidelity polymerase chain reaction (PCR) kit for amplification of cDNAs for cloning/expression and for production of templates for dsRNA synthesis.

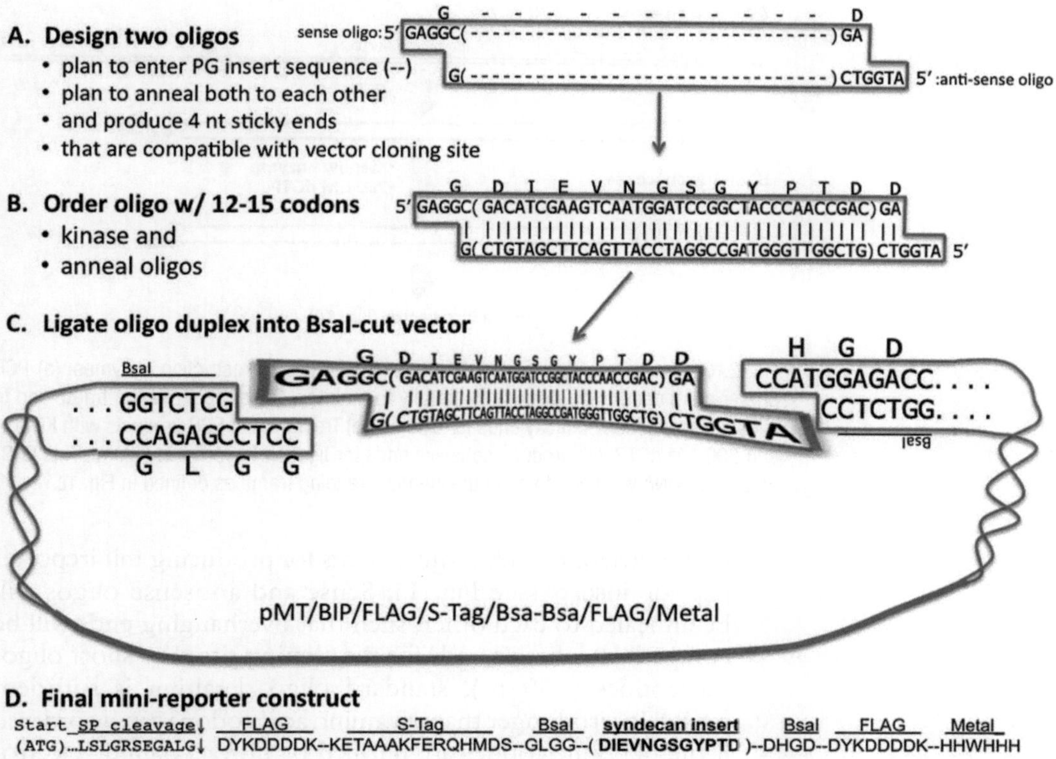

Fig. 1. Rapid cloning of defined proteoglycan modification sites into a minireporter protein expression construct. (**a**) A cDNA insert is created by designing a pair of complementary oligonucleotides with overhanging, "sticky" ends for ligation into a plasmid vector. (**b**) An example of annealed sense and antisense oligonucleotides that encode for a syndecan proteoglycan modification site with 12 amino acid codons (between the parenthesis, and labeled above the DNA in single-letter amino acid code). A BamHI restriction site (GGATCC) is in the insert coding region and used for restriction mapping/screening of expression clones. (**c**) Complementary cohesive ends in the vector are produced by digestion of the vector with BsaI, for directional cloning of the insert. The inserted cDNA needs to be in-frame with two glycine codons upstream in the vector and with the His–Gly–Asp codons in the downstream vector sequence. (**d**) The configuration of the insert (*in parenthesis*) relative to the BIP signal peptide cleavage site, purification/detection tags, and amino acid sequences that flank the proteoglycan insert. *Dashed lines* represent variable length spacer sequences between the specific epitope tags.

6. T7 RNA polymerase (Megascript Kit, Ambion) for production of double-stranded RNA from a PCR product that is flanked with T7 promoters on both ends.
7. Klenow (Large fragment of DNA polymerase I) (New England Biolabs) for creating 4-bp sticky ends at the end of PCR products, when used in the presence of dGTP and the PCR primers described in this study (see Fig. 2).
8. pMT/V5-His (Invitrogen): A plasmid for cloning and expression of full-length glycosyltransferases in S2 cells, like the pLC-S2-FL-CE-pXTase plasmid for the *C. elegans* pXTase rescue plasmid.
9. pMT/BIP/V5-His (Invitrogen): A parent plasmid for construction of a reporter with FLAG, S-Tag, and BsaI sites (see pLC-S2-SP-BsaI in ref. (3)).

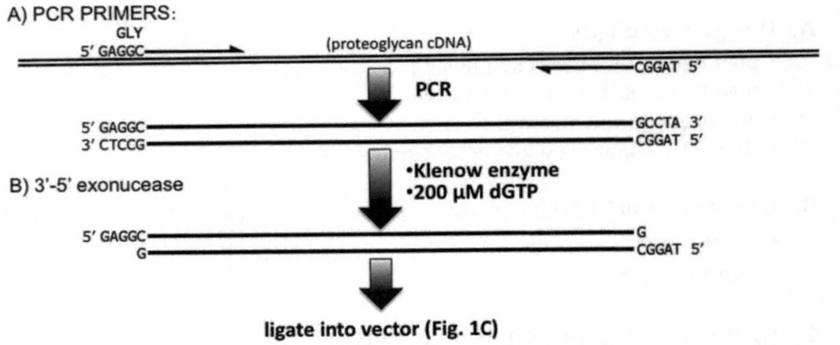

Fig. 2. Cloning of proteoglycan cDNAs without the use of restriction enzymes. (**a**) PCR primer pairs need to be designed carefully to maintain the correct reading frame and to allow for production of sticky ends for cloning. (**b**) Treatment of PCR products with Klenow and 200 μM dGTP will produce cohesive ends for ligation to vector in Fig. 1c. This PCR-amplified cDNA will need to be in the identical reading frame as defined in Fig. 1c.

10. Oligonucleotides are requirements for producing minireporter peptide inserts (see Fig. 1). Sense and antisense oligos will be annealed to each other, such that overhanging ends will be compatible cohesive ends for the cloning site. For short oligonucleotides (<50 nt), standard oligo desalting is sufficient purity. Inserts longer than 15 amino acid codons can be ordered if the oligonucleotides are purified by polyacrylamide electrophoresis. The sense oligo must begin with 5′ GAGGC, prior to the cDNA sequence and end with a GA, while the antisense oligo must begin with 5′ ATGGTC and end with a 3′ "G" (see Fig. 1 and Note 1 for sequence requirements).
11. Requirements for large proteoglycan or full-length cDNA inserts (see Fig. 2). Sense strand PCR primers need to begin with GAGGC immediately before the first codon of the proteoglycan insert; therefore, a glycine codon (GGC) will precede the insert. The antisense oligo must begin with the sequence ATGGC at its 5′end, such that the downstream 3′end of the PCR product ends with GCCAT-3′ (on the sense strand), so that a histidine codon (CAT) is in frame with the proteoglycan cDNA.
12. QuikChange Mutagenesis Kit (Stratagene) includes Turbo-PFU DNA polymerase and Dpn-1 for removal of the parent template after PCR.

2.2. Cell Culture

1. *Drosophila* S2 Cells (Invitrogen) (see Note 2).
2. Schneider's *Drosophila* Medium (SDM) supplemented with 10% fetal bovine serum (FBS) (Invitrogen).
3. Fugene-6 (Roche Applied Science) will be diluted in serum-free SDM.
4. 100 mM $CuSO_4$ (Sigma) filter sterilize and plan to dilute 1:166 to a final concentration of 600 μM $CuSO_4$.

2.3. Sample Preparation

1. Anti-FLAG M2 antibody agarose (Sigma)—do not confuse with the FLAG M1 antibody which has a slightly different sequence requirement for binding.
2. IP Wash Buffer (20 mM Tris, 200 mM NaCl, 5% glycerol, pH 7.4).
3. Glycosidases from the Glycoprotein Deglycosylation Kit; Calbiochem:
   - *N*-glycanase (500 mU of peptide *N*-glycanase F).
   - *O*-glycanase (0.125 mU of endo-α-*N*-acetylgalactosaminidase).
   - Sialidase (0.5 mU of α 2–3,6,8,9-neuraminidase).
4. Amylase (Sigma): control for deglycosylation of N-linked protein. The amylase protein is about 50% unmodified and 50% *N*-glycosylated and thus runs as a dimer on SDS-PAGE.

2.4. Western Blotting

1. SDS-PAGE on Bis-Tris 12% SDS-PAGE with MOPS running buffer (NuPAGE; Invitrogen).
2. Nitrocellulose membranes (this is more sensitive than PVDF membranes).
3. TBS (Tris-buffered Saline): 20 mM Tris, 100 mM NaCl, pH 7.6.
4. TBS-T: TBS with 0.05% Tween-20.
5. S-Protein HRP-conjugate (horseradish peroxidase) (Novagen): use at a 1:5,000 dilution.
6. SuperSignal West Femto Maximum Sensitivity substrate (Pierce) for detection of HPR.

3. Methods

Mapping the position of a proteoglycan site is accomplished by mutating a suspected amino acid that defines the modification site and then conducting Western blot analysis of the wild-type and mutated protein. If a single amino acid point mutation results in the loss of an attached carbohydrate chain, then there will be a mobility shift in the electrophoretic gel run. Carbohydrate chains may be attached to single or multiple serine residues of proteoglycans; therefore, mapping experiments are designed either on full-length proteins, when there is high confidence of a single proteoglycan attachment site, or on isolated glycosylation sites in a minireporter protein containing a short segment of 10–30 amino acids. The latter recombinant minireporter system is useful and lends to more rapid analysis and interpretation, when there are many candidate proteoglycan modification sites in a large protein.

The expression of glycosylated reporter proteins requires careful consideration with regard to the repertoire of the endogenous glycosyltransferase machinery that is expressed in the cell culture system. Generally, all four glycosyltransferases that attach and build the tetrasaccharide linker (GlcAβ1-3Galβ1-3Galβ1-4Xyl-) to a protein serine side chain are ubiquitously expressed in all cell types. However, the glycosyltransferases that express the terminal sugars chondroitin and heparan may not be expressed in all cell lines and this will impact the use of anti-carbohydrate antibodies for mapping glycosylation sites. Should there be any question about the ability to probe for specific terminal sugars, then it is desirable to map the glycosylation site in a minireporter system, as the addition of the tetrasaccharade linker to a small reporter protein/peptide is sufficient for mobility shift in SDS-PAGE, which eliminates the need to use expensive anti-carbohydrate antibodies.

It is also possible to probe the proteoglycan sequence requirements in a species-specific manner, using a single cell culture system, if the endogenous initiating glycosyltransferase is knocked down by RNA interference and an exogenous polypeptide xylose transferase is coexpressed with the reporter protein in the RNAi-treated culture cells. These molecular tools to knock down and to reintroduce glycosylation activity are also critical for setting up controls and interpreting the gel mobility shift and Western blot assay data (see Fig. 3).

3.1. Molecular Cloning of Single Modification Sites, Large Proteoglycans, and Mutants

1. *Expression plasmid.* Both full-length cDNAs or small DNA segments that encode for single proteoglycan modification sites can be introduced into the same type of expression plasmid with a directional, sticky-end cloning strategy, using sites generated by the restriction enzyme BsaI in the expression plasmid pLC-S2-SP-BsaI (3). Cleavage of this plasmid with BsaI creates upstream and downstream sites for directional cloning, as depicted in Fig. 1c, for insertion of a cDNA fragment generated by either oligoduplexes or PCR products. Prior to the ligation of insert, the cleaved plasmid is treated with calf intestinal phosphatase to remove the 5′ phosphates of the cloning sites and prevent vector self-ligation.
2. *Proteoglycan minigene reporter.* Oligonucleotide duplexes that encode for a single proteoglycan modification site are designed with upstream and downstream overhanging ends, which are used for ligation into the vector (see Fig. 1a, b). It is essential that these oligonucleotides be treated with T4 polynucleotide kinase and 1 mM rATP prior to ligation into the dephosphorylated vector. Ligation reactions use 10–40 ng of vector with 0.4 ng of the oligonucleotide duplex that was previously annealed by heating to 90°C for 1 min and cooling gradually in a temperature gradient to room temperature over 30 min in a thermal cycler instrument. All oligonucleotide inserts contain a restriction site that is used for screening for positive expression

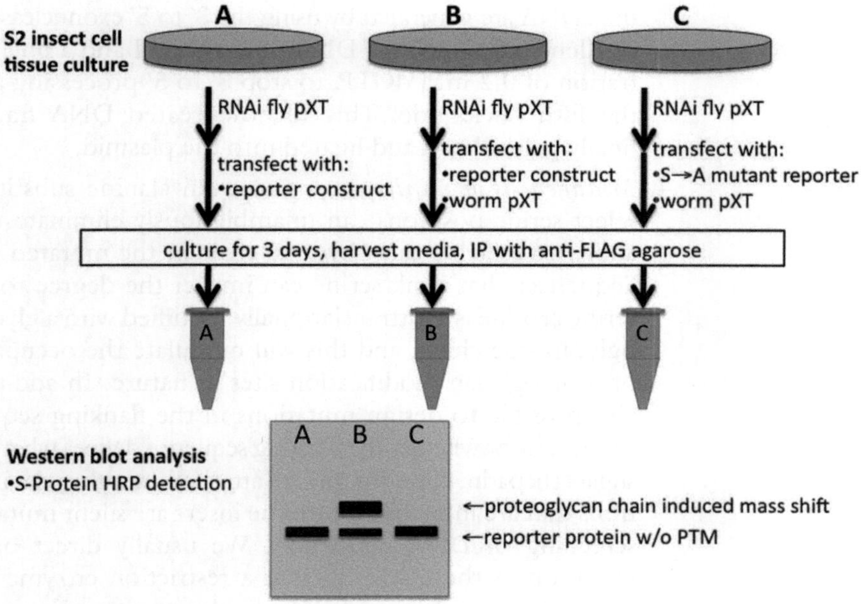

Fig. 3. Expression of reporter glycoproteins in insect cells. All cells are RNAi-treated to knock down endogenous insect cell polypeptide xylose transferase (pXT) activity, producing cell background that is devoid of glycosyltransferase activity that initiates proteoglycan attachment. Reporter proteins are immuno-precipitated and examined by SDS-PAGE and Western blot analysis. A hypothetical western blot of three samples (*A*, *B*, *C*) demonstrates how a proteoglycan is detected and definitely mapped. (*A*) Transfection with a wild-type reporter construct in the absence of pXT activity will produce an unmodified protein, which serves as a reference maker for this reporter. (*B*) Cotransfection of a *C. elegans* pXT expression plasmid with the report protein construct will reconstitute pXT activity and allow for glycosylation at the proteoglycan attachment site, producing a larger MW reporter that can be observed at a slow mobility by SDS-PAGE. (*C*) Confirmation of the glycosylation site is obtained by coexpressing the pXT with a reporter construct, harboring a point mutation at the suspected proteoglycan modification site. In this insect cell model system, both the pXT expression construct and proteoglycan reporter can be derived from any metazoan organism, to examine species-specific glycosylation. See Wang et al. (3) for many examples of reporter gel-shift assays.

constructs, which are always sequenced to confirm reading frame and insert sequence integrity. This cloning scheme is very efficient, allowing for the production of minireporters in batches of 10 or more unique reporter constructs at the same time.

3. *Large proteglycan expression constructs.* To enable insertion of a proteoglycan cDNA into the cloning site above, full-length secreted regions of the proteoglycan need to be PCR-amplified with an upstream primer that starts with the sequence 5′GAGGC contiguous with frame 1 of the cDNA. The downstream antisense primer needs to begin with the sequence 5′ATGGC. PCR products of the proteoglycan cDNA with these engineered ends are cleaned up with a phenol extraction and then ethanol-precipitated by addition of 50 μL of 7.5 M ammonium acetate and 330 μL of 100% ethanol to a 100-μL DNA sample. 5′ protruding sticky ends with 4 nt overhangs for cloning

the cDNA are generated by using the 3′ to 5′ exonuclease activity of Klenow fragment of DNA polymerase I and a final concentration of 0.2 mM dGTP, to stop 3′ to 5′ processing at the G, the fifth nucleotide. This Klenow-treated DNA fragment is finally gel-isolated and ligated into the plasmid.

4. *Mutagenesis of minireporter genes.* An alanine substitution of select serine positions can unambiguously eliminate the cell's ability to attach a carbohydrate chain at the mutated position. Sequences that flank serine can impact the degree to which a serine residue is posttranslationally modified with a glycosaminoglycan side chain, and this will modulate the occupancy rate of proteoglycan modification sites in nature. In addition, it is also possible to design mutations in the flanking sequence to probe how or whether the flanking sequence defines what terminal sugars (heparin, chondroitin) are attached. Another class of mutations that are introduced into the insert are silent mutations for screening of DNA constructs. We usually direct one point mutation in the insert, so that a restriction enzyme cleavage site is created or destroyed. This enables rapid restriction mapping of mutant reporter constructs and selection of positive clones. All positive clones are then submitted for DNA sequencing.

5. *Mutagenesis of large cDNA inserts.* PCR is used in an oligonucleotide-directed mutagenesis method, based from the QuikChange Mutagenesis Kit (Stratagene) to alter the sequence of 25 ng of the target plasmid. Sense and antisense oligonucleotides (about 40 nt long, with >12 nt long segments that flank the directed base change) at 0.3 μM final concentration are used with 200 μM dNTPs, 2.5 U of Turbo PFU in a 25-μL reaction. First, five thermal cycle reactions are conducted at 95°C for 30 s, 45°C for 30 s, and 68°C for 8 min. Next, 13 cycles are immediately conducted at 95°C for 30 s and 68°C for 8 min. After the thermal cycling reaction is complete and cooled, the amplified samples are treated with 3.5 U of DpnI, which cleaves the methylated parental DNA strand and leaves the newly synthesized strands intact. After a 1-h digestion at 37°C, the sample is ethanol-precipitated by adding 12.5 μL of 7.5 M ammonium acetate and 60 μL of 100% ethanol, placing the tube on ice for 10 min and centrifuging at 4°C for 30 min. The pellet is washed by addition of 50 μL of ice-cold 70% ethanol, immediate centrifugation for 2 min, and finally removal of the supernatant. After drying under vacuum in a microcentrifuge for 2 min, the mutated DNA pellet is resuspended in 2.5 μL of water and all of it is used in a single transformation of chemical competent *E. coli*. Individual colonies are screened (four per construct) for the restriction enzyme ablated or created by mutagenesis (see Subheadings 3 and 3.1, step 4) and one restriction enzyme positive clone is selected for DNA sequence analysis for each mutant construct.

3.2. Expression of Recombinant Proteoglycans in Tissue Culture

1. Expression of the peptide reporters in insect cells—*Drosophila* Schneider (S2) cells was cultured in SDM medium (Gibco) supplemented with 10% (v/v) FBS at 27°C. Cells were passaged every 3–4 days. One million S2 cells were added to fresh media in each 35-mm dish for transfections, incubated for 1 day, and then treated with a DNA-FuGENE-6 mixture. Briefly, 3 μL of FuGENE-6 (Roche Applied Science) transfection reagent was diluted into 97 μL of serum-free SDM medium and incubated at room temperature for 5 min. Then 1 μg of reporter plasmid DNA was added with mixing to the diluted FuGENE-6, and this mixture was incubated for 20 min at room temperature. Then the DNA-FuGENE-6 mixture was added in a drop-wise manner to cells. Vigorous agitation/swirling was used to mix the cells and transfection reagent. After a 1-day incubation period, cells are induced to express the reporter protein by the addition of $CuSO_4$ to a final concentration of 600 μM. After three additional days in cell culture, the culture medium is harvested for the secreted reporter, clarified by centrifugation at 2,000 rpm in a SH3000 rotor (Sorval) for 10 min, aliquoted to three tubes, and stored at −80°C.
2. Inhibiting the expression of endogenous polypeptide xylose transferase-RNA interference was used to eliminate the endogenous initiating enzyme. This is performed for two goals: (a) to produce a control reporter lacking an attached proteoglycan sugar chain, which is important in the gel mobility shift assay for establishing the electrophoretic position of the negative control (unmodified) peptide, and (b) to generate an endogenous glycosyltransferase knockdown coupled with an induced expression of a polypeptide xylose transferase from a different species or isoform. RNA interference in S2 cells is easily achieved by exposure of cells to dsRNA, prior to the transfection step. Briefly, on the day of transfection, cells are first washed once with serum-free SDM media, resuspended with 1 mL serum-free SDM, treated with 15 μg of fly pXT dsRNA (>800 bp in length) template in 1 mL SDM media for 2 h, and then immediately transfected with plasmid, without washing the cells, using the procedure described above.
3. For coexpression of worm (or any other) polypeptide xylosyltransferase with the reporter proteins, 0.05–0.1 μg of pLC-S2-FL-CE-pXTase plasmid, harboring a species- or isoform-specific polypeptide xylose transferase cDNA, is cotransfected at the same time with 0.95 μg of the reporter plasmid construct. Theoretically, a glycosyltransferase cDNA from any metazoan (worm, insect, frog, mouse, human, etc.) can be cotransfected with a reporter plasmid to examine how proteoglycan modification sites or machinery is evolutionarily conserved.

3.3. Sample Preparation for the Identification of Proteoglycan Modification Sites

1. *Immuno-precipitation of proteoglycan reporter proteins.* 25 μL (packed resin volume) of anti-FLAG M2 agarose is added directly to 1 mL of clarified cell culture medium. The affinity resin is rocked at 4°C overnight and washed twice in IP Wash Buffer. Reporter proteins bound to beads are treated on-bead, with 0.125 mU of *O*-glycanase, 0.5 mU of sialidase, and 500 mU of *N*-glycanase in a final volume of 48 μL for 5 h at 37°C, according to the manufacturer's specifications. For SDS-PAGE analysis, 16 μL of 4xNuPage sample loading buffer is added to the glycosidase-treated samples.
2. *Controls for glycosidase treatment.* To demonstrate that the glycosylhydrolases are removing N- and O-linked sugar chains, control *N*-glycanase digestions are run using human salivary amylase, which migrates as a doublet of *N*-glycosylated and nonglycosylated isoforms. An *O*-glycanase control protein requires a mucin-type O-glycosylated domain, rich in serine/threonine/proline that is not larger than 80 kDa, so that an electrophoretic mobility shift is easily visible after deglycosylation treatment.

3.4. Gel-Shift Assays for Proteoglycan Attachment

1. Detection of glycosylated reporter proteins—in this cell culture system, we have never observed a minireporter protein that is 100% posttranslationally modified with carbohydrate chains. There is always a nominal percentage of protein that is secreted in its unmodified state, and this unmodified protein migrates faster on SDS-PAGE, by virtue of its lower molecular mass. Western blot analysis is used to determine the percent of glycosylation of the reporter protein, by calculating the ratio of the glycosylated protein (higher MW) over the unmodified (lower MW) isoform in the same lane (see Fig. 3b).
2. Important controls for Western blot analysis of glycoproteins—to verify that the unmodified form of a protein doublet is the lower MW protein species, the reporter construct is transfected in a parallel set of cells with and without RNAi-treatment to knock down the glycosylation machinery, to demonstrate the mobility of the same peptide sequence in the absence of glycosylation (see Fig. 3a). Parallel expression of a serine to alanine point mutant will definitively identify the modification site and is especially useful if multiple serine residues are present in the reporter.
3. Western blot analysis must be of very high quality and very sensitive. Because recombinant proteins are immuno-precipitated with anti-FLAG M2 antibody agarose, blots are probed for a second tag in the reporter protein, the S-Tag, or they are probed with anti-carbohydrate antibodies (see Note 3). High-quality nitrocellulose membranes are blocked in 0.5% casein in Tris-buffered Saline with 0.05% Tween for 2 h at room temperature. S-protein conjugated to horseradish

peroxidase is added to this blocked membrane at a 1:5,000 dilution for an overnight incubation. On the next morning, the membranes are washed 5 times with TBS-T and 1 time with TBS, before HRP detection, using the Super Signal West Femto Maximum Sensitivity substrate. All gels are run in triplicate to obtain an average densitometry scan of the fraction of the protein in the glycosylated state.

4. Notes

1. We avoid designing peptide minireporter proteins that contain cysteine residues, as they tend to multimerize in SDS-PAGE, even in the presence of reducing agents. Multimerized reporters are difficult to interpret in the Western blot analysis.
2. We favor the use of *Drosophila* S2 insect cells as our tissue culture system, as they do not require a CO_2 incubator, are very sensitive to RNA interference (using dsRNA), transfect well, and can be scaled up to large 1 L suspension culture with stable selection for transfected cells. Theoretically, cells from other insects (sf9, sf21, Hi-5) could also be used if they are available, but RNAi knockdowns will need to be tailored to that organism's glycosyltransferase cDNAs.
3. *Probing for proteoglycan sugars directly.* Such probing can be performed on duplicate Western blots, using anti-chondroitin or anti-heparin stub antibodies after chondroitin ABC or heparin lyase treatment of the anti-FLAG M2 antibody immunoprecipitates (see Subheadings 3 and 3.3, step 1). However, not all cell lines produce these terminal sugars with high efficiency, so this could be problematic for detection. Also, probing directly for terminal sugars is usually considered as a qualitative experiment, as it lacks an internal control for quantification. Therefore, Western blotting for terminal sugars may be used for confirmation of proteoglycan modification sites, but ratios of unmodified and glycosylated reports are determined by probing with the epitope tags in the fusion protein.

References

1. Brunner, A., Kolarich, D., Voglmeir, J., Paschinger, K., and Wilson, I. B. (2006) Comparative characterisation of recombinant invertebrate and vertebrate peptide O-Xylosyltransferases. *Glycoconj J.* **23**, 543–554.
2. Dong, S., G.J. Cole, and Halfter, W. (2003) Expression of collagen XVIII and localization of its glycosaminoglycan attachment sites. *J Biol Chem.* **278**, 1700–1707.
3. Wang, H., Julenius, K., Hryhorenko, J., and Hagen, F. K. (2007) Systematic Analysis of proteoglycan modification sites in Caenorhabditis elegans by scanning mutagenesis. *J Biol Chem.* **282**, 14586–14597.
4. Brinkmann, T., Weilke C. and Kleesiek, K. (1997) Recognition of acceptor proteins by UDP-D-xylose proteoglycan core protein beta-D-xylosyltransferase. *J Biol Chem.* **272**, 11171–11175.

5. McCormick, D., van der Rest, M., Goodship, J., Lozano, G., Ninomiya, Y., and Olsen, B. R. (1987) Structure of the glycosaminoglycan domain in the type IX collagen-proteoglycan. *Proc Natl Acad Sci USA.* **84**, 4044–4048.
6. Hagen, F.K., Van Wuyckhuyse B. and Tabak, L.A. (1993) Purification, cloning, and expression of a bovine UDP-GalNAc: polypeptide N-acetyl-galactosaminyltransferase. *J Biol Chem.* **268**, 18960–18965.
7. Julenius, K., Mølgaard, A., Gupta, R., and Brunak, S. (2005) Prediction, conservation analysis, and structural characterization of mammalian mucin-type O-glycosylation sites. *Glycobiology* **15**, 153–164.
8. Shakin-Eshleman, S.H., Spitalnik, S.L., and Kasturi, L. (1996) The amino acid at the X position of an Asn-X-Ser sequon is an important determinant of N-linked core-glycosylation efficiency. *J Biol Chem.* **271**, 6363–6366.
9. Nehrke, K., Ten Hagen, K. G., Hagen, F. K., and Tabak, L. A. (1997) Charge distribution of flanking amino acids inhibits O-glycosylation of several single-site acceptors in vivo. *Glycobiology* 7, 1053–1060.

Chapter 3

Mapping of the Wnt/β-Catenin/TCF Response Elements in the Human Versican Promoter

Maziar Rahmani, Jon M. Carthy, and Bruce M. McManus

Abstract

Versican, a chondroitin sulfate proteoglycan, is one of the main components of the extracellular matrix and is considered to be crucial to several key cellular processes involved in development and disease. There is differential temporal and spatial expression of versican by multiple cell types and in different developmental and pathological timeframes. In order to fully appreciate the functional roles of versican as it relates to changing patterns of expression in development and disease, an in-depth knowledge of versican's biosynthetic processing is necessary. We have recently shown that β-catenin/T-cell factor (TCF) complex formation at the versican promoter site is essential for activation of versican transcription. The transcriptional activator β-catenin is the key mediator of the canonical Wnt signaling pathway. However, β-catenin does not itself bind DNA and thus functions via interaction with TCF/Lymphoid-enhancing factor (LEF) transcription factors. These proteins contain a high-mobility group (HMG) box that binds DNA in a sequence-specific manner. Thus, in the case of active Wnt signaling, β-catenin activates, in cooperation with proteins of the TCF/LEF family, the expression of a wide variety of target genes. The goal of this chapter is to describe the techniques used to elucidate the transcriptional control of versican by the β-catenin/TCF response elements in its promoter site and to demonstrate how this signaling may be assayed experimentally. These approaches provide insight into the transcriptional regulation of the versican gene and provide the basis for the identification of novel Wnt/β-catenin/TCF-regulated genes that are part of the signaling machinery regulating early embryogenesis, neoplasia, and cardiovascular remodeling.

Key words: Versican, Wnt pathway, β-catenin, TCF/LEF recognition site, Luciferase, Promoter activity, EMSA

1. Introduction

Versican is the most versatile hyalectin in regard to its structure and tissue distribution. Recent data emerging from in vitro and in vivo studies suggest versican modulates cell adhesion, proliferation, and migration, and hence play a central role in tissue development and maintenance as well as in a number of pathologic processes. Many

Françoise Rédini (ed.), *Proteoglycans: Methods and Protocols*, Methods in Molecular Biology, vol. 836,
DOI 10.1007/978-1-61779-498-8_3, © Springer Science+Business Media, LLC 2012

aspects of versican biology have recently been reviewed, including the biosynthesis of versican (1, 2), the role of versican in vascular disease (3–5), cancer (6), the central nervous system (7, 8), and the interaction of versican with its binding partners (9). However, our understanding of the regulation of versican expression in different developmental stages and diseases is mainly limited (review in ref. (2)).

β-catenin is a 92-kDa protein which links E-cadherin to α-catenin and the actin microfilament network of the cytoskeleton (see Fig. 1) (10). However, another important function of β-catenin is its transcriptional activity in the nucleus (11). Conventional Wnt signaling causes β-catenin to accumulate intracellularly where it forms a complex with the TCF/LEF transcription factors to regulate target gene expression (see Fig. 1). Briefly, the key factors in β-catenin signaling are its stabilization and accumulation in the cytoplasm, whereupon it translocates to the nucleus and in cooperation with TCF/LEF mediates the expression of a wide variety of genes. Many of the direct Wnt target genes are already described (a listing is available at: http://www.stanford.edu/~rnusse/pathways/targets.html). For a more profound understanding of the different effects of Wnt signaling, it is necessary to define additional β-catenin targets. We have recently shown that PI3K and its downstream effector PKB mediate serum stimulation of versican transcription in vascular smooth muscle cells (SMC) (12). Further, we showed that cytoplasmic accumulation of β-catenin via the inhibition of GSK-3β stimulated transcription of versican in these cells. Finally, we demonstrated that a TCF/LEF binding site in the versican promoter is involved in β-catenin-induced versican transcription and that at least one of the members of the TCF family of transcription factors, namely TCF-4, binds to the TCF/LEF site in the human versican promoter. In addition, we and our collaborators have recently shown that versican is transcriptionally regulated by androgens in androgen receptor (AR)-positive human prostate epithelial and stromal cancer cells (13). In this study, we demonstrated a strong and specific binding of recombinant AR to the DNA sequences located in the +75 region of the proximal human versican promoter. Interestingly, we demonstrated that β-catenin is required for AR-mediated transcription in both a ligand-dependent and ligand-independent manner in prostate stromal fibroblast cells. These observations may support a new concept for the pathophysiological role of PI3K/PKB/GSK-3β signaling through the β-catenin/TCF transactivation complex to modify the gene expression patterns observed in provisional matrix formation during development and disease (3, 4, 14–19). Our data also suggest a novel role for β-catenin in nuclear hormone receptor-mediated transcription in prostate stromal cells, which may augment tumor–stromal interactions and may contribute to prostate cancer progression.

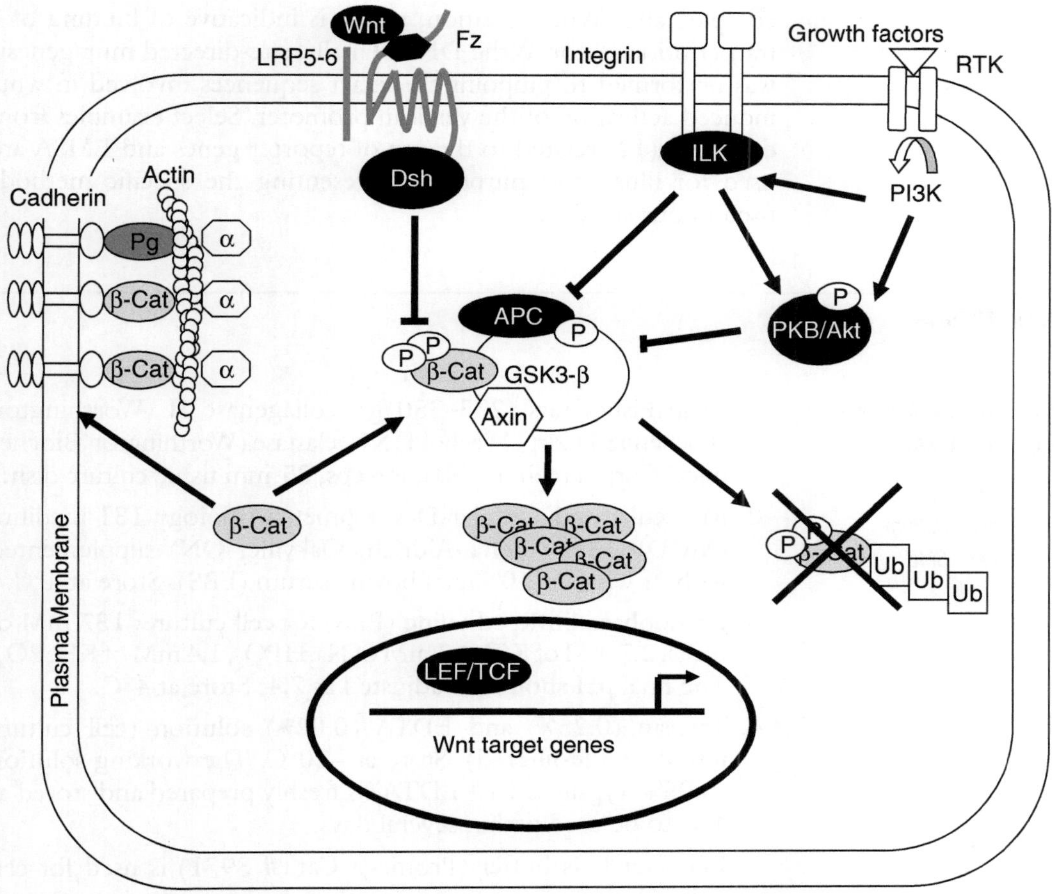

Fig. 1. Schematic representation of Wnt/β-catenin/TCF signaling. In cells devoid of a Wnt and/or growth factor-mediated signal, β-catenin is held by the GSK-3β/APC/axin-conductin complex, resulting eventually in degradation of β-catenin. TCF target genes are repressed by TCFs interacting with nuclear corepressors. Upon Wnt stimulation (through Disheveled) and/or growth factor-mediated signals (such as PI3K/PKB), GSK-3β is inactivated, resulting in disruption of the quarternary complex and β-catenin accumulation. A transcriptionally active complex with TCFs is formed that leads to target genes transcription. The abbreviations used are: *APC* adenomatosis polyposis coli; β-*cat* β-catenin; *Dsh* Disheveled; *Fz* frizzled receptor; *GSK-3*β glycogen synthase kinase-3β; *ILK* integrin link kinase; *LRP5-6* LDL receptor-related protein (LRP) 5-6; *Pg* plakoglobin; *PI3K* phosphatidylinositol 3-kinase; *PKB* protein kinase B; *RTK* receptor tyrosine kinase; *TCF/LEF* T-cell factor/lymphoid enhancer factor; *Ub* ubiquitin; *Wnt* wingless.

The methods described herein were used to determine the transcription factor that mediates the β-catenin/TCF-induced activation of the versican promoter and include luciferase-based promoter reporter assays and electrophoretic mobility shift assay (EMSA) analysis of the TCF sequence in the versican promoter site. More precisely, luciferase-based reporter assays were used to define the sequence within the versican promoter that is responsive to Wnt signaling. Electrophoretic mobility shift assays were employed to determine whether the β-catenin/TCF promoter sequence identified with luciferase-based reporter assays formed a nucleoprotein complex after Wnt activation, which is indicative of binding of a transcription factor to the DNA. Finally, site-directed mutagenesis was performed to pinpoint the exact sequences involved in wnt-induced activation of the versican promoter. Select examples from our work (12) related to the use of reporter genes and EMSA are used for illustrative purpose in presenting the specific methods used in our assays.

2. Materials

2.1. Luciferase-Based Reporter Assay

2.1.1. Isolation and Primary Culture of Rat Aortic SMC

1. Male Fisher rats (275–350 g), collagenase II (Worthington Biochemical Corp, Freehold, NJ), elastase (Worthington Biochemical Corp, Freehold, NJ), forceps, 35-mm tissue culture dish.
2. Molecular and Cellular Developmental Biology 131 medium (MCDB-131) (Sigma-Aldrich, Oakville, ON) supplemented with or without 10% fetal bovine serum (FBS). Store at 4°C.
3. Phosophate-buffered saline (PBS) for cell culture: 137 mM of NaCl, 2.7 mM of KCl, 4.2 mM of Na_2HPO_4, 1.4 mM of KH_2PO_4. The final pH should be adjusted at 7.4. Store at 4°C.
4. Trypsin (0.25%) and EDTA (0.02%) solution (cell culture tested, sterile-filtered). Store at −20°C. The working solution (0.05% trypsin, 0.01% EDTA) is freshly prepared and stored at 4°C to be used within several days.
5. Reporter lysis buffer (Promega Cat. # 3971) is used for cell lysis. However, the manufacturing company does not reveal the composition of the buffer. So, if you do not want to use this buffer, the following buffer will also work: 25 mM of glycylglcine (pH 7.8), 15 mM of $MgSO_4$, 4 mM of Ethylene glycol bis(2-aminoethyl ether)tetraacetic acid (EGTA), 1%(v/v) of Triton X-100, 1 mM of dithiothreitol (DTT, add immediately before use). Store at room temperature.
6. Cell scrapers.
7. 24-Well cell culture dishes. Larger culture dishes (6-well or 12-well) may be needed if a lower transfection efficiency is observed.

2.1.2. Transfection and Cell Lysis

1. Reporter vectors. Versican promoter (−634/+118) and a shorter fragment (−438/+118) corresponding to the versican promoter sequences (20) were generated by PCR from the human genomic DNA with appropriate sets of primers (available upon request) (see Figs. 2 and 3). These inserts were cloned into a pGL3 basic vector (Promega) by standard molecular biology techniques and called wt-versican-Luc and −438del-versican-Luc, as depicted in Fig. 3. The putative TCF

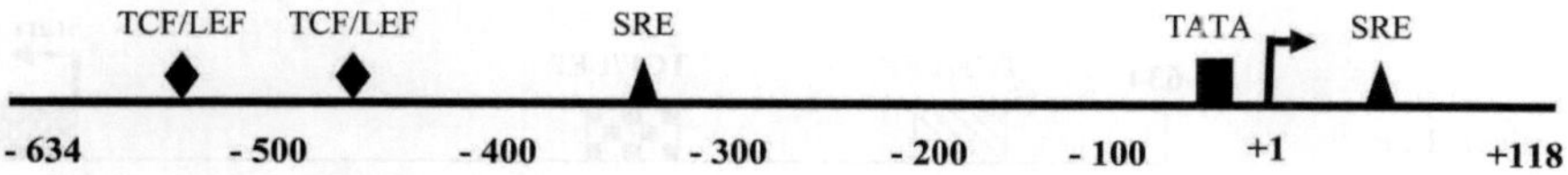


```
-632 GAATTCTTACACTTTCCCTCTAGGTCCCCGAAGGATCTCGTTTCTCAGT
                                               TCF/LEF
-582 GTCTCTTTCAGGTTGGCAGGAGCCTTGAGCCTGACACTTCCCTTTGATGG
                                                 TCF/LEF
-532 GACAGGCAAGCTCTGTGGGCGCGTAAACACGCTGTAACCAAGTTCTTTGC
-482 TGATTTTACAGTTTTGTGTGCTCCCGAGAAGAAGTGATCGTACTCAATTG
-432 TCTATTGCTGGCCTGCCCCCTAAGAGCCTGGGGCTCCTTTCCCCTAACCC
            SRE
-382 AGAACTAGCTGCACGGGGGGCGGGGAAGTGGGGGTGGGGAAGGAGTGGGA
-332 GGGCAGTGGTTTCCGCGAGCAGAGCATGTTACTGAGTGAGTCCCTGAATG
-282 GGGAGCGCTGCTGTCCCAAGCCGATTGGTACTTCTTGTCAGGAAGAAACG
-232 CCAAGAGGTGGGAGTGCCTGGGGAGGGAGGCAGGCGGTCCCTACCGCAGG
-182 CGCGGGGAGCTGCCTTTCCCCCCTCCGCCTGCTTTCCAAGCCTGGACTCT
-132 TAGGAGTGGCTGAAGCTGCGGAGCGCTTTTGGAGCCTGTGAATGAACCCT
-82  CCTCCTCTCCCTCCTCCTTCTTCTCGCTGAGTCTCCTCCTCGGCTCTGAC
          (TATA-box)            +1
-32  GGTACAGTGATATAATGATGATGGGTGTCACAACCCGCATTTGAACTTGC
18   AGGCGAGCTGCCCCGAGCCTTTCTGGGGAAGAACTCCAGGCGTGCGGACG
                SRE
68   CAACAGCCGAGAACATTAGGTGTTGTGGACAGGAGCTGGGACCAAGATCT
118  TCGGCCAGCCCCGCATCCTCCCGCATCTTCCAGCACCGTCCCGCACCCTC
```

Fig. 2. Sequence analysis of the versican gene promoter and exon 1. Graphic representation of the nucleic acid sequence of the proximal versican promoter from position −632 to +118 relative to the transcription start site. The versican proximal promoter was searched against the MatInspector and ConSite databases to detect potential T-cell factor/lymphoid enhancer factor (TCF/LEF) and steroid response elements (SREs). Previously and newly described protein-binding sites indicated are the T-cell factor/lymphoid enhancer factor (TCF/LEF) and SRE.

binding sites, −546 bp TCCCTTTGATGG and −492 bp TTCTTTGCTGAT contained in the wt-versican-Luc, were mutated as denoted in the Fig. 3 by site-directed mutagenesis using a Quick Change mutagenesis kit from Stratagene as described before (21). The mutated inserts were generated by PCR and then inserted into the promoterless luciferase vector pGL3-Basic. All constructs were verified by sequencing. PSVβ-gal (promega), an expression vector for β-galactosidase under control of the SV40 promoter, is used for normalization of transfection. pEGFP-C1 (Clontech), an expression vector for green fluorescent protein, is used for the confirmation of transfection efficiency before the reporter assay (optional). It is essential to obtain high-quality plasmid DNAs for transfection assays. We use a plasmid purification kit from QIAGEN. Each plasmid DNA is dissolved in 0.1× TE (10 mM of Tris–HCl, pH 7.6 and 1 mM of EDTA), which is carefully prepared using endotoxin-free tissue culture grade water. The concentration of each DNA solution is kept at 1 mg/mL and aliquots are stored at −80°C (for long-term storage) or at 4°C (for short-term storage).

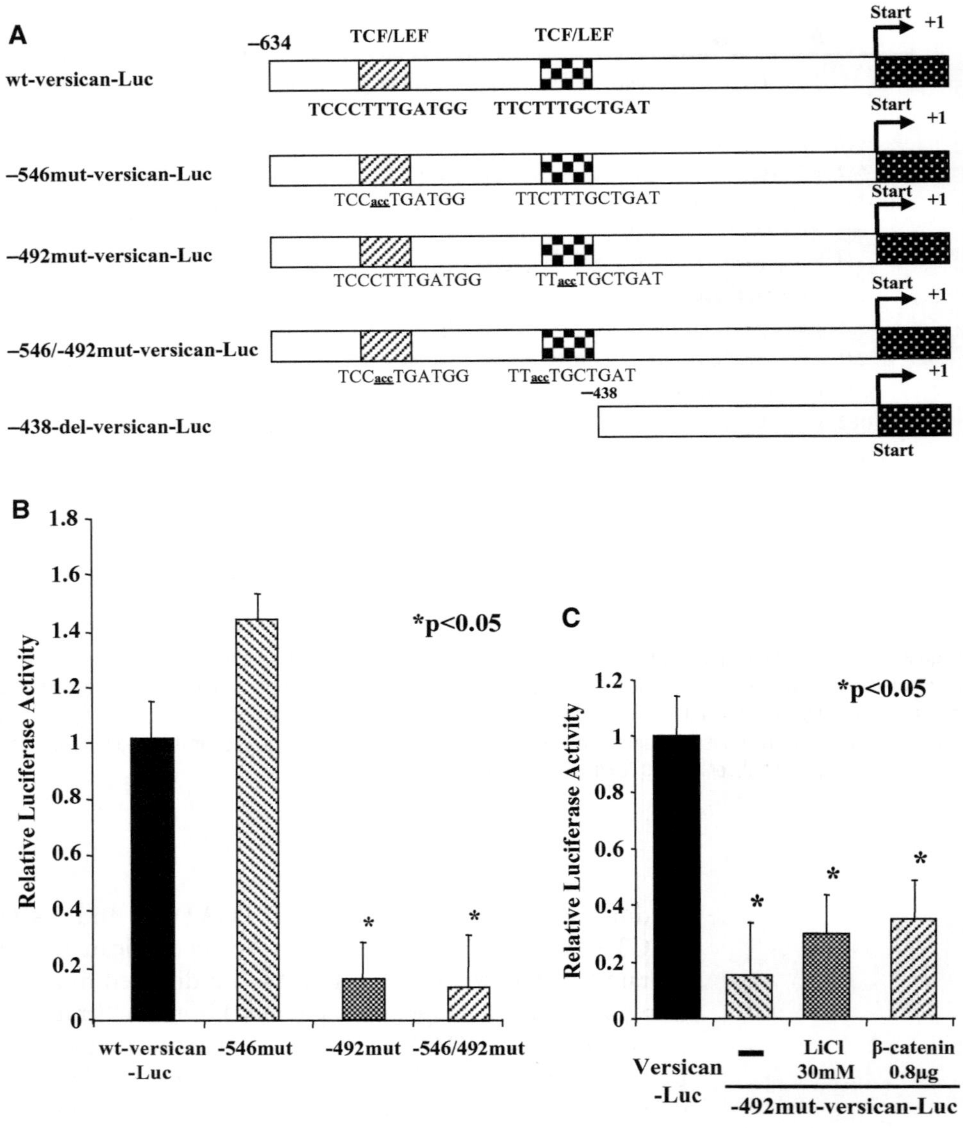

Fig. 3. The TCF/LEF binding sequence at position –492 is crucial for activation of the versican promoter by the β-catenin-TCF complex. (**a**) Schematic representation of the sequence for TCF/LEF binding sites in wild-type human versican promoter and mutant constructs. Two putative TCF/LEF binding sites are localized at position –546 and –492 bp from the transcription start site. Mutations known to abolish TCF binding were introduced into the TCF/LEF recognition sites to make the mutant constructs –546mut-versican-Luc, –492mut-versican-Luc, and –546/–492-mut-versican-Luc in the context of the wt-versican-Luc promoter. A deletion construct lacking the TCF/LEF binding sites, –438del-versican-Luc, was also generated. (**b**) Mutation at position –492 of the versican promoter inhibited the response of luciferase promoter activity to LiCl treatment and β-catenin transactivation in SMC. SMC containing normal endogenous β-catenin were transiently transfected with 0.5 μg of wt-versican-Luc and treated with 30 mM LiCl. The cells were cotransfected with 0.8 μg of β-catenin expression vector or equal amount of empty vector and harvested 24 h after transfection. (**c**) A deletion construct lacking the TCF/LEF binding sequences showed no β-catenin-mediated versican promoter transacativation. SMC were transfected with wt-versican-Luc or –438del-versican-Luc reporter, along with β-catenin (0.8 μg) or empty vector as indicated in the figure. Total amounts of plasmid DNA were kept constant by adding the empty pCDNA3.1 vector. All experiments were performed in duplicate and repeated at least 3 times. Mean ± S.D. is presented (reproduced from ref. (12) with permission from *J Biol Chem*).

2. FuGENE® 6 Transfection Reagent (Roche Applied Science) (Cat.No.11814443001). This transfection reagent is suitable for use with several primary and immortalize cell lines (see Note 1).
3. Sterile Polystyrene Tubes (12 × 75 mm).

2.1.3. Reporter Assays

1. Luciferase assay reagent (Promega E3971). Store aliquots at −80°C, but avoid multiple freeze/thaw cycles by determining the appropriate aliquot size for your experiments.
2. β-galactosidase assay reagent: 80 mM of sodium phosphate butter, pH7.3, 102 mM of 2-mercaptoethanol, 9 mM of $MgCl_2$, 8 mM of CPRG (β-D-galactopyranoside, add immediately before use). Store this reagent without CPRG at room temperature. CPRG is dissolved at 400 mM in water and stored as a stock solution for aliquoting at −80°C.

2.2. Protein-DNA-Binding Reactions

2.2.1. Preparation of Nuclear Extracts

1. Dithiothreitol (DTT) (Invitrogen): 100 mM in Buffer D (20 mM HEPES, pH 7.9, 100 mM KCl, 10% glycerol (v/v)).
2. Buffer A: 10 mM HEPES, pH 7.9, 1.5 mM $MgCl_2$, 0.5 mM DTT added immediately prior to use.
3. Buffer C: 20 mM HEPES, pH 7.9, 25% (v/v) glycerol, 0.42 M NaCl, 1.5 mM $MgCl_2$, 0.2 mM disodium ethylenediamine tetraacetate (EDTA), 0.5 mM phenylmethanesulphonylfluoride (PMSF, optional), 0.5 mM DTT added immediately prior to use.
4. A prechilled, clean glass Dounce homogeniser (Kontes, B type pestle).
5. Teflon cell scraper(s).
6. Tissue culture materials appropriate to the cell line of interest.

2.2.2. Generation of Labeled DNA Probe for EMSA Experiment

1. Synthesized dephosphorylated complimentary single-stranded DNA oligomers (typically 18–35 bases) corresponding to the region of interest dissolved in nuclease-free Tris-EDTA buffer (TE) to a concentration of 20 μg/mL or 3–5 μg of plasmid containing the cloned region of interest.
2. Tris-EDTA buffer: 10 mM Tris base, pH 8.0, 1 mM EDTA.
3. 30% acrylamide: *N*,*N*-methylene-bis-acrylamide solution (37.5:1) (BioRad Laboratories).
4. Ammonium persulfate (APS): 10% (w/v) prepared in dH_2O.
5. *N*,*N*,*N*,*N*-Tetramethyl-ethylenediamine (TEMED) (Invitrogen).
6. 5× Tris-borate buffer (5× TBE): 0.445 M Tris base, pH 8.0, 0.445 M boric acid, 10 mM EDTA.
7. Vertical gel electrophoresis unit of ~30 cm in height with corresponding glass plates, 1.5-mm spacers and combs with 14 wells.
8. Power supply capable of delivering at least 400 V/100 mA.

9. DNA loading buffer (5×): 50% glycerol (v/v), 0.04% (w/v) Bromophenol Blue, 0.04% (w/v) xylene cyanol in TBE.
10. Deoxynucleotide triphosphates (dNTP's) (dATP, dTTP, dGTP, dCTP) (Invitrogen): diluted to 10 mM with sterile dH_2O.
11. Klenow enzyme (for plasmid sourced DNA probes) (New England Biolabs) and the necessary restriction enzymes to generate a single-end-labeled DNA on either the upper or lower strand.
12. T4 polynucleotide kinase (PNK) (for oligomer-based DNA probes) (New England Biolabs).
13. [α-^{32}P]dCTP (3,000 Ci/mmole) (Perkin Elmer) (used for Klenow labeling of plasmid sourced DNA probes with 5′ overhangs containing guanines).
14. [α-^{32}P]dATP (6,000 Ci/mmole) (Perkin Elmer) (used for T4 PNK labeling of oligomers).
15. PCR cleanup kit (Qiagen).
16. 95 and 70% ethanol.
17. Buffer D: 20 mM HEPES, pH 7.9, 100 mM KCl, 10% glycerol (v/v).
18. Autoradiography films and cassettes, darkroom facilities for film processing.
19. MicroSpin Sephadex G50 spin columns (GE Healthcare).
20. Gel elution buffer: 0.6 M ammonium acetate (NH_4OAc), 0.1% SDS, 1 mM EDTA.

2.2.3. EMSA

1. Radiolabeled DNA probe (see Subheading 2.2.2).
2. Poly(dI-dC).(dI-dC) (GE Healthcare) is made up to 2 mg/mL in 100 mM KCl. Heat to 43°C and cool to room temperature to get duplex strands, aliquot, and store at −20°C.
3. 100 mM DTT in Buffer D.
4. DNA-binding buffer (DBB): Buffer D supplemented with 1 mM DTT.
5. 0.5× TBE: 1/10 dilution of 5× recipe in Subheading 2.2.2, item 6.
6. 30% Acrylamide: *N,N*-methylene-bis-acrylamide (37.5:1) solution (BioRad Laboratories).
7. Ammonium persulfate (APS) (see Subheading 2.2.2, item 4).
8. *N,N,N,N*-Tetramethyl-ethylenediamine (TEMED).
9. Vertical gel electrophoresis unit of ~30 cm in height with corresponding glass plates, 1.5-mm spacers, and combs with 14 wells.
10. 5× DNA loading buffer (see Subheading 2.2.2, item 9).

11. Gel drier (heat and vacuum system with vapor trap).
12. Autoradiography films and cassettes, darkroom facilities for film processing or phospho-imager system.

2.2.4. Supershift EMSA

Antibodies including test antibodies directed either against the protein of interest or an attached tag molecule and a control non-specific antibody (e.g., normal IgG or an antibody directed against an unrelated target). Control and test antibodies should be resuspended in similar buffers and at similar concentrations.

3. Methods

3.1. Luciferase-Based Reporter Assay

This technique is commonly known as the reporter gene assay. The firefly luciferase gene used in this assay is cloned downstream of the chosen DNA promoter sequence in an otherwise promoterless DNA vector. The luciferase enzyme is then synthesized after transient transfection of the DNA vector into cells only if the upstream DNA promoter sequence drives transcription of the luciferase gene. The ability of the cloned DNA promoter sequence to be activated by a certain stimulus is proportional to the amount of light produced by the oxidation of luciferin by luciferase in the presence of ATP in an in vitro reaction performed using cell extracts. Cotransfection with β-galactosidate expression vector is commonly used to compensate for variations in transfection efficiency/sample manipulation. By generating a series of deletions and mutations of the versican promoter fused to the luciferase gene, we have identified one sequence within the versican promoter that responded to activation of Wnt/β-catenin/TCF signaling.

3.1.1. Isolation and Primary Culture of Rat Aortic SMCs

1. A rat aortic SMC culture was established by a modification of the enzymatic dispersion technique (22, 23). Briefly, four adult male Fisher rats (275–350 g) were euthanized. The thoracic aorta was then removed and immediately washed in MCDB-131 medium.
2. Enzyme I (0.5 mg/mL collagenase II) was applied to the exposed media and the tissue was incubated for 20 min at 37°C to loosen the media from the underlying adventitia.
3. Medial strips were removed with sterile forceps, taking care not to reach the adventitial layer, transferred to a 35-mm tissue culture dish containing 500 μL of Enzyme II (0.5 mg/mL collagenase II and 0.2 mg/mL elastase), and minced.
4. Medial tissue from the aortas of all four rats were pooled, additional Enzyme II solution was added to the dish, and the tissue mixture was incubated at 37°C for 2.5 h with pipetting at regular intervals to disperse cells.

5. Liberated cells were subsequently pelleted at 1,000 rpm for 5 min, resuspended in 1 mL MCDB-131 containing 20% FBS, and seeded into a 35-mm tissue culture dish.
6. Cells were grown to confluence, released by trypsinization and subcultured at a density of 1.0×10^4 cells/cm^2 in MCDB-131 supplemented with 5% FBS.

3.1.2. Transfection and Cell Lysis

1. Cells were seeded in 6-well plates at a concentration of 2×10^4 cells per well in 3 mL of MCDB-131 medium containing 10% FBS. Incubate the cells at 37°C in a CO_2 incubator until the cells are 80% confluent. This will usually take 18–24 h.
2. Cells are transiently transfected in 6-well plates using up to 2 μg of the versican promoter plasmid DNA or the luciferase reporter construct alone and a β-galactosidae expressing vector and FuGENE6 reagent (Roche Applied Science) according to the manufacturer's recommendations (see Notes 2 and 3). In brief, a ratio of 3:1 for FuGENE reagent (μL): plasmid (μg) was incubated for 30 min at room temperature in serum-free medium before addition to 70–80% subconfluent cells in medium for 24 and 48 h (see Note 4).
3. Twenty-four or 48 h posttransfection, remove the growth medium from the wells. Wash the cells once with ice-cold PBS, being careful not to dislodge any of the cells. Remove as much of the final PBS as was possible.
4. Add 100 μL per well of reporter lysis buffer to cover the cells. Rock the dish slowly several times to ensure complete coverage of the cells.
5. Scrape all areas of the dish, then tilt the dish, and scrape the cell lysate to the lower edge of plate. Take care to scrape down all visible cell debris. Transfer the cell lysate to a microcentrifuge tube with a pipette and place the tube at −80°C for at least 15 min.
6. After thawing the cell lysate at room temperature, vortex the tube for 15 s, then centrifuge in a microcentrifuge ($11,000 \times g$) for 15 s at room temperature. Transfer the supernatant to a fresh tube (see Note 5).

3.1.3. Reporter Assays

1. Mix 10 μL of cell extract with 50 μL of Luciferase Assay Reagent in sample tubes at room temperature and measure the light produced by luminometer (see Notes 6 and 7).
2. Mix 2 μL of cell extract (room temperature) with 100 μL of β-galactosidase assay reagent on a microtiter plate. As a control, mix 2 μL of Reporter Lysis Buffer (room temperature) with 100 μL of β-galactosidase assay reagent. Incubate samples at room temperature for 5–30 min. Read the absorbance at 580 nm with a microtiter plate reader.

3. Calculate the normalized luciferase activity by data obtained from microtiter plate reader and express data as ration of luciferase relative light unit to β-galactosidase light unit (see Fig. 3).

3.2. Protein-DNA-Binding Reactions, EMSA, and Supershift EMSA

Electrophoretic Mobility Shift Assay, or EMSA, allows for detection of sequence-specific DNA-transcription factor interactions. It is based on the premise that free DNA will run quicker than DNA bound to a protein when resolved on a polyacrylamide gel. When an antibody against the protein of interest is introduced into the mix, it further adds to the retardation of the complex in the gel matrix and allows the identification of a transcription factor, which binds the DNA sequence under investigation. This is deemed a supershift. Incubation with an excess of unlabeled, double-stranded oligonucleotides representing the consensus sequence for the investigated transcription factor prevents the formation of the labeled DNA–protein complex and is often used to corroborate supershift data. Although ^{32}P-labeled DNA has been the cornerstone for EMSA studies, there are alternative ways to label DNA, including the biotin-based method. Here, by using EMSA analysis, we found that activation of Wnt/β-catenin signaling promotes the binding of the TCF-4 transcription factor to the TCF/LEF consensus elements within the versican promoter.

3.2.1. Preparation of Nuclear Extracts

The following protocol was used for preparing nuclear extracts for DNA-binding studies. The reagent volumes suggested are for a single 15-cm tissue culture plate, approximately 75% confluent (~2×10^7 cells).

1. Wash cells twice with ice-cold PBS to remove media (care should be taken not to dislodge the cells).
2. Scrape cells into 10 mL of ice-cold PBS, collect in a 15-mL conical tube.
3. Pellet cells by centrifuging at $800 \times g$ for 10 min at 4°C.
4. Remove supernatant and resuspend cells (gentle vortexing) in 5 mL ice-cold PBS, centrifuge as before.
5. Remove supernatant and resuspend cells (by gentle vortexing) in 5 mL ice-cold Buffer A and incubate on ice for 10 min.
6. Centrifuge at $800 \times g$ for 10 min at 4°C, remove supernatant, and resuspend in 1.5 mL ice-cold Buffer A. Incubate on ice for 10 min.
7. Disrupt cells with 10 strokes of a prechilled, clean glass Dounce homogeniser (Kontes, B type pestle).
8. Place cell lysate into a chilled 1.5-mL microcentrifuge tube and centrifuge at $18,000 \times g$ at 4°C for 1 min.
9. Remove supernatant (cytoplasmic fraction) and discard. Repeat centrifugation and remove any remaining residue.

10. Resuspend nuclei in 200 μL chilled Buffer C. Incubate at 4°C for 30 min with mixing (gentle vortexing is best).
11. Centrifuge at maximum speed (microcentrifuge) for 20 min at 4°C. Recover and aliquot supernatant (25 μL volumes work well) (see Note 8).

3.2.2. Generation of Labeled DNA Probe for EMSA Experiment

Appropriate safety measures and adherence to local and national rules should be followed when working with ionizing radiation sources and discarding waste materials.

Using Synthetic Complimentary Oligomers

1. Radiolabel 100 ng of one complimentary oligomer using T4 polynucleotide kinase and [α^{-32}P]dATP in the following reaction:

   100 ng oligomer.

   5 μL 10× T4 kinase buffer (supplied with the enzyme by the manufacturer).

   10 U T4 polynucleotide kinase.

   2.5 μL [α^{-32}P]dATP (25 μCi).

   Sterile dH_2O to 50 μL.
2. Incubate at 37°C for 1 h.
3. Remove the unicorporated nucleotides by passing through a G25 spin column according to the manufacturer's instructions.
4. Anneal the radiolabeled oligomer to its complimentary molecule by adding an equimolar amount of the second oligomer to the labeled oligomer and heating to 95°C for 5 min followed by slowly cooling to room temperature.
5. Gel purify the DNA probe as described in Subheading 3.2.2, step 2.

Gel Purification of Labeled DNA Probes

1. Assemble a 1 mm vertical nondenaturing 5% acrylamide in 0.5× TBE gel, using the appropriate spacers and comb (see Note 9).
2. Prerun the acrylamide gel at 400 V for 3 min in 0.5× TBE.
3. Add 1/5 volume of 5× DNA loading dye to the labeled DNA reaction (described in Subheading 3.2.2, step 1) and load it into one well of the gel.
4. Run the gel at 400 V until the bromophenol blue (dark blue) dye-front has migrated approximately 2/3 of the way down the gel.
5. Stop the reaction and disassemble the gel running apparatus. Lay the gel flat and remove the top glass plate, leaving the gel adhered to the bottom plate.

6. Wrap the gel and glass plate in plastic wrap and place it into an X-ray film cassette. Use a method of registering the film to the gel during exposure such that the registration can be reproduced on the bench after the film is exposed. You can typically use phosphorescent stickers on the plastic wrap to leave registration exposures on the film following developing.
7. In the dark room, expose wet gel to film for 1 min and develop film.
8. Reorientate the film on top of the plastic wrapped-gel. Using a razor blade or scalpel, cut out the section of the gel containing the DNA of interest and transfer it to a 1.5-mL microcentrifuge tube (see Note 10).
9. Elute the DNA fragment in 500 μL gel elution buffer by rotating overnight (see Note 11).
10. Carefully transfer the elution buffer containing the DNA probe to a new tube (do not transfer any fragments of the acrylamide) and add 1 mL of 95% ethanol and mix thoroughly by inverting the tube (see Note 12).
11. Precipitate the DNA probe by incubating on ice for 30 min and centrifuge at maximum rpm in a microcentrifuge, at 4°C for 30 min.
12. Remove the supernatant and wash twice with 500 μL 70% ethanol, centrifuge at maximum rpm for 5 min, and again, aspirate the supernatant. The location of the probe DNA through these manipulations can be followed by Geiger counter to ensure the pellet is not lost.
13. Air-dry the pelletted DNA and resuspend in 20 μL of buffer D
14. Count 1 μL of the radiolabeled DNA using a scintillation counter and dilute the DNA probe to 10,000 dpm/μL in buffer D.

3.2.3. EMSA

Assemble a 1.5 mm vertical 0.5× TBE nondenaturing 5% acrylamide gel using the appropriate spacers and combs.

1. Thaw either nuclear extracts or recombinant protein and 2 μg/μL poly(dI-dC). (dI-dC) aliquots on ice (see Note 13).
2. Prepare a master mix containing 7 μL of DNA-binding buffer and 1 μL of 2 μg/μL poly(dI-dC).(dI-dC) for each reaction (include one extra tube for free probe control).
3. Aliquot 8 μL of the master mix for each reaction and add 2 μL of nuclear extract or recombinant protein. Gently mix by pipetting up and down and avoid introducing air bubbles.
4. For the free probe control, add 2 μL of buffer D.
5. Incubate the EMSA reaction tubes at room temperature for 15 min (see Note 14).

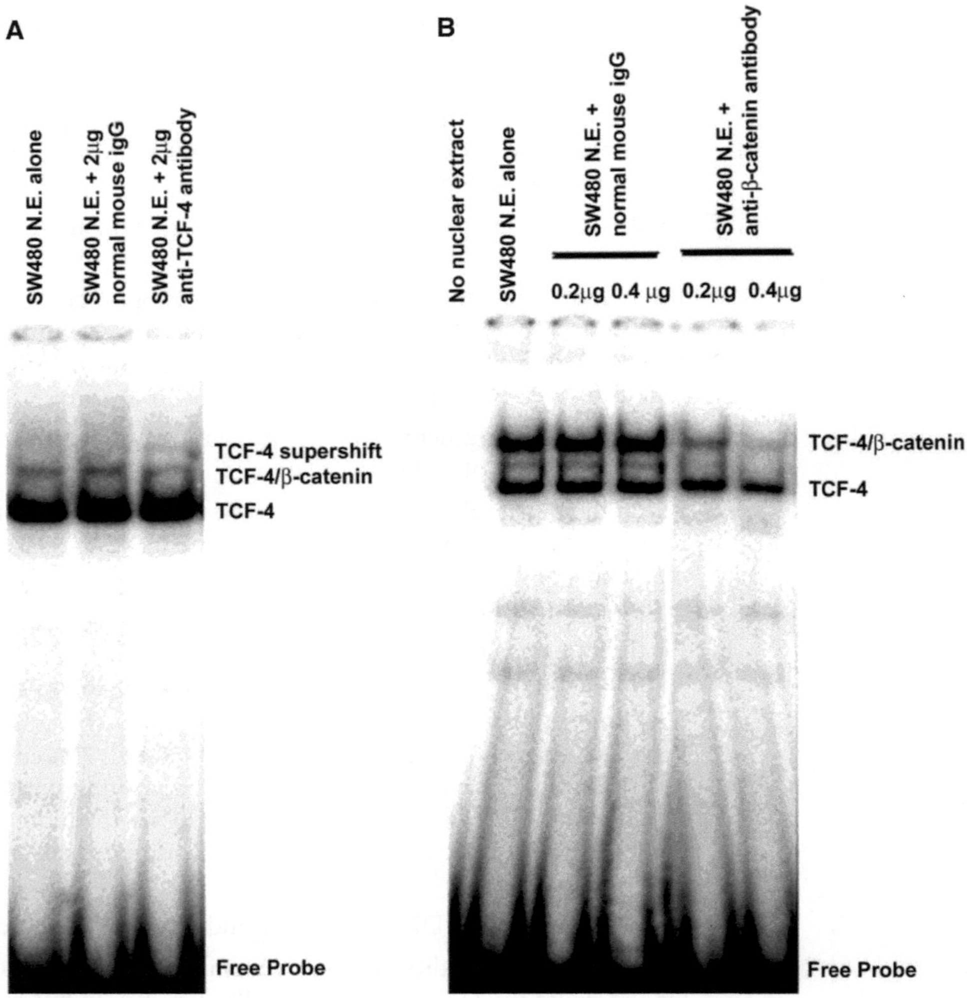

Fig. 4. EMSA and supershift assays of a potential TCF-4 binding site in the versican promoter. A double-stranded oligonucleotide corresponding to a potential TCF-4 binding site within the versican promoter was radiolabeled and incubated with nuclear extracts derived from the SW480 cell line. The indicated antibodies were added to the DNA-binding reaction prior to electrophoresis. (**a**) Anti-TCF-4 antibodies generated a supershifted complex while (**b**) anti-β-catenin antibodies interfered with the formation of the higher molecular weight complex present in the control reactions (reproduced from ref. (12) with permission from *J Biol Chem*).

6. Add 2 µL of radiolabeled DNA probe at 10,000 dpm/µL to each tube, again gently by pipetting up and down and avoiding air bubbles.
7. Incubate for a further 15 min.
8. Prerun the gel at 300 V for 10 min in 0.5× TBE.
9. Stop the gel and load 1× DNA loading buffer into and outside well. Load the free probe control and all other reactions into the appropriate wells without DNA loading buffer.

10. Run the gel approximately 1.5 h at 300 V.
11. Dry the gel on filter paper in a vacuum gel drier and expose to a phosphor-imaging screen or autoradiography film overnight (see Fig. 4).

3.2.4. Supershift EMSA

1. Assemble the EMSA reactions as described in Subheading 3.2.3, steps 1–4.
2. To one supershift EMSA reaction, add 0.5–1 μL (usually 102 μg) specific antibody to supershift reaction and nonspecific antibody to the control reaction (see Note 15).
3. Incubate at room temperature for 15 min, then add 2 μL of radiolabeled DNA probe at 10,000 dpm/μL to each tube and incubate for a further 15 min.
4. Load and electrophorese the reactions on a vertical 0.5× TBE nondenaturing 5% acrylamide gel as described for EMSA in Subheading 3.2.3, steps 8–11 (see Fig. 4; Notes 16 and 17).

4. Notes

1. There are many suitable transfection reagents available that can be used for these experiments. We used FuGENE®6 Transfection Reagent (Roche Applied Science) for the experiment described here. Using standard experimental conditions, 1 mL of FuGENE®6 Transfection Reagent will perform up to 300 transfections in 35-mm tissue culture dishes that contain HeLa, NIH3T3, COS-1, COS-7, or CHO-K1 cells. This is equivalent to over 1,200 transfections in 24-well plates.
2. At this time, you should check the activity without the added stimulants because versican promoter may be activated by the reagents themselves due to their potential damaging effects on transfected cells.
3. If you cotransfect an expression vector, such as β-catenin, with the versican reporter vector, the total amount of plasmid DNA should remain constant. If you investigate the effects of an expression vector for X protein on versican promoter, you should use the same expression vector without the gene for X protein as a negative control.
4. Incubation time before adding stimulants should be determined by each preliminary experiment. In the case of SMC (12), an incubation of 24–48 h before adding stimulants is needed to achieve an acceptable induction of the luciferase expression. According to our experience, a longer incubation time is better for measuring the induced luciferase activities, but not for measuring the basal luciferase activities.

5. Luciferase in the cell extract is not stable even when stored at −80°C. On the other hand, one freeze/thaw cycle provides better extraction from the cells. We recommend measuring the luciferase activity in freshly prepared samples after one freeze/thaw cycle. If stored samples are used, all samples in the set should be used because a measurement of luciferase activity on one day is different from that measured on another day.
6. Measuring the luciferase activity is sensitive to temperature. It is important to keep the samples at room temperature. The luminometer should be turned on at least 15 min before use.
7. Luciferin is light-sensitive and should be protected from light by wrapping container in foil.
8. It is important not to disturb the pellet; it is preferable to sacrifice yield to ensure the pellet is not disturbed and contaminates the extract.
9. Using narrower gel spacers improves elution efficiency; however, if the gel is too thin, it is difficult to work with and can easily tear.
10. Bands can be curved because of the relatively high salt content of the labeling reaction. If no column cleanup was performed, there may also be a "cloud" at the bottom of the gel due to the presence of unincorporated radionucleotides. Cut out the portion of the gel corresponding to the DNA fragment of interest. To maximize the recovery of DNA, macerate the gel fragment (we use a pipette tip) prior to the elution step.
11. The duration of the elution can be shortened by gently heating the elution mixture with rotation for 3 h at 45°C. If available, a hybridization oven is suitable.
12. If the gel fragment was macerated after being excised, filter the elution mixture through a spin filter (800 × *g* for 2 min) to remove any fine acrylamide fragments prior to adding ethanol. Samples contaminated with acrylamide will give inaccurate estimation of the amount of radioactivity and will also affect the electrophoresis.
13. The overall salt concentration in the EMSA reaction should be in the range of 100–120 mM (including all buffer salts, NaCl, KCl, etc.). The optimal salt concentration for a given protein must be determined empirically; however, concentrations outside this range are nonphysiological.
14. Incubation temperature may influence the binding reaction, particularly for supershift reactions. Performing the incubations on ice may stabilize the bound antibody-protein complex.
15. Care should be taken to ensure that the added antibody does not significantly alter the overall salt concentration of the reaction.

16. Antibody/protein reactions in supershift experiments may cause interference rather than a shifted band; if antibody binding sterically interferes with DNA–protein interactions, a shifted band normally seen in the absence of antibody may be attenuated or absent.

17. Shifts using nuclear extracts, in particular, may generate a number of different bands representing a variety of DNA/protein species. Binding reaction specificity can be determined by using a cold DNA competition; 10- to 100-fold excess unlabeled DNA probe is added to the binding reaction prior to addition of the labeled DNA probe. Bands that persist under these conditions are likely the result of nonspecific DNA/protein interactions.

Acknowledgments

This work was supported in part by a Grant-in-Aid from the Heart and Stroke Foundation of British Columbia and Yukon.

References

1. Kinsella, M.G., Bressler, S.L., and Wight, T.N. (2004) The regulated synthesis of versican, decorin, and biglycan: extracellular matrix proteoglycans that influence cellular phenotype. *Crit Rev Eukaryot Gene Expr.* **14**, 203–234.
2. Rahmani, M., Wong, B.W., Ang, L., Cheung, C.C., Carthy, J.M., Walinski, H., and McManus, B.M. (2006) Versican: signaling to transcriptional control pathways. *Can J Physiol Pharmacol.* **84**, 77–92.
3. Wight, T.N. and Merrilees, M.J. (2004) Proteoglycans in atherosclerosis and restenosis: key roles for versican. *Circ Res.* **94**, 1158–1167.
4. Rahmani, M., McDonald, P.C., Wong, B.W., and McManus, B.M. (2004) Transplant vascular disease: Role of lipids and proteoglycans. *Can J Cardiol.* **20**: 58B–65B.
5. Wight, T.N. (2008) Arterial remodeling in vascular disease: a key role for hyaluronan and versican. *Front Biosci.* **13**, 4933–4937.
6. Ricciardelli, C., Sakko, A.J., Ween, M.P., Russell, D.L., and Horsfall, D.J. (2009) The biological role and regulation of versican levels in cancer. *Cancer Metastasis Rev.* **28**, 233–245.
7. Morgenstern, D.A., Asher, R.A., and Fawcett, J.W. (2002) Chondroitin sulphate proteoglycans in the CNS injury response. *Prog Brain Res.* **137**, 313–332.
8. Zimmermann, D.R. and Dours-Zimmermann, M.T. (2008) Extracellular matrix of the central nervous system: from neglect to challenge. *Histochem Cell Biol.* **130**, 635–653.
9. Wu, Y.J., Lapierre, D., Wu, J., Yee, A.J., and Yang, B.B. (2005) The interaction of versican with its binding partners. *Cell Res.* **15**, 483–494.
10. Cowin, P. (1994) Unraveling the cytoplasmic interactions of the cadherin superfamily. *Proc Natl Acad Sci USA* **91**, 10759–10761.
11. Korinek, V., Barker, N., Morin, P.J., van Wichen, D., de Weger, R., Kinzler, K.W., et al. (1997) Constitutive transcriptional activation by a beta-catenin-Tcf complex in APC−/− colon carcinoma. *Science* **275**, 1784–1787.
12. Rahmani, M., Read, J.T., Carthy, J.M., McDonald, P.C., Wong, B.W., Esfandiarei, M., et al. (2005) Regulation of the versican promoter by the beta-catenin-T-cell factor complex in vascular smooth muscle cells. *J Biol Chem.* **280**, 13019–13028.
13. Read, J.T., Rahmani, M., Boroomand, S., Allahverdian, S., McManus, B.M., and Rennie, P.S. (2007) Androgen receptor regulation of the versican gene through an androgen response element in the proximal promoter. *J Biol Chem.* **282**, 31954–31963.
14. Farb, A., Kolodgie, F.D., Hwang, J.Y., Burke, A.P, Tefera, K., Weber, D.K., et al. (2004) Extracellular matrix changes in stented human coronary arteries. *Circulation* **110**, 940–947.

15. Lin, H., Ignatescu, M., Wilson, J.E., Roberts, C.R., Horley, K.J., Winters, G.L., et al. (1996) Prominence of apolipoproteins B, (a), and E in the intimae of coronary arteries in transplanted human hearts: geographic relationship to vessel wall proteoglycans. *J Heart Lung Transplant.* **15**, 1223–1232.

16. Lin, H., Wilson, J.E., Roberts, C.R., Horley, K.J., Winters, G.L., Costanzo, M.R., and McManus, B.M. (1996) Biglycan, decorin, and versican protein expression patterns in coronary arteriopathy of human cardiac allograft: distinctness as compared to native atherosclerosis. *J Heart Lung Transplant.* **15**, 1233–1247.

17. Lin, H., Kanda, T., Hoshino, Y., Takase, S., Kobayashi, I., Nagai, R., and McManus, B.M. (1998) Versican, biglycan and decorin protein expression patterns in coronary arteries: Analysis of primary and restenotic lesions. *Cardiovasc Pathol.* 7, 31–37.

18. Williams, K.J. and Tabas, I. (1995) The response-to-retention hypothesis of early atherogenesis. *Arterioscler Thromb Vasc Biol.* **15**, 551–561.

19. Williams, K.J. (2001) Arterial wall chondroitin sulfate proteoglycans: diverse molecules with distinct roles in lipoprotein retention and atherogenesis. *Curr Opin Lipidol.* **12**, 477–487.

20. Naso, M.F., Zimmermann, D.R., and Iozzo, R.V. (1994) Characterization of the complete genomic structure of the human versican gene and functional analysis of its promoter. *J Biol Chem;* **269**, 32999–33008.

21. Yeung, L.H., Read, J.T., Sorenson, P., Nelson, C.C., Gleave, M., Jia, W., and Rennie, P.S. (2003) Identification and characterization of a prostate-specific androgen independentprotein binding site in the probasin promoter. *Biochem J.* **371**, 843–855.

22. Thyberg, J. (1996) Differentiated properties and proliferation of arterial smooth muscle cells in culture. *Int Rev Cytol.* **169**, 183–265.

23. Campbell, J.H. and Campbell, G.R. (1993) Culture techniques and their applications to studies of vascular smooth muscle. *Clin Sci (Lond)* **85**, 501–513.

Chapter 4

Gene Silencing in Mouse Embryonic Stem Cells

Norihiko Sasaki and Shoko Nishihara

Abstract

Embryonic stem cells (ESCs) are promising tools for regenerative medicine as well as for biotechnological research. However, to exploit ESCs for clinical purposes, a better understanding of the molecular mechanisms that control the pluripotency and differentiation of ESCs is required. Several extrinsic signaling pathways contribute to the maintenance of pluripotency, as well as induction of differentiation, in ESCs. However, the mechanisms that regulate extrinsic signaling in ESCs are largely unknown. Heparan sulfate (HS) is present ubiquitously as a component of cell surface proteoglycans and is known to play crucial roles in the regulation of several signaling pathways. We have validated that RNA interference (RNAi) is a useful method for the functional analysis of some target genes in mouse ESCs (mESCs). Indeed, we have investigated the functions of HS in mESCs by using RNAi and have demonstrated that HS on mESCs is involved in regulating signaling pathways that are important for the maintenance of mESCs. In this chapter, we describe detailed methods for the gene silencing of proteoglycan-related genes in mESCs by RNAi.

Key words: RNA interference, Short hairpin RNA, Mouse embryonic stem cells, EXT1, Heparan sulfate

1. Introduction

RNA interference (RNAi) is an evolutionarily conserved phenomenon in which double-stranded RNA induces posttranscriptional gene silencing (1, 2). Recently, RNAi has been utilized to study gene function in many model organisms, including plants, *C. elegans*, and *D. melanogaster* (3, 4). The use of long double-stranded RNA molecules to induce RNAi can result in nonspecific silencing and a fatal interferon response in mammalian cells. However, short interfering RNA (siRNA) can be used to avoid such effects (5, 6). Recently, a method for the selection of highly effective siRNA sequences has been reported (7). Moreover, RNAi can be easily induced by endogenous expression of short hairpin RNAs (shRNAs)

Françoise Rédini (ed.), *Proteoglycans: Methods and Protocols*, Methods in Molecular Biology, vol. 836,
DOI 10.1007/978-1-61779-498-8_4,

using a DNA-based vector system. The shRNAs are processed intracellularly into siRNA (8). As a consequence, RNAi is used widely for the functional analysis of target genes in mammalian cells.

RNAi has advantages over gene knockout because, in addition to the direct effects of knocking out a gene, there might also be secondary effects that are caused by adaptation of the cells during long-term culture. For example, gene knockout might unexpectedly induce the expression of an unrelated gene that has secondary effects on the cells. In fact, the knockout of some genes that are related to the sulfation of heparan sulfate (HS), such as *6-O-endosulfatase*, *C5-epimerase*, and *HS2ST*, leads to a number of unexpected changes in the structure of sulfated glycosaminoglycans, presumably due to secondary effects (9–11). In addition, if the target gene is essential for cell survival and proliferation, it will be impossible to analyze the knockout cells due to cell death.

Embryonic stem cells (ESCs) (12, 13) are promising tools for biotechnology and possess key features that should allow their exploitation in clinical applications (14). Extrinsic signaling pathways are key mechanisms for determining the fate of ESCs, and sulfated glycans, such as HS, are well-known regulators of signal transduction (15). Previously, functional roles of HS chains in mouse ESCs (mESCs) have been demonstrated by using vector-based RNAi to either reduce HS chains or reduce their sulfation (16, 17). Therefore, RNAi is a useful method for the functional analysis of HS-related genes in mESCs. In this chapter, we take the *EXT1* gene, which is required for HS chain elongation (18), as an example and describe methods for the transient and stable knockdown of this gene in ESCs by vector-based RNAi.

2. Materials

2.1. Cell Culture

1. Mitomycin C (Sigma) is dissolved in phosphate-buffered saline (PBS) at 1 mg/mL and stored in single-use aliquots at –20°C.
2. Dulbecco's Modified Eagle's Medium (DMEM; Gibco) is stored at 4°C.
3. Fetal bovine serum (FBS; HyClone) for the culture of mESCs is stored at –20°C.
4. Penicillin/streptomycin (Gibco) is stored at 4°C.
5. 2-Mercaptoethanol (Gibco) is stored at 4°C.
6. Nonessential amino acids (Gibco) are stored at 4°C.
7. Leukemia inhibitory factor (LIF; Chemicon) is stored at 4°C.
8. Gelatin (Sigma) is dissolved in ultra pure water at 0.1%, autoclaved, and stored at room temperature.
9. 0.25% Trypsin-EDTA (Gibco) is stored at –20°C.

10. FBS (Gibco) for the culture of feeder cells and PLAT-E cells is stored at –20°C.
11. Brastcidine (Sigma) is dissolved in PBS at 1 mg/mL and stored in single-use aliquots at –20°C.
12. ESC medium : DMEM supplemented with 15% FBS, 1% penicillin/streptomycin, 0.1 mM 2-mercaptoethanol, and 0.1 mM nonessential amino acids.
13. PLAT-E medium : DMEM supplemented with 10% FBS, 1% penicillin/streptomycin, 10 μg/mL brastcidine, and 1 μg/mL puromycin.

2.2. Transient Transfection

1. pSilencer 3.1-H1 (Ambion) is stored at –20°C.
2. Lipofectamine 2000 (Invitrogen) is stored at 4°C.
3. Puromycin (Sigma) is dissolved in PBS at 1 mg/mL and stored in aliquots at –20°C.

2.3. Stable Transfection

1. pSUPER.retro.puro (OligoEngine) is stored at –20°C.
2. Polybrene (Sigma) is dissolved in PBS at 8 mg/mL and stored at 4°C.

2.4. Real-Time PCR

1. TRIZOL Reagent (Invitrogen) is stored at 4°C.
2. Oligo-dT primer (Invitrogen) is stored at –20°C.
3. The Superscript II First Strand Synthesis Kit (Invitrogen) is stored at –20°C.
4. Double-dye probe (Nippon EGT) is dissolved in ultra pure water at 100 μM and stored in aliquots at –20°C.
5. pGEM®-T Easy Vector (Promega) is stored at –20°C.

2.5. Fluorescence-Activated Cell Sorting (FACS)

1. EDTA (Sigma) is dissolved in PBS at 0.05% and stored at room temperature.
2. FACS buffer: 0.5% BSA and 0.1% sodium azide in PBS; stored at 4°C.
3. Primary antibodies: mouse IgM isotype control (Chemicon) and the anti-HS antibody 10E4 (Seikagaku Corp.).
4. Secondary antibody: FITC-conjugated anti-mouse IgM antibody (Sigma).

3. Methods

In general, the efficiency of transfection in mESCs is much lower than that of other mammalian cell lines. Therefore, it is necessary to select only transfected cells for accurate functional analysis and this can be achieved by using a vector that contains a drug-resistance gene. mESCs grow very rapidly; therefore a vector that has been introduced transiently into mESCs will decrease in abundance relatively soon after transfection. This will be followed by a decrease in the effects of the knockdown. For this reason, in the case of transient transfection, short-term functional analysis should be performed within 7 days.

mESCs are pluripotent cells and can be differentiated into three germ layers: endoderm, mesoderm, and ectoderm. This differentiation process takes many days. As a consequence, for functional analysis of the differentiation of mESCs, it is necessary to perform stable transfection. Stable transfection can be achieved quickly and efficiently by using a retroviral vector.

3.1. Transient Transfection of shRNA into mESCs

1. R1 (19) and E14TG2a (20) mESC lines are maintained on feeder cells in ESC medium with 1,000 U/mL LIF. Feeder cells are prepared from mouse embryonic fibroblasts (MEFs) (see Note 1) by treatment with 10 μg/mL mitomycin C and cultured on gelatin-coated 35-mm tissue culture dishes. Gelatin-coated dishes are prepared by adding sufficient 0.1% gelatin to cover the dish and then incubating the dish at room temperature for at least 10 min. The ESC medium is changed every day and mESCs are passaged every other day.
2. shRNA vectors are prepared as follows. The shRNA sequences used for RNAi are designed as described previously (7) using the online design site siDirect (http://sidirect.jp/esd/modules/modsiperfect/), which is based on an accelerated off-target search algorithm (21) (see Note 2). shRNA vectors are produced by inserting the appropriate dsDNAs between the BamHI and HindIII sites of pSilencer 3.1-H1. The resulting vectors are transformed into *Escherichia coli* and purified by using QIAGEN Plasmid Purification Kits (QIAGEN) in accordance with the manufacturer's protocol (see Note 3).
3. Prior to transfection, the mESCs are harvested after trypsinization with 0.25% Trypsin-EDTA, replated at a density of 1×10^6 cells/plate on gelatin-coated feeder-free 60-mm tissue culture dishes in ESC medium with LIF, and incubated for 16 h. Gelatin-coated dishes are prepared as described above (see Subheading 3.1, step 1).
4. On day 1, the cells are transfected with an shRNA expression vector (2 μg per culture dish) using Lipofectamine 2000 in accordance with the manufacturer's protocol.
5. On day 2 (24 h after transfection), the cells are harvested after trypsinization with 0.25% Trypsin-EDTA and replated at a density of 3×10^6 cells/plate on gelatin-coated feeder-free

60-mm tissue culture dishes in ESC medium with LIF and 2 μg/mL puromycin (see Note 4).

6. On day 3 (2 days after transfection), the transfected cells are harvested and analyzed as described below (see Note 5).

3.2. Stable Transfection of shRNA into mESCs

1. To produce retrovirus, the pSUPER.retro.puro constructs are transfected into ecotropic virus-packaging (PLAT-E) cells using Lipofectamine 2000 in accordance with the manufacturer's protocol. The pSUPER.retro.puro constructs are prepared in the same manner as the pSilencer 3.1-H1 constructs (see Subheading 3.1, step 2 and Note 3).
2. One day after transfection, the PLAT-E medium is changed to ESC medium without LIF. Prior to infection, the mESCs are harvested after trypsinization with 0.25% Trypsin-EDTA, replated at a density of 2.5×10^5 cells/plate on gelatin-coated feeder-free 60-mm tissue culture dishes in ESC medium with LIF, and incubated for 16 h.
3. After the PLAT-E cells have been cultured for 24 h, supernatants derived from the cultured cells, which contain virus, are mixed with 8 μg/mL polybrene and then mESCs are incubated with the virus/polybrene mixtures in medium that contains LIF.
4. Twenty-four hours after infection, the medium is replaced with fresh ESC medium that contains LIF.
5. From the next day, 2 μg/mL puromycin is added to the medium and the cells are cultured for 5–7 days for selection.
6. After selection, cells are used for analysis or stored until usage (see Note 6).

3.3. Real-Time PCR Analysis of the EXT1 Transcript in EXT1-Knockdown mESCs

1. Total RNA is isolated from transfected cells with TRIZOL Reagent and subsequently reverse-transcribed using an oligo-dT primer and a Superscript II First Strand Synthesis Kit in accordance with the manufacturer's protocol.
2. Standard plasmids that contain the *EXT1* or *β-actin* coding sequence are prepared by inserting the approximately 1-kb *EXT1* coding sequence or *β-actin* sequence into the pGEM®-T Easy Vector.
3. Real-time PCR is performed using an ABI PRISM® 7700 Sequence Detection System (Applied Biosystems). The absolute amount of *EXT1* or *β-actin* transcript in each prepared sample is calculated from a standard curve of *EXT1* or *β-actin* transcript, which is obtained by using the appropriate standard plasmid. The relative amounts of *EXT1* transcript are normalized to the *β-actin* transcript. Examples of this analysis in *EXT1*-knockdown mESCs are shown in Fig. 1. The single-stranded DNA sequences that were used to construct the shRNA vector used in this study are listed in Table 1. Primer sets and probes for real-time PCR are listed in Table 2.

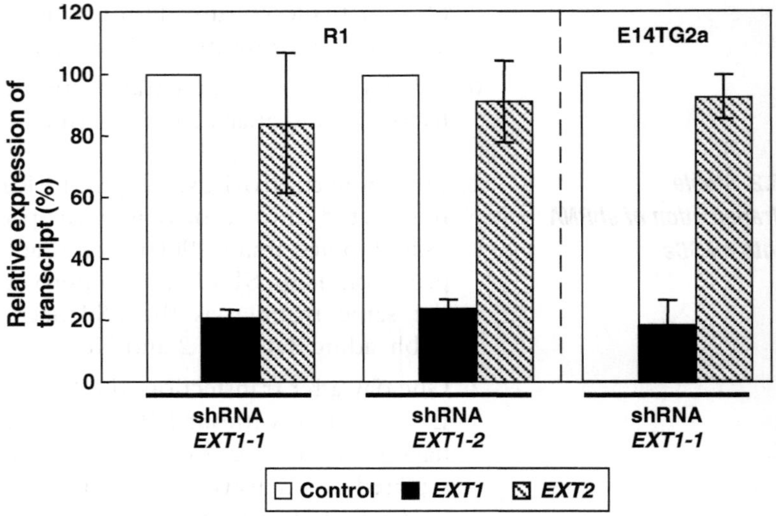

Fig. 1. *EXT1* shRNA induces efficient knockdown of the *EXT1* transcript in mESCs. The results of real-time PCR analysis of cells 2 days after transfection are shown. The results are shown as the amount of expression of the *EXT1* or *EXT2* transcript in the knockdown cells as a proportion (%) of that observed in control cells. The control cells are transfected with shRNA expression plasmids that target *EGFP*. The results show efficient knockdown of *EXT1* in mESCs. The level of the *EXT2* transcript was unaffected by the *EXT1* shRNAs, which confirmed the specificity of the targeting sequences. The values shown are the means ± SD of three independent experiments. We used two constructs expressing different shRNAs that targeted *EXT1* (*EXT1-1* and *EXT1-2*).

Table 1
The single-stranded DNA oligonucleotide sequence

Gene	
EGFP	5′-GATCCCGCCACAACGTCTATATCATGG**GGAAAAT**CCATGATATAGACGTTGTGGCTTTTTTGGAAA-3′
EXT1-1	5′-GATCCCGTCCTACCGCAGTATTCATCT**GCTTCCTGTCAC**AGATGAATACTGCGGTAGGACTTTTTTGGAAA-3′
EXT1-2	5′-GATCCCGGTCTATTCATCAGGATAAAA**GCTTCCTGTCAC**TTTTATCCTGATGAATAGACCTTTTTTGGAAA-3′

The *underlined* 21 nucleotides are sense and antisense siRNA target sequences, respectively. The *boldface* nucleotides are loop sequences

3.4. FACS Analysis to Quantitate HS on the Surface of EXT1-Knockdown mESCs

1. After washing with PBS, 0.05% EDTA is added to the transfected cells and the cells are incubated at 37°C for 20 min (see Note 7).
2. The cells are harvested and the resulting cell suspension is incubated with primary antibodies diluted in FACS buffer on ice for 30 min.

Table 2
The list of gene-specific primers and probes for real-time PCR

Gene	Forward primer	Reverse primer	Probe[a]
EXT1	CAGGCTTGGGT CCTTCAGATT	CCATCCGTTGCT GAGCATT	CTTTGCAGGCTGCCTGTGT CCCTG
EXT2	CCCACAGAGGC AGATTGAAGA	TGGCTTTAATGGAC TGGAAGTATG	CAGAACCACCGTGCCTGTCTC TGCAT

[a]The probe was labeled at the 5′-end with the reporter dye, 3FAM, and at the 3′-end with the quencher dye, TAMRA (applied biosystems)

3. After washing with FACS buffer, the cell suspension is incubated with FITC-conjugated secondary antibody diluted in FACS buffer on ice for 30 min in the dark.
4. After washing with FACS buffer, cell sorting and analysis are performed using a FACSAria Cell Sorter (Becton Dickinson). Examples of this analysis in *EXT1*-knockdown mESCs are shown in Fig. 2.

4. Notes

1. MEFs should be isolated from E14.0 mouse embryos.
2. This site enables highly effective targeting sequences that will exclude off-target effects to be designed easily. In general, a few sequences should be tested in order to avoid off-target effects. In this chapter, we used two targeting sequences for the *EXT1* gene that had been designed using this site.
3. Purified shRNA vectors are stored in single-use aliquots at −20°C.
4. In general, puromycin selection of transfected cells is carried out for 24 h. The transfection efficiency of the cells is approximately 60%, as confirmed by FACS analysis after transfection of an expression vector for enhanced green fluorescent protein (EGFP; data not shown), but only transfected cells survive after selection with puromycin.
5. After puromycin selection, almost all cell types may show occasional cell death. In the case of cell death, the purity of the shRNA vector may be affected. Therefore, the shRNA vector should be reprepared from retransformed *Escherichia coli*.
6. After frozen cells have been thawed, the cells should be passaged 2–3 times before analysis.

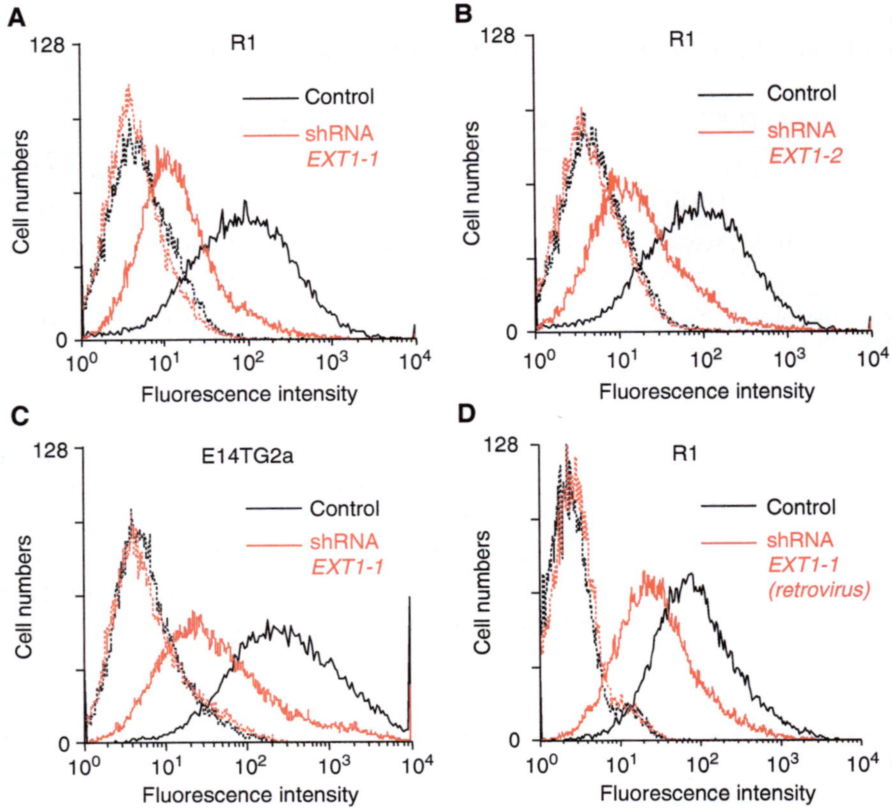

Fig. 2. *EXT1* shRNA induces a reduction in the amount of HS on mESCs. The results of FACS analysis of cells 2 days after transfection (**a**–**c**) or 7 days after infection (**d**) using an anti-HS antibody are shown (the *dotted lines* represent IgM isotype controls for the control and *EXT1*-knockdown cells, respectively). (**a**–**d**) Show the results obtained in transiently transfected cells and stably transfected cells, respectively. Control cells are transfected with shRNA expression plasmids that target *EGFP*. These flow cytometry plots show clearly the effect of knockdown of *EXT1* by RNAi on the synthesis of HS in mESCs.

7. mESCs are usually harvested after trypsinization with 0.25% Trypsin-EDTA. However, trypsinization reduces the expression of some cell surface antigens. Therefore, for FACS analysis, the cells are harvested after treatment with only EDTA.

Acknowledgments

We thank Prof. Toshio Kitamura and Prof. Kumiko Ui-Tei for gifting experimental materials. Our research was partially supported by funds from the MEXT, a Grant-in-Aid for Scientific Research (B) to SN, 20370051, 2008–2010, and from MEXT, the Matching Fund for Private Universities, S0901015, 2009–2014.

References

1. Fire, A., Xu, S., Montgomery, M. K., Kostas, S. A., Driver, S. E., and Mello, C. C. (1998) Potent and specific genetic interference by double-stranded RNA in Caenorhabditis elegans. *Nature* **391**, 806–811.
2. Almeida, R. and Allshire, R. C. (2005) RNA silencing and genome regulation. *Trends Cell Biol.* **15**, 251–258.
3. Giordano, E., Rendina, R., Peluso, I., and Furia, M. (2002) RNAi triggered by symmetrically transcribed transgenes in Drosophila melanogaster. *Genetics* **160**, 637–648.
4. Ueda, R. (2001) Rnai: a new technology in the post-genomic sequencing era. *J. Neurogenet.* **15**, 193–204.
5. Elbashir, S. M., Harborth, J., Lendeckel, W., Yalcin, A., Weber, K., and Tuschl, T. (2001) Duplexes of 21-nucleotide RNAs mediate RNA interference in cultured mammalian cells. *Nature* **411**, 494–498.
6. Stark, G. R., Kerr, I. M., Williams, B. R., Silverman, R. H., and Schreiber, R. D. (1998) How cells respond to interferons. *Annu. Rev. Biochem.* **67**, 277–264.
7. Ui-Tei, K., Naito, Y., Takahashi, F., Haraguchi, T., Ohki-Hamazaki, H., Juni, A., et al. (2004) Guidelines for the selection of highly effective siRNA sequences for mammalian and chick RNA interference. *Nucleic Acids Res.* **32**, 936–948.
8. Brummelkamp, T. R., Bernards, R., and Agami, R. (2002) A system for stable expression of short interfering RNAs in mammalian cells. *Science* **296**, 550–552.
9. Li, J. P., Gong, F., Hagner-McWhirter, A., Forsberg, E., Abrink, M., Kisilevsky, R., et al. (2003) Targeted disruption of a murine glucuronyl C5-epimerase gene results in heparan sulfate lacking L-iduronic acid and in neonatal lethality. *J. Biol. Chem.* **278**, 28363–28366.
10. Merry, C. L., Bullock, S. L., Swan, D. C., Backen, A. C., Lyon, M., Beddington, R. S., et al. (2001) The molecular phenotype of heparan sulfate in the Hs2st−/− mutant mouse. *J. Biol. Chem.* **276**, 35429–35434.
11. Lamanna, W. C., Baldwin, R. J., Padva, M., Kalus, I., Ten Dam, G., van Kuppevelt, T. H., et al. (2006) Heparan sulfate 6-O-endosulfatases: discrete in vivo activities and functional co-operativity. *Biochem. J.* **400**, 63–73.
12. Evans, M. J. and Kaufman, M. H. (1981) Establishment in culture of pluripotential cells from mouse embryos. *Nature* **292**, 154–156.
13. Martin, G. R. (1981) Isolation of a pluripotent cell line from early mouse embryos cultured in medium conditioned by teratocarcinoma stem cells. *Proc. Natl. Acad. Sci. U.S.A.* **78**, 7634–7638.
14. Keller, G. (2005) Embryonic stem cell differentiation: emergence of a new era in biology and medicine. *Genes. Dev.* **19**, 1129–1155.
15. Bernfield, M., Gotte, M., Park, P. W., Reizes, O., Fitzgerald, M. L., et al. (1999) Functions of cell surface heparan sulfate proteoglycans. *Annu. Rev. Biochem.* **68**, 729–777.
16. Sasaki, N., Okishio, K., Ui-Tei, K., Saigo, K., Kinoshita-Toyoda, A., Toyoda, H., et al. (2008) Heparan Sulfate Regulates Self-renewal and Pluripotency of Embryonic Stem Cells. *J. Biol. Chem.* **283**, 3594–3606.
17. Sasaki, N., Hirano, T., Ichimiya, T., Wakao, M., Hirano, K., Kinoshita-Toyoda, A., et al. (2009) The 3′-phosphoadenosine 5′-phosphosulfate transporters, PAPST1 and 2, contribute to the maintenance and differentiation of mouse embryonic stem cells. *PLoS One* **4**, e8262.
18. Lind, T., Tufaro, F., McCormick, C., Lindahl, U., and Lidholt, K. (1998) The putative tumor suppressors EXT1 and EXT2 are glycosyltransferases required for the biosynthesis of heparan sulfate. *J. Biol. Chem.* **273**, 26265–26268.
19. Nagy, A., Rossant, J., Nagy, R., Abramow-Newerly, W., and Roder, J. C. (1993) Derivation of completely cell culture-derived mice from early-passage embryonic stem cells. *Proc. Natl. Acad. Sci. USA* **90**, 8424–8428.
20. Smith, A. G. and Hooper, M. L. (1987) Buffalo rat liver cells produce a diffusible activity which inhibits the differentiation of murine embryonal carcinoma and embryonic stem cells. *Dev. Biol.* **121**, 1–9.
21. Yamada, T. and Morishita, S. (2005) Accelerated off-target search algorithm for siRNA. *Bioinformatics* **21**, 1316–1324.

Chapter 5

A Novel Strategy for a Splice-Variant Selective Gene Ablation: The Example of the Versican V0/V2 Knockout

María T. Dours-Zimmermann and Dieter R. Zimmermann

Abstract

The complete knockout of genes that give rise to alternative splice products can often provide only an integral view of the dominant function(s) of all the isoforms they encode. If one of these isoforms is indispensable for life, a constitutive and complete inactivation may even preclude any in vivo studies of later expressed splice-variants in mice. To explore function of the tissue-restricted versican V2 isoform during central nervous system maturation, for instance, we had to circumvent the early embryonic lethality of the complete knockout by employing a novel splice-variant-specific gene ablation approach. For this purpose, we introduced a preterm translational stop codon preceded by an ER-retention signal (KDEL) into the alternatively spliced exon 7 of the VCAN gene. This way the synthesis of the V2- and the V0-forms of the proteoglycan was entirely abolished in the mutant mice, most likely mediated by a KDEL-promoted intracellular degradation of the mutant fragment and by a nonsense-mediated decay mechanism. The expression of the vitally important V1-isoform and the smallest V3-variant remained, however, unaffected. Here we provide the details of our targeting strategy, the screening procedure, the generation of isoform-specific antibodies, and the transcript analysis and we supply the experimental protocols for the biochemical and immunohistological examinations of the mutant mouse strain Vcan$^{tm1.1Dzim}$.

Key words: Knock-in, Alternative splicing, Gene inactivation, Proteoglycan, Versican, Hyaluronan, Hyaluronidase, Antibodies, Immunofluorescence, PCR

1. Introduction

Most of the extracellular matrix (ECM) proteins are multitask molecules that fulfill their specific functions on the basis of their modular compositions. Some of these modules may participate in the supramolecular assembly providing tissue stability; others might be required for cell adhesion and migration or control proliferation and differentiation by gathering specific signaling molecules. It is therefore not surprising that several ECM proteins are indispensable for

Françoise Rédini (ed.), *Proteoglycans: Methods and Protocols*, Methods in Molecular Biology, vol. 836,
DOI 10.1007/978-1-61779-498-8_5, © Springer Science+Business Media, LLC 2012

normal development (1). Consequently, a full elimination of their expression, e.g., by a constitutive gene ablation, leads frequently to pre- or perinatal lethality. Although certain information about the roles of a particular protein may be drawn from such a knockout mouse model, more subtle approaches are necessary to gain deeper functional insights beyond the critical developmental stage. Among these strategies, conditional and/or inducible knockouts or knock-downs allowing a spatially and timely limited suppression of specific gene expressions have recently been described in detail in this book series (2). Alternatively, knock-in experiments targeting only a small portion or an alternatively spliced exon may circumvent an early lethal phenotype and provide detailed knowledge about particular domains and isoforms.

For example, the complete knockout of all four splice-variants of the large aggregating chondroitin sulfate proteoglycan versican leads to early intrauterine death of the embryos around E10.5 due to a heart defect (3). To be anyhow able to analyze the role of the central nervous system-specific variant V2 during peri- and postnatal brain development (4), we have designed a knock-in strategy for an isoform-restricted gene ablation (5). We anticipated that by only eliminating versican V2 (and alongside V0 owing to the gene constellation) (see Fig. 1), we would not interfere with normal heart development, which seems to depend mainly on the V1 variant. To omit a compensatory up-regulation of the remaining isoforms (V1 and V3) and to reduce the risk of potential splicing errors, we aimed primarily at the translation of the V2 core protein while trying to affect the overall transcription of the versican gene (*VCAN*) as little as possible. For this purpose, we have introduced an ER-retention signal (KDEL) and an early translational termination codon into the center of the alternatively spliced exon 7. The rational for incorporating the ER-retention signal was to prevent the potential secretion of an artificial N-terminal proteoglycan-rudiment and to facilitate the intracellular degradation of such a putative product of the mutant allele. Since the markers used for ES clone selection could be subsequently removed by CRE recombination, the gene modification in the mutant mice was minimal at the end.

To test the effectiveness of our splice-variant-specific gene ablation, we established analyses of the different versican transcripts and generated domain-specific polyclonal antibodies to distinguish between the alternative core proteins in comparative Western blotting and immunohistological staining experiments. Some of these antibodies have been devised to detect exclusively the V0 and V2 isoforms and to potentially recognize also the putative truncation product of the mutated gene. Here, we have now detailed the strategy and protocols for the generation and analysis of the versican V0/V2 knockout mouse. For more general information about the work with ES cells, the blastocyst injection, and the generation of mouse chimera, we refer the reader to a previous article of *Methods in Molecular Biology* (6).

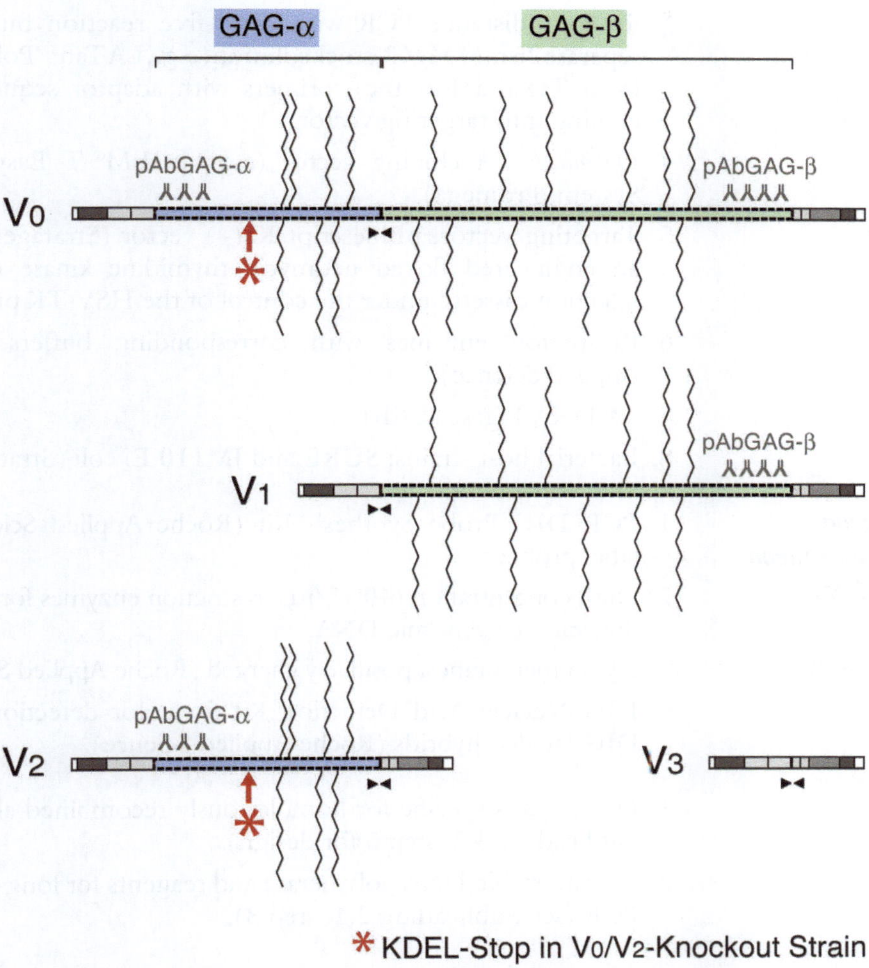

Fig. 1. Structural models of the isoform V0 to V3 of mouse versican. *Wavy lines* represent glycosaminoglycan (GAG) side chains. The alternatively spliced GAG attachment domains are delimited on the top. The site of modification and translational termination in the V0/V2 knockout strain is marked with an *asterisk*. The binding positions of domain-specific polyclonal antibodies are depicted above the core proteins. *Arrowheads* show the location of the primers used for isoform-specific RT-PCR.

2. Materials

2.1. Preparation of Splice-Variant-Specific Targeting Construct

1. OLIGO Primer Analysis Software (Molecular Biology Insights, Inc) for primer design.
2. Thermocycler (e.g., GeneAmp® PCR System 9700, Applied Biosystems).
3. PCR reagents: DNase/RNase-free Water (e.g., UltraPure™ from Invitrogen), nucleotide mix (e.g., 10 mM PCR Nucleotide Mix, Roche Applied Science), thermostable DNA polymerase

for long-distance PCR with Mg^{2+}-free reaction buffer and separate 25 mM $MgCl_2$ stock solution (e.g., LA Taq™ Polymerase from TaKaRa Bio Inc), primers with adaptor sequence for cloning into targeting vector.

4. *Optional*: TA cloning vector (e.g., pGEM®-T Easy Vector System, Promega).
5. Targeting vector: pBluescript KS(+) vector (Stratagene) with an engineered floxed neomycin–thymidine kinase (neo–tk) selection cassette under the control of the HSV-TK promoter.
6. Restriction enzymes with corresponding buffers (Roche Applied Science).
7. T4 DNA Ligase (USB).
8. Bacterial host strains: SURE and JM110 E. coli (Stratagene).

2.2. Screening for Target Integration in ES Cell Clones and Mice

2.2.1. Southern Blotting

1. PCR DIG Probe Synthesis Kit (Roche Applied Science) to label probe.
2. High concentration (40 U/μL) restriction enzymes for efficient digestion of genomic DNA.
3. Nylon membranes, positively charged (Roche Applied Science).
4. DIG Nucleic Acid Detection Kit for color detection of the DIG-labeled hybrids (Roche Applied Science).

2.2.2. Long-Distance PCR

1. Primer pairs specific for homologously recombined allele (see Subheading 3.2, step 5 for design).
2. Thermostable DNA polymerase and reagents for long-distance PCR (see Subheading 2.1, step 3).

2.2.3. Standard PCR (to Detect the Mutant Allele After Removal of Selection Cassette)

1. Primers specific for gene sequences flanking the unique loxP site.
2. Hot start TaqDNA polymerase and reagents (e.g., AmpliTaq Gold® with PCR Buffer II and $MgCl_2$ Solution, Applied Biosystems; 10 mM PCR Nucleotide Mix, Roche Applied Science; UltraPure™ water, Invitrogen).
3. PCR-compatible extraction buffer to isolate DNA from mouse biopsies: 50 mM Tris pH 8.5, 1 mM EDTA, 0.5% Tween-20 containing proteinase K, sterile-filtered.

2.3. Analysis of Alternatively Spliced Transcripts in KO Mouse Strain by Quantitative RT-PCR

2.3.1. Isolation of Total RNA

1. Tissue homogenizer with disposable probes (e.g., Omni *TH* with Omni Tips™).
2. RNeasy Protect kit (Qiagen).
3. RNase-Free DNase Set (Qiagen).

2.3.2. Quantitative RT-PCR (qRT-PCR)

1. Real-time PCR system (e.g., ABI PRISM 7900 Sequence Detection System).
2. RT-PCR kit for qRT-PCR (e.g., QuantiTect® SYBR® Green RT-PCR kit; Qiagen).
3. Primers covering specific splice junctions of mRNA of interest.
4. Primers specific for transcript of house-keeping gene.

2.4. Analysis of Alternatively Spliced Transcripts by Nonradioactive Northern Blotting

2.4.1. Isolation of Total RNA

1. Tissue homogenizer.
2. Trizol® reagent or acid phenol method (7).

2.4.2. Preparation of Riboprobes

1. Thermocycler, PCR reagents (see Subheading 2.1).
2. TA cloning vector (e.g., pGEM®-T Easy Vector System, Promega).
3. Primers to amplify cDNA template for riboprobe preparation (lower primer supplemented with inverse complementary T7-promoter sequence for generation of anti-sense riboprobes).
4. DIG RNA Labeling Kit (SP6/T7) for RNA labeling with digoxigenin-UTP by in vitro transcription with T7 RNA polymerase (Roche Applied Science).

2.4.3. Northern Blotting

1. 0.24–9.5 kb RNA ladder.
2. Nylon membranes, positively charged.
3. Formaldehyde gel and all standard reagents for RNA blotting.
4. Methylene Blue staining solution: 0.04% methylene blue, 500 mM sodium acetate, pH 5.2.
5. DIG Easy Hyb Granules (Roche Applied Science) for membrane hybridization using DIG-labeled probe.
6. DIG Nucleic Acid Detection Kit for color detection of the DIG-labeled hybrids (Roche Applied Science).

2.5. Generation of Domain-Specific Antibodies for Splice-Variant Identification

2.5.1. Preparation of Recombinant Antigen

1. Modified primers to amplify cDNA of selected antigen.
2. PCR or RT-PCR reagents, restriction enzymes, and T4 ligase used for cloning.
3. Bacterial expression vector (e.g., pQE-30 Xa, Qiagen) with a 6xHis tag (*and optionally, the codon for the Factor Xa protease recognition sequence IEGR*) 5′ to the cloning site (N-terminal tag).
4. Bacterial expression system with bacterial host strain, Ni^{2+}-NTA agarose, columns and buffers for purification, etc. (e.g., QiaExpress Kit, Qiagen).

2.5.2. Affinity-Purification of Polyclonal Antibodies

1. Two HiTrap™ NHS-activated HP Columns (G&E Healthcare, formerly Pharmacia) for antigen coupling.
2. Coupling buffer: 200 mM $NaHCO_3$, 500 mM NaCl, pH 8.3.
3. Blocking buffers:
   (a) 500 mM ethanolamine, 500 mM NaCl, pH 8.3.
   (b) 100 mM acetic acid, 500 mM NaCl, pH 4.
4. High salt buffer: 10 mM Tris, 500 mM NaCl, pH 7.5.
5. Antibodies elution buffer I (acid): 100 mM glycine, pH 2.5.
6. Antibodies elution buffer II (basic): 100 mM triethylamine, pH 11.5, prepare fresh!
7. Neutralization buffers: 10 mM Tris pH 8.8 and pH 7.5.
8. Saturated ammonium sulfate solution to concentrate eluates.
9. 10% BSA in PBS (10× stock).
10. 2% NaN_3 stock solution.

2.6. Protein Analysis of Tissue Extracts from Knockout Mice by SDS-PAGE and Immunoblotting

1. Tissue homogenizer (e.g., Polytron® PT2100 from Kinematica).
2. Ultracentrifuge (e.g., Beckman® TL100 table top ultracentrifuge).
3. Protein extraction buffer (8): 0.5 M NaCl, 50 mM Tris pH 7.5, 25 mM EDTA, 0.5% Igepal® (NP-40), 1 mM Pefabloc (Pentapharm or Roche Applied Science), plus 1 tablet of Complete® protease inhibitors per 50 mL buffer (Roche Applied Science).
4. BCA Protein Assay (Thermo Scientific Pierce).
5. Chondroitinase buffer: 40 mM Tris–HCl, 40 mM Na acetate, 10 mM EDTA, 1 mM Pefabloc, pH 8.0, plus 1 tablet of Complete® protease inhibitors per 50 mL buffer.
6. Chondroitinase ABC Protease-Free (Seikagaku No. 100332, or MP Biomedicals No. 08320302). Dissolve lyophilized enzyme (stored at −80°C) at a concentration of 10 mU/μL in 50 mM Tris pH 8.0. After reconstitution, aliquot and keep at −20°C. Use each time freshly thawed aliquot.
7. Leupeptin (Roche Applied Science): prepare a 1,000× stock solution in water (2 mg/mL), aliquot and store at −20°C.
8. Pepstatin (Roche Applied Science) prepare a 1,000× stock solution in methanol (1 mg/mL), aliquot and store at −20°C.
9. PhastSystem™ Separation-Control Unit with 4–15% PhastGels®, PhastGel Buffer Strips-SDS, and sample applicators (G&E Healthcare).
10. Immobilon-P blotting membrane (Millipore) or other PVDF membrane, 0.45 μm.
11. Gel loading buffer, 10× stock solution: 100 mM Tris–HCl, 10 mM EDTA, 20% SDS, 0.1% bromophenol blue, pH 8.0.

12. Precision Plus Protein dual color standards (Bio-Rad).
13. TBST buffer: 10 mM Tris–HCl, 150 mM NaCl, 0.05% Tween20, pH 8.0.
14. Blocking buffer: 3% dry low-fat milk in TBS buffer.
15. Alkaline phosphatase (AP)-conjugated secondary antibodies (Jackson Immuno Research):
    - Goat anti-Mouse IgG (H + L) AP-conjugated.
    - Goat anti-Rabbit IgG (H + L) AP-conjugated.
    - Donkey anti-Guinea Pig IgG (H + L) AP-conjugated
16. Alkaline phosphatase AP-buffer: 10 mM Tris–HCl, 10 mM NaCl, 5 mM $MgCl_2$, pH 9.5.
17. Western Blue® Stabilized Substrate for Alkaline Phosphatase (Promega).

2.7. Collection and Fixation of Tissues for Histological Characterization

1. Fixation buffer: 4% PFA (paraformaldehyde) in PBS (prepare always freshly before use). Dissolve 4 g PFA in 50 mL distilled H_2O, add 2 drops of 10 N NaOH; stir with magnetic stirrer while heating to 60°C; continue until solution becomes clear; add 50 mL 2× PBS buffer; adjust pH to 7.3.
2. Ethanol series (70, 96, and 100% ethanol in water), xylene and paraffin (Paraplast® Tissue Embedding Medium, Leica No. 39501006).

2.8. Preparation of Paraffin Sections for Immunostaining

1. Rotary Microtome (e.g., Leica RM2235).
2. Positively charged glass slides (Super Frost Plus®) and cover slips (Menzel-Gläser, Thermo Scientific).
3. Staining dishes with glass racks for histological slides.
4. Controlled antigen retrieval device (see Note 1) or a medium-size (2 L) stainless steel household pressure-cooker and electrical heating plate.
5. Antigen retrieval buffers, prepared freshly from 10× stocks:
   (a) 10 mM tri-sodium citrate buffer, pH 6.0.
   (b) 2 mM Tris, 1.7 mM EDTA, 1 mM tri-sodium citrate, pH 7.8 (TEC buffer).

2.9. Preparation of Frozen Sections for Immunostaining

1. Cryostat (e.g., Leica CM3050 S cryostat).
2. Positively charged glass slides and cover slips (same as for paraffin sections).

2.10. Characterization of KO Mouse Strain by Immunohistochemistry

1. Slide staining tray with lid (e.g., StainTray™ Simport Scientific or Immuno slide staining tray from Thermo Scientific).
2. PBS buffer.

3. Blocking buffer for immunohistochemistry: 3% Normal Donkey Serum (Jackson ImmunoResearch) in PBS.
4. *Optional for immunohistochemistry*: Mouse-On-Mouse Immunodetection Kit, M.O.M.™ Kit (Peroxidase) (Vector Laboratories).
5. Hydrogen peroxide 30% (H_2O_2) (*harmful! avoid contact to eyes or skin*) (Merck).
6. Secondary antibodies for immunohistochemistry (all from Jackson ImmunoResearch):
   - Biotin-SP-conjugated donkey anti-rabbit IgG (H + L).
   - Biotin-SP-conjugated donkey anti-goat IgG (H + L).
   - Biotin-SP-conjugated donkey anti-guinea pig IgG (H + L).
7. VECTASTAIN® Elite ABC (Avidin-Biotin Complex) Reagent (Vector Laboratories).
8. Peroxidase substrate solution: dissolve 30 mg 3,3′-diaminobenzidine tetrahydrochloride hydrate (DAB, *harmful! avoid contact to eyes or skin*; also available as ready-to-use tablets from Sigma) in 50 mL of 50 mM Tris pH 7.5 and add 100 μL H_2O_2.
9. Mayer's hematoxylin solution, filtered before use (*toxic! protect eyes, respiratory tract and skin*).
10. Glycergel® Mounting Medium (Dako) (warm up to 37°C before use).

2.11. Characterization of KO Mouse Strain by Immunofluorescence

1. Slide staining tray with lid (e.g., StainTray™ Simport Scientific or Immuno slide staining tray from Thermo Scientific). Use light-protective dark cover for immunofluorescence stainings!
2. Fluorescence microscope equipped with digital camera and image analysis software (we generally use an Olympus BX61 microscope, an F-View camera, and the AnalySISPro software).
3. PBS.
4. Blocking buffer for immunofluorescence: 0.5% bovine serum albumin (BSA), 0.2% gelatine, 0.02% sodium azide in PBS, sterile-filtered.
5. *Optional for immunofluorescence*: Mouse Ig Blocking Reagent (Vector Laboratories).
6. Secondary antibodies for double immunofluorescence (all from Invitrogen; see Note 2):
   - Alexa® Fluor-488 goat anti-rabbit IgG (H + L) highly cross-adsorbed, combined with.
   - Alexa® Fluor-594 goat anti-mouse IgG (H + L) highly cross-adsorbed.

- For triple immunofluorescence stainings also including primary antibodies generated in goat:
- Alexa® Fluor 488 donkey anti-rabbit IgG (H + L), combined with
- Alexa® Fluor 555 donkey anti-goat IgG (H + L)
- Alexa® Fluor 647 donkey anti-mouse IgG (H + L)

7. Hoechst H33258 pentahydrate (bis-benzimide) solution.
8. Fluorescent Mounting Medium (Dako) (stored at 4°C, warm up to room temperature before use).

2.12. Cleavage of Hyaluronan in Histological Sections

1. Slide staining tray with lid (e.g., StainTray™ Simport Scientific or Immuno slide staining tray from Thermo Scientific).
2. Hyaluronidase digestion buffer: 30 mM sodium acetate trihydrate, 125 mM sodium chloride, pH 5.2. Just before using, add one Complete® Protease Inhibitor cocktail tablet (Roche Applied Science) per 50 mL buffer.
3. Hyaluronate Lyase from Streptomyces hyalurolyticus (Sigma; see Note 3). Prepare a 100 U/mL stock solution in digestion buffer without protease inhibitors, aliquot and keep frozen. Prior to application on the slides, dilute 1:10 in digestion buffer supplemented with Complete® protease inhibitors.

2.13. Staining of Hyaluronan in Histological Sections

1. Slide staining tray with lid (e.g., StainTray™ Simport Scientific or Immuno slide staining tray from Thermo Scientific).
2. PBS.
3. Blocking buffer for immunohistochemistry: 3% Normal Donkey Serum (Jackson ImmunoResearch) in PBS.
4. Hydrogen peroxide 30% (H_2O_2) (*harmful! avoid contact to eyes or skin*) (Merck).
5. Biotinylated Hyaluronic Acid-Binding Protein (b-HABP; Seikagaku). Prepare fresh solutions of b-HABP in PBS (0.5 μg/mL).
6. VECTASTAIN® Elite ABC (Avidin-Biotin Complex) Reagent (Vector Laboratories).
7. Peroxidase substrate solution: dissolve 30 mg 3,3′-diaminobenzidine tetrahydrochloride hydrate (DAB, *harmful! avoid contact to eyes or skin*; also available as ready-to-use tablets from Sigma) in 50 mL of 50 mM Tris pH 7.5 and add 100 μL H_2O_2.
8. Mayer's hematoxylin solution, filtered before use (*toxic! protect eyes, respiratory tract and skin*).
9. Glycergel® Mounting Medium (Dako) (warm up to 37°C before use).

3. Methods

3.1. Preparation of Splice-Variant-Specific Targeting Construct

To keep the modifications at the knock-in site minimal, we use a targeting vector containing a floxed neomycin–thymidine kinase (neo–tk) selection cassette under the control of the HSV-TK promoter (see Fig. 2). This combination of marker genes allows positive selection with G418 for DNA integration and subsequent negative selection with ganciclovir for CRE-loxP-mediated removal of the marker genes in homologously recombined embryonic stem (ES) cell clones (see Note 4).

1. Choose position in target exon, where ER-retention signal and early translational stop codon will be introduced.
2. Design primer pairs to amplify two long genomic DNA fragments (usually 4–5 kb) extending upstream and downstream from this position within this exon.
3. For the amplification of the upstream fragment, include at the 5′ end of the upper strand primer (a) unique restriction site(s) suitable for cloning and later linearization of the targeting construct (in our example NotI; Fig. 2). Add to the 5′ end of the lower primer the (inverse complementary) codons for KDEL-sequence, the translational stop and a restriction site compatible with the 5′-cloning site of the targeting vector (in our example BstZI).
4. For the amplification of the downstream fragment, supplement the primers with the corresponding restriction site sequences (in our example ClaI).
5. Amplify these fragments by long-distance PCR from genomic DNA of mice syngeneic to the stem cell origin.
6. Clone the fragments via a TA cloning vector or directly into the targeting vector following restriction with the corresponding enzymes.
7. Verify targeting construct by sequencing.
8. Linearize construct and electroporate ES cells.

3.2. Strategy to Screen for Target Integration in ES Cell Clones and Mice (General Considerations)

The Southern blot analysis is still the most reliable method to screen for homologously recombined alleles in ES clones and mice (see Fig. 3). For this purpose, application of a nonradioactive detection system with digoxigenin (DIG)-labeled probes is highly recommended. Alternatively, long-distance PCR approaches are suitable, which allow the detection of recombined alleles, even when both of the homologous sequences of the targeting construct exceed lengths of 4–5 kb (see Note 5).

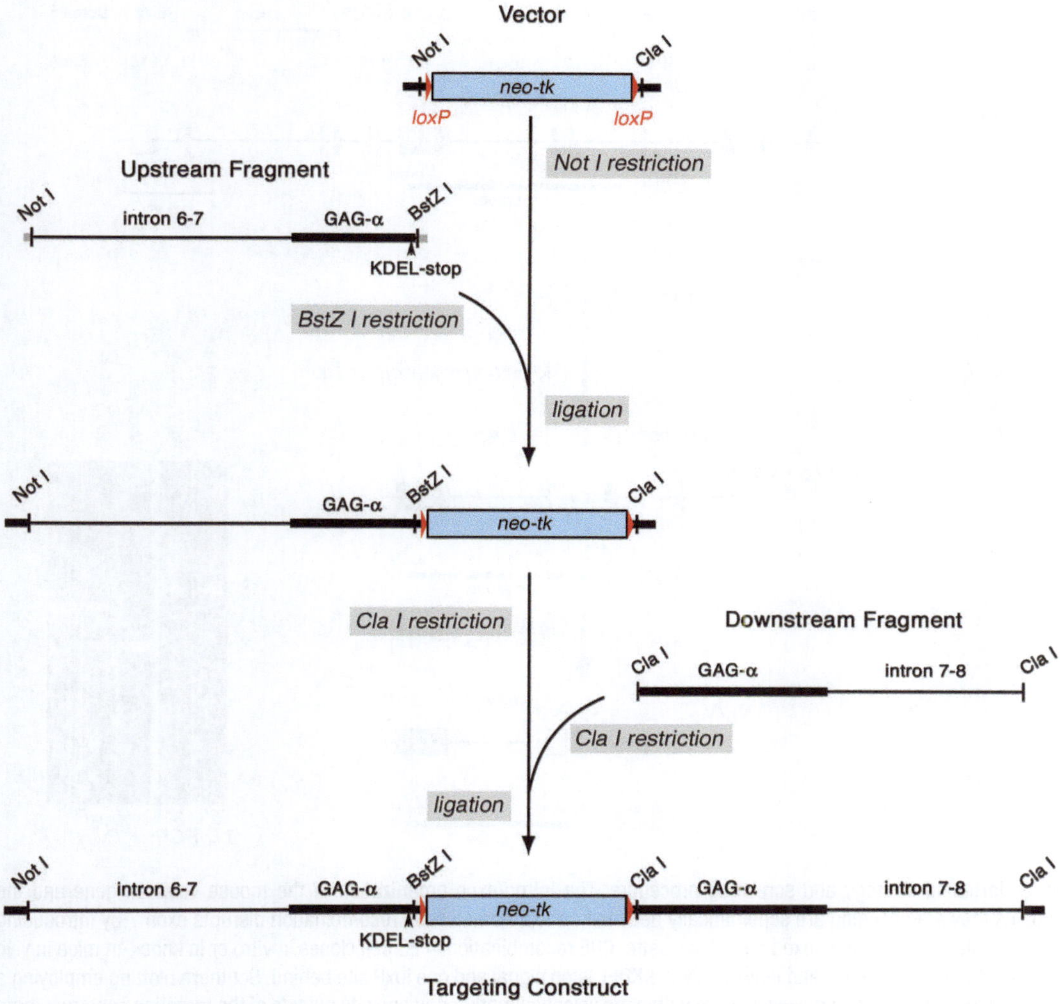

Fig. 2. Preparation of the targeting construct. LD-PCR fragments extending upstream and downstream from the modification site in exon 7 (GAG-α) are introduced into the corresponding positions of the cloning vector containing a neomycin–thymidine kinase (neo–tk) marker cassette flanked by loxP sequences (vector only partly shown). The codons for the ER-retention signal (KDEL) and the translational termination (stop) at the 3′-end of the upstream fragment as well as the specific restriction sites at the ends of both fragments are incorporated by primer adaptation.

Consider following points for screening:

1. High-quality genomic DNA dissolved in TE pH 8 required for both screening techniques. Use 3–5 μg genomic DNA per restriction for Southern blotting or 100–200 ng per reaction for LD-PCR (1 well of ES culture in a 24-well dish yields about 20–25 μg genomic DNA).
2. *Southern blot*: For screening, choose restriction with one site within and one site outside of the sequence present in the targeting construct. Use commonly used methylation-insensitive restriction enzymes preferentially available at high concentration.

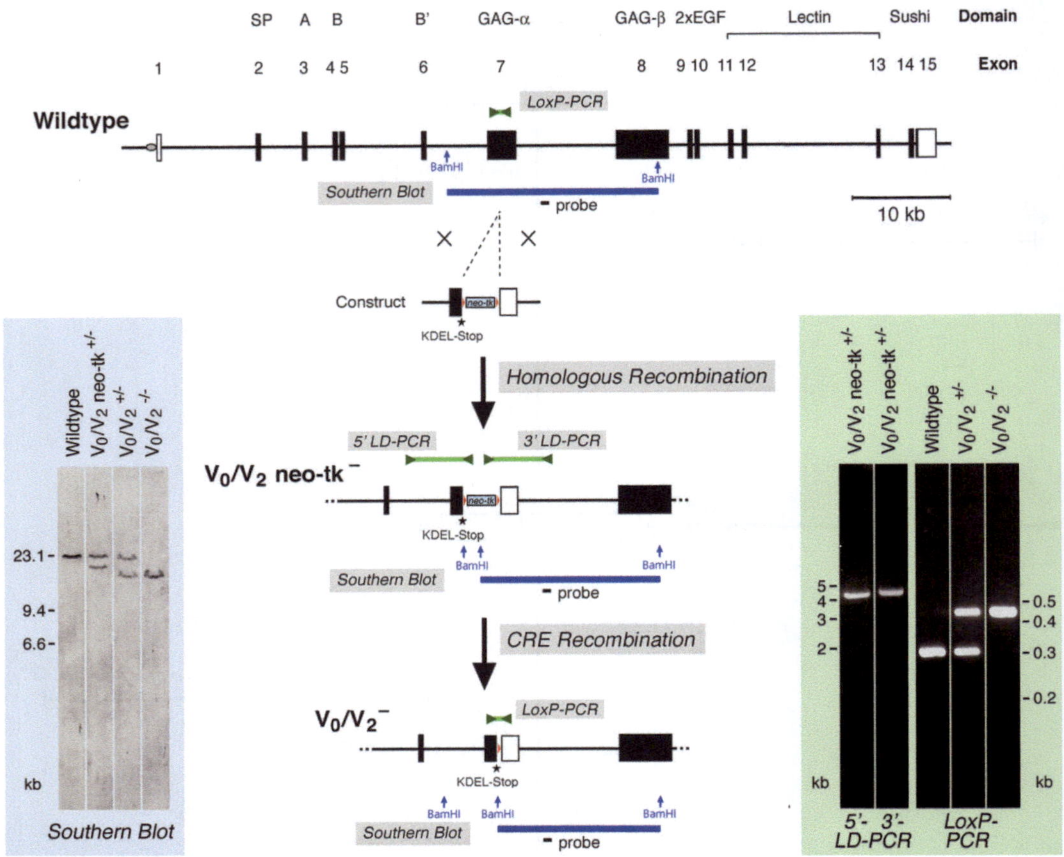

Fig. 3. Targeting strategy and screening procedure. The intron/exon organization of the mouse versican gene and the domains they are encoding are schematically depicted on top. Homologous recombination disrupts exon 7 by introducing the KDEL-stop codon and a floxed neo–tk cassette. CRE recombination in ES cell clones in vitro or in knock-in mice in vivo removes the neo–tk marker and leaves only the KDEL-stop signal and one loxP site behind. Southern blotting employing a restriction enzyme that recognizes one site within the selection marker and one site outside of the targeting construct (here Bam HI) is used for screening for homologous recombination. The probe, which should hybridize close to but not within the region covered by the targeting constructs, labels the characteristic fragments of the wild-type and the targeted alleles before and after CRE recombination. Alternatively, screening with long-distance PCRs (5′ and 3′ LD-PCR) with one primer within the marker sequence and one primer outside of the target construct displays the corresponding amplification product only in case of a target integration. A standard PCR across the remaining loxP site (LoxP-PCR) is sufficient to distinguish between the wild-type and the mutant allele after CRE recombination.

3. *Southern blot probe*: PCR-amplify an about 500 bp-long sequence included in the corresponding restriction fragment, but not present within the sequence of the targeting construct.
4. *Southern blot probe*: Label probe with digoxigenin-dUTP by incorporation via probe-specific PCR for sensitive detection.
5. *LD-PCR*: Choose one primer within the sequence of the targeting construct and one primer adjacent to the homologous targeting sequence. Keep the amplicon as small as possible (<10 kb).

6. *LD-PCR*: Verify specificity of the amplification in homologously recombined clones by sequencing.
7. *PCR to screen for the mutant allele after CRE-mediated removal of selection cassette*: Design primers up- and downstream of the singly remaining loxP site.
8. Use diluted DNA-extract of ES cells or biopsies directly for PCR analysis (see Note 6).

3.3. Analysis of Alternatively Spliced Transcripts in KO Mouse Strain by Quantitative RT-PCR

Isolation of total RNA:

1. Harvest tissue, weigh, and immediately submerge in RNA*later* RNA Stabilization Reagent from RNeasy Protect kit (10 μL Reagent/1 mg tissue) (see Note 7).
2. Disrupt tissue in lysis buffer (included in the kit) using the homogenizer.
3. Isolate total RNA with the RNeasy Protect kit. To eliminate residual genomic DNA, process sample on the Qiagen column with DNase.

Quantitative RT-PCR (general considerations):

1. Design primer pairs for RT-PCR amplification across splice-junction. To differentiate between alternative transcripts, choose isoform-specific splice junctions (example versican; Fig. 1).
2. Test assay specificity in conventional RT-PCR.
3. Prepare and keep reactions at all times on ice before starting the cycler program.
4. Use 100 ng total RNA per reaction for one-step RT-PCR.
5. Analyze each sample in triplicate from at least three animals per genotype and per developmental time point.
6. Include control reactions: (a) without reverse transcription and (b) without template.
7. Normalize against an endogenous control mRNA (e.g., mRNA from HPRT, GAPDH, RPLP0 genes).

3.4. Analysis of Alternatively Spliced Transcripts by Nonradioactive Northern Blotting (General Considerations)

1. To generate the PCR template for riboprobe preparation, select a cDNA fragment hybridizing to all transcripts of the gene of interest (e.g., G1 domain-encoding sequences of mouse versican). This allows an estimation of relative contributions of alternatively spliced gene products later on the Northern blot.
2. T/A-clone the PCR template and perform another PCR-round now using a lower primer supplemented with the inverse complementary sequence of the T7-promoter at its 5′ end. Purify this second PCR-product and use 1 μg as template for T7-transcription incorporating DIG-UTP. This indirect approach minimizes unspecific hybridizations with ribosomal RNAs.

Determine the efficiency of labeling by comparison with DIG-labeled RNA standards in a dot blot (as indicated in the DIG RNA Labeling Kit).

3. Isolate total RNA with the acid phenol method (7) to obtain the highest quality of RNA, required for the detection of very large transcripts (e.g., for mouse versican isoforms up to 13 kb). Load 5–10 μg per slot in a formaldehyde gel.
4. After standard Northern procedure (gel electrophoresis, capillary transfer, and UV fixation), stain blot with methylene blue (9) by sequential immersion in 5% acetic acid (15 min) and methylene blue staining solution (10 min). Destain in RNase-free water, photograph, and mark positions of the size markers. Clearly distinct 28S- and 18S-rRNA bands serve as indicator of RNA quality and quantity. This reversible staining will not interfere with subsequent hybridization.
5. Prehybridize and hybridize in DIG Easy Hyb buffer (Roche Applied Science) at 68°C for 1 h and overnight, respectively. Use 100 ng/mL riboprobe for hybridization.
6. Apply high stringent conditions for blot washing (twice for 5 min in 2× SSC/0.1% SDS at RT and twice for 15 min in 0.1× SSC/0.1% SDS at 68°C).
7. Visualize DIG-labeled bands with a DIG Nucleic Acid Detection Kit. Allow color reaction to proceed in the dark overnight.

3.5. Generation of Domain-Specific Antibodies for Splice-Variant Identification (General Considerations)

The generation of monospecific antibodies against proteoglycans is not a trivial task, since extensive similarities of some of the core protein domains and carbohydrate structures frequently cause strong cross-reactivity among different gene products due to epitope-sharing. To circumvent this problem, we have pursued following strategy (10):

1. Prepare carbohydrate-free core protein fragments in a bacterial expression system.
2. Restrict antigen production to unique sequence portions of the core protein (see Note 8).
3. Choose regions with low probability for glycosaminoglycan-attachment to elude epitope-masking in the intact proteoglycan.
4. *Optional*: Select fragment from alternatively spliced domains for antigen production to enable isoform-specific detection.
5. Prepare recombinant antigens fused to a short tag (e.g., N-terminal His-tag).
6. Use affinity-purified antigens for immunization.
7. Immuno-absorb antisera first on an affinity column coupled with bacterial control extract to remove antibodies against contaminant proteins (11).

8. Purify run-through by binding to column coupled with recombinant antigen.
9. Wash with high salt buffer.
10. Elute adsorbed antibodies by consecutive application of acid and basic elution buffers.
11. Immediately neutralize pH of acid- and basic-eluates.
12. Precipitate antibodies by slowly adding an equal volume of a saturated ammonium sulfate solution.
13. Resuspend in PBS containing 1% BSA and 0.02% Na-azide as stabilizer.
14. Aliquot and store at −20°C. Keep aliquot in use at 4°C (stable for about 1 year).

3.6. Protein Analysis of Tissue Extracts from Knockout Mice by SDS-PAGE and Immunoblotting

1. Weigh tissue and disrupt in the homogenizer with 4 volumes per weight of ice-cold extraction buffer.
2. Continue extraction by stirring overnight at 4°C.
3. Centrifuge homogenate at 100,000 RCF for 1 h at 4°C (see Note 9).
4. Discard pellet and dialyze supernatant extensively against chondroitinase buffer at 4°C.
5. Determine protein concentration with the BCA Protein Assay.
6. To analyze core glycoproteins of proteoglycans, digest the samples with chondroitinase ABC (1–2 mU/μg proteoglycan) in the presence of Leupeptin (2 μg/mL) and Pepstatin (1 μg/mL) overnight at 37°C.
7. Add 1:10 volume of SDS-loading buffer stock solution.
8. Optional: to analyze samples under reducing conditions, add β-mercaptoethanol (3% final concentration) just before denaturation.
9. Denature protein samples for 5–7 min at 95°C.
10. Run samples on 4–15% PhastGel polyacrylamide gels for 80–220 Vh (see Note 10). Temperature setting 15°C.
11. Remove buffer strips, but leave the gel on the separation bed for diffusion blotting (see Note 11).
12. Shortly wet Immobilon-P membrane in methanol and then in H_2O.
13. Place the prewetted membrane on the gel avoiding air bubbles.
14. Change the standby temperature of the PhastSystem to 70°C (temperature increases gradually from 15°C).
15. Place buffer strip holder back on top to prevent electrodes from touching the gel.

16. Close the lid and blot proteins onto Immobilon-P membrane by diffusion at 70°C for 20–30 min.
17. Gently detach membrane from gel by shaking in TBST buffer for 10 min.
18. Incubate blots in blocking buffer for at least 30 min.
19. Bind first antibodies (diluted usually 1:1,000–1:2,000 in blocking buffer) overnight.
20. Wash membranes in TBST buffer for 5 min. Repeat twice.
21. Bind AP-conjugated secondary antibody (diluted 1:10,000–1:30,000 in TBST) for 2 h at room temperature.
22. Wash in TBST buffer for 5 min. Repeat twice.
23. Rinse for 5 min with AP-Buffer.
24. Place membranes in the dark for chromogenic detection using the Western blue® substrate.
25. Let staining proceed until bands are visible (maximally overnight).

3.7. Collection and Fixation of Tissues for Histological Characterization

Excellent morphological preservation of tissues is achieved by applying the two following fixation procedures (see Notes 12–14).

The first protocol relies on standard PFA fixation conditions routinely used in surgical pathology:

1a. Sacrifice mice by CO_2-inhalation and decapitation (use firm cage lid with CO_2-connector).

2a. Rapidly dissect tissues and fix immediately in 4% PFA in PBS overnight at room temperature.

3a. Dehydrate tissues in an ascending ethanol series (70, 96, and 100%; each step 1.5 h at 42°C), then place in xylene (2 h; 37°C) and paraffin (3 h; 62°C) before finally embedding in paraffin.

The alternative procedure is recommended when the histological structure of interest is highly sensitive to rapid autolytic processes:

1b. Anesthetize mice with a lethal dose of pentobarbital (100–150 mg/kg i.p.) and transcardially perfuse with PBS, followed by perfusion-fixation with 4% PFA in PBS (perfusion pressure of about 100 mbar either generated by gravity or a peristaltic pump).

2b. Brain, spinal cords and other tissues are quickly removed, postfixed overnight at 4°C in the same fixative and either paraffin-embedded or cryo-protected, and snap-frozen in Tissue-Tek® O.C.T. compound.

3.8. Preparation of Paraffin Sections for Immunostaining

1. Cut 2–4 μm thick sections, extend on water surface, and mount on positively charged glass slides. Let section attach for 10 min on heating plate at 42°C and allow to dry at room temperature overnight or in a drying cabinet at 60°C for 30 min.

2. Dewax and hydrate tissue sections by successive immersion in a decreasing series of xylene/ethanol/water (3 times 5 min in xylene, 3 min in absolute ethanol, 5 min in absolute ethanol, 5 min in 96% ethanol, 5 min in running deionized water).
3. *Optional* (steps 3–5): If antigen retrieval is necessary, place the slides in a pressure-cooker or a controlled antigen retrieval device filled with 1 L of the suitable buffer (citrate pH 6.0 or TEC buffer; to be tested for each antigen).
4. Retrieve epitopes for 2–3 min in the steam-cooker or for 2 min in a retrieval device set to 105–120°C. Cool down until normal pressure is reached (in device at 70°C).
5. Submerge the slides shortly in distilled water and then rinse twice for 5 min in PBS.
6. Continue with the immunohistochemistry or the immunofluorescence protocol (see Note 15).

3.9. Preparation of Frozen Sections for Immunostaining

1. Cut 5–7-μm thick sections in a cryostat and mount on positively charged glass slides. Let air-dry for at least 1 h. Store at −80°C (see Note 12).
2. Remove slides with cryosections from −80°C freezer and return them to ambient temperature inside the sealed bag, in which they were stored. Then, open bag and let sections air-dry for 1 h.
3. *Optional*: If not fixed before, submerge slides in cold acetone for 10 min at −20°C. Let air-dry for 1 h.
4. Dip fixed sections shortly in PBS and continue with staining procedure.

3.10. Characterization of KO Mouse Strain by Immunohistochemistry

Except for the binding of primary antibodies, all steps are performed at room temperature (see also Note 16).

1. Block unspecific binding of antibodies by incubating at least for 30 min with blocking buffer (3% normal donkey serum in PBS). Alternatively, when using mouse monoclonal antibodies (mAbs) on mouse tissues, employ the M.O.M.™ Kit (Peroxidase) to avoid nonspecific immunoreactions (follow the manufacturer's protocol).
2. Carefully absorb rests of blocking buffer and cover slides with primary antibodies diluted in 1% normal donkey serum in PBS; incubate overnight at 4°C in a humid chamber (see Note 17).
3. Remove unbound antibodies by immersing sections 3 times for 5 min in PBS.
4. Incubate with secondary antibodies (diluted 1:800–1:1,000 in PBS) for 1–2 h.
5. Wash unbound antibodies away by submersing 3 times for 5 min in PBS.

6. Prepare the ABC reagent for step 9 by mixing 18 μL of each solution (A and B) from VECTASTAIN® Elite ABC Kit per 1 mL of PBS. Vortex and let stand for about 30 min before use.
7. Quench tissue-endogenous peroxidase by incubating slides in 3% H_2O_2 in PBS for 20 min.
8. Wash 3 times for 5 min in PBS.
9. Cover slides with previously prepared ABC reagent for 40 min.
10. Wash 3 times for 5 min in PBS.
11. Incubate sections in peroxidase substrate solution until desired signal intensity is reached (usually 5–10 min).
12. Stop reaction by rinsing sections for 5 min in tap water.
13. Counterstain cell nuclei by dipping slides shortly (1 min or less) into Mayer's hematoxylin solution and then immediately into warm tap water (reddish colored sections after hematoxylin will turn blue in water).
14. Mount the sections in Glycergel® Mounting Medium. Let the slides dry at least overnight before light microscopical examination (particularly when using oil immersion objectives) (see Fig. 4).

3.11. Characterization of KO Mouse Strain by Immunofluorescence

Important: Protect slides from strong light exposure starting from step 5.

1. When using mouse monoclonal antibodies (mAbs) on mouse tissues, incubate first for 1 h with the Mouse Ig Blocking Reagent Vector® (dilution according to company's instructions) to avoid nonspecific reactions. Otherwise proceed directly to step 2.
2. Block for at least 30 min with BSA/gelatine blocking buffer.
3. Carefully remove blocking buffer, cover sections with primary antibodies diluted in the BSA/gelatine blocking buffer, and incubate overnight at 4°C in a humid chamber. For immunostainings of multiple antigens, incubate primary antibodies (originating from different species) simultaneously (see Notes 16 and 17).
4. Wash unbound antibodies away by immersing the slides 3 times for 5 min in PBS.
5. Incubate with Alexa® Fluor secondary antibodies (diluted 1:200 in PBS) for 1 h at room temperature.
6. Counterstain cell nuclei for 2 min with Hoechst H33258 pentahydrate (*harmful! known mutagen*).
7. Remove unbound antibodies and nuclear dye by immersing 3 times for 5 min in PBS.

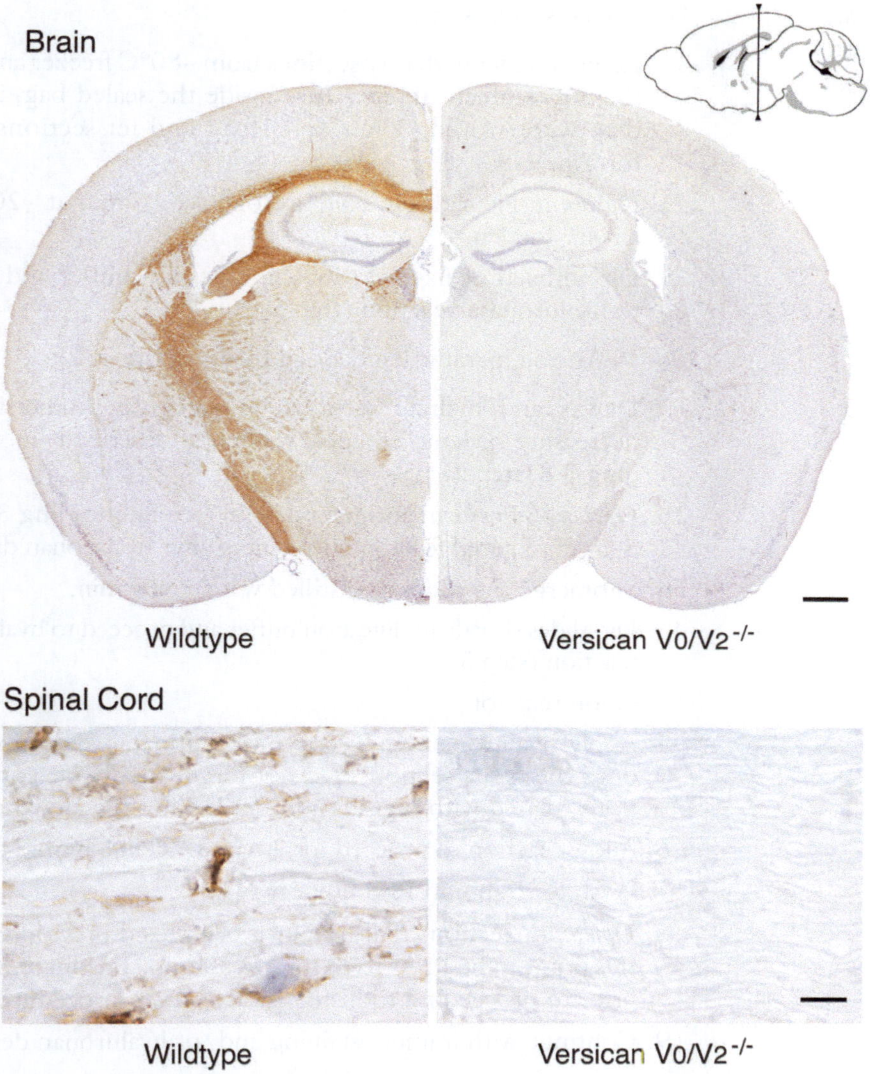

Fig. 4. Immunohistochemical staining of brain and spinal cord of adult mice. The sections from wild-type and versican V0/V2-deficient mice treated with antibodies against the GAG-α domain demonstrate the complete absence of versican V2 in the mouse mutants. Versican V2 is normally present in the white matter and particularly prominent at the nodes of Ranvier (visible here in the spinal cord image). The position of the brain section is depicted at the top right. Scale bars: brain 0.5 mm, spinal cord 5 μm.

8. Mount sections in fluorescent mounting medium. Let slides dry at least overnight before analyzing by fluorescence microscopy.
9. Acquire separate black and white images with filters appropriate for Alexa® Fluor 488, Alexa® Fluor 555, Alexa® Fluor 594, Alexa® Fluor 647, and DAPI dyes. Superimpose pseudo-colored pictures using an image editing software (e.g., Adobe Photoshop).

3.12. Cleavage of Hyaluronan in Histological Sections

For frozen sections (a):

1a. Remove slides with cryosections from −80°C freezer and return them to ambient temperature inside the sealed bag, in which they were stored. Then, open bag and let sections air-dry for 1 h.

2a. *Optional*: Fix slides in cold acetone for 10 min at −20°C. Let air-dry for 1 h.

3a. Dip unfixed or fixed sections in digestion buffer and proceed to hyaluronase-reaction (step 5).

For PFA-fixed, paraffin-embedded tissue sections (b):

1b. Dewax and hydrate sections by successive immersion in a decreasing xylene/ethanol/water as described in Subheading 3.8, step 2.

2b. *Optional*: Perform antigen retrieval (see Subheading 3.8, steps 3–5), if required for immunostaining after hyaluronan digestion.

3b. Submerge the slides in distilled water for 5 min.

4b. Dip slides shortly in digestion buffer and proceed to hyaluronase-reaction (step 5).

Digestion reaction:

5. Apply hyaluronate lyase enzyme solution in digestion buffer onto the tissue sections. Do not forget to include control sections treated with digestion buffer lacking the enzyme.
6. Allow reaction to proceed for 2 h at 37°C in a humid chamber.
7. Wash for 5 min in PBS. Repeat twice.
8. *Optional*: Sections that have not been fixed previously have at this point to be PFA- fixed for 10–30 min at room temperature prior to proceeding to histological staining procedures.
9. Continue with immunostaining and/or hyaluronan detection.

3.13. Staining of Hyaluronan in Histological Sections

1. Dewax and hydrate paraffin sections as described (see Subheading 3.8, step 2; antigen retrieval is not required for hyaluronan staining). Alternatively, use acetone- or PFA-fixed cryosections (see Subheadings 3.12, steps 1a and 2a or 3.7, steps 1b and 2b, respectively).
2. Submerge slides in PBS for 5 min (see Note 15).
3. Block unspecific binding by incubating at least 30 min with blocking buffer (3% normal donkey serum in PBS).
4. Remove rests of blocking buffer by washing slides 3 times in PBS for 5 min (see Note 16).
5. Cover slides with b-HABP and incubate overnight at 4°C in a humid chamber.
6. Remove unbound b-HABP by immersing sections 3 times for 5 min in PBS.

7. Proceed with ABC (Avidin-Biotin Complex) staining as described for immunohistochemistry (see Subheading 3.10, steps 6–14).

4. Notes

1. Unfortunately, the antigen retrieval device FSG 120-T/T from Milestone, which we use routinely in our lab, is not sold anymore. Nevertheless, descriptions of other suitable equipment can be found in the protocol database of the website www.ihc-world.com/protocol_database.
2. These combinations of secondary antibodies have been successfully used in our analysis. However, depending on the species origin of the primary antibodies, different combinations of secondary antibodies may be required. Also the combination of fluorophores may be changed, but their spectral compatibility must be verified and matched to the filter sets available in the fluorescence microscope.
3. Hyaluronate lyase from Streptomyces hyalurolyticus cleaves exclusively hyaluronic acid, while other hyaluronidases (e.g., from bovine or sheep testis) are also reactive with chondroitin and chondroitin sulfate.
4. The floxed selection cassette can be removed by transient expression of the CRE-recombinase in the homologously recombined ES cell clones. Alternatively, the cassette may later be removed in vivo by crossing the heterozygous KO-mice with a CRE-deleter strain (more risky approach due to potential fertility problems of male mice expressing the HSV-thymidine kinase marker gene).
5. Utmost care has to be taken to avoid cross-contamination during PCR screening. A strict division of working areas for pre- and post-PCR processing steps and the use of separate pipettes and filter-tips in the pre-PCR area are highly recommended to avoid the enormous extra work and frustrations associated with false positive results.
6. When using a PCR-compatible extraction buffer containing Tween-20 instead of SDS as detergent (see Subheading 2.2, step 9), PCR can be performed directly from the heat-inactivated extract without further purification (120–150 μL extraction buffer per biopsy; incubation overnight at 55°C in a thermomixer; proteinase K inactivation for 10 min at 95°C; short centrifugation to clear extract). In general, 1 μL of a 1:20 dilution of this DNA-extract per 20 μL PCR gives optimal results.
7. The conservation of tissue in RNA*later* is only required, when samples are not immediately extracted (e.g., extraction delayed

by dissection of multiple tissue samples for parallel processing or when sample has to be intermediately stored). Otherwise, proceed directly to the tissue lysis step.

8. Unfortunately, most of the core protein region fulfilling all the criteria for a specific antibody-binding is frequently poorly species-conserved. It therefore is necessary to verify cross-species immunoreactivity and to generate eventually new antibodies for the detection of orthologous core proteins.
9. For the analysis of proteoglycans secreted by cells in cultures, centrifuge conditioned medium at 10,000 RCF for 10 min to remove cellular debris and dialyze supernatant extensively against chondroitinase buffer before proceeding to chondroitinase digestion.
10. By extending the length of the electrophoresis from 80 to 220 Vh, the smaller size protein fragments will run out of the gel. However, larger core proteins (e.g., the large isoforms of versican or other lecticans) will be better separated in 220 Vh-runs. In any case, optimize the running conditions for your particular needs by using high and/or low molecular size standards.
11. Diffusion blotting of the ultra-thin PhastGels has proven to be very efficient even for the large core glycoproteins of versican V0, V1, and V2. Nevertheless, the transfer should be monitored by processing in parallel a prestained high molecular weight marker and a positive control sample (e.g., chondroitinase-treated conditioned medium from the human U251MG glioma cell line or corresponding preparations of primary embryonic fibroblast medium from other species for versicans V0 and V1 isoforms).
12. The morphology is best preserved in PFA-fixed, paraffin-embedded tissues. However, some epitopes are affected by this treatment. In this case, tissue conservation by direct snap freezing is recommended. Adequate fixation (e.g., acetone, methanol) can be performed later on the tissue sections, shortly before immunostaining. Frozen tissues and tissue sections are usually stored at −80°C. Seal sections in paper bags commonly used for sterilizing instruments (e.g., stericlin®).
13. Option for cryoprotection: to avoid potential artifacts caused by ice crystal formation, immerse the fixed tissue in 10% sucrose (in PBS) for 3–12 h and then soak in 30% sucrose until it sinks to the bottom of the solution (24–48 h). Only then snap-freeze.
14. If the sample includes bone, decalcify tissue before paraffin-embedding by an EDTA-based ultrasound procedure (12).
15. Once rehydrated, tissue sections must never dry out during the entire immunostaining procedure! They should always be incubated in a humid chamber containing a wet tissue paper at its base.
16. All washings steps are done in histology-staining dishes with racks for slides. Submerge the slides completely. Before applying

antibody solutions, rapidly dry area around the sections with a tissue paper, but maintain the section moist. Take profit of the surface tension to keep the volume of the solution at the minimum necessary to cover the tissue (50–70 μL, depending on size of the sections).

17. Always prepare control stainings lacking primary antibodies in the incubation solution in parallel. Alternatively, block primary antibody-binding by including the corresponding antigen in the solution.

Acknowledgments

We thank Reinhard Fässler and Uwe Rauch for crucial inputs for the targeting strategy and invaluable advices for the generation of the mouse strain, Bernhard Odermatt for many helpful hints regarding the immunhistological stainings, and Philipp U. Heitz and Holger Moch for their continuous support of the project. Our work has been financed in part by grants from the Swiss National Science Foundation and the Velux-Foundation to D.R.Z.

References

1. Aszodi, A., Legate, K. R., Nakchbandi, I., and Fassler, R. (2006) What mouse mutants teach us about extracellular matrix function. *Annu Rev Cell Dev Biol.* **22**, 591–621.
2. Kuhn, R. and Wurst, W. (2009) Overview on mouse mutagenesis. *Methods Mol Biol.* **530**, 1–12.
3. Mjaatvedt, C., Yamamura, H., Capehart, A., Turner, D., and Markwald, R. (1998) The *cspg2* gene, disrupted in the *hdf* mutant, is required for right cardiac chamber and endocardial cushion formation. *Dev. Biol.* **202**, 56–66.
4. Zimmermann, D. R. and Dours-Zimmermann, M. T. (2008) Extracellular matrix of the central nervous system: from neglect to challenge. *Histochem. Cell Biol.* **130**, 635–653.
5. Dours-Zimmermann, M. T., Maurer, K., Rauch, U., Stoffel, W., Fassler, R., and Zimmermann, D. R. (2009) Versican V2 assembles the extracellular matrix surrounding the nodes of ranvier in the CNS. *J Neurosci.* **29**, 7731–7742.
6. Talts, J. F., Brakebusch, C., and Fassler, R. (1999) Integrin gene targeting, *Methods Mol. Biol.* **129**, 153–187.
7. Chomczynski, P. and Sacchi, N. (2006) The single-step method of RNA isolation by acid guanidinium thiocyanate-phenol-chloroform extraction: twenty-something years on. *Nat. Protoc.* **1**, 581–585.
8. Yamagata, M., Shinomura, T., and Kimata, K. (1993) Tissue variation of two large chondroitin sulfate proteoglycans (PG-M/versican and PG-H/aggrecan) in chick embryos. *Anat. Embryol. Berl.* **187**, 433–444.
9. Herrin, D. L. and Schmidt, G. W. (1988) Rapid, reversible staining of Northern blots prior to hybridization. *BioTechniques* **6**, 196.
10. Zimmermann, D. R., Dours-Zimmermann, M. T., Schubert, M., and Bruckner-Tuderman, L. (1994) Versican is expressed in the proliferating zone in the epidermis and in association with the elastic network of the dermis. *J. Cell Biol.* **124**, 817–825.
11. Harlow, E. and Lane, D. (1988) *Antibodies, a laboratory manual*, Cold Spring Harbor Laboratory, Cold Spring Harbor.
12. Reineke, T., Jenni, B., Abdou, M. T., Frigerio, S., Zubler, P., Moch, H., and Tinguely, M. (2006) Ultrasonic decalcification offers new perspectives for rapid FISH, DNA, and RT-PCR analysis in bone marrow trephines. *Am J Surg Pathol.* **30**, 892–896.

Chapter 6

Detection of Neurocan in Cerebrospinal Fluid

Uwe Rauch

Abstract

Cerebrospinal fluid (CFS) is the most easily accessible component of the human central nervous system and has been successfully used for the analysis of disease-associated molecular imbalances, particularly for extracellular matrix components. Alterations in the presence of the nervous system-associated chondroitin sulfate proteoglycan neurocan had been reported from active multiple sclerosis lesions. Neurocan could be detected as a component of human CFS after enrichment of proteoglycans by anion exchange chromatography from pooled liquor as well as individual 300 μL samples by Western blot. However, a general alteration in neurocan levels in CFS sample with high immunoglobulin content could not be demonstrated. To further reduce the sample size, the development of a PG capturing assay based on polybrene-coated 96-well plates was initiated. This approach could be an interesting alternative option for the analysis of PGs in biological fluid and tissue samples.

Key words: Neurocan, Chondroitin sulfate proteoglycan, Cerebrospinal fluid, Anion exchange chromatography, Polybrene

1. Introduction

Neurocan is a chondroitin sulfate proteoglycan (CSPG), which was initially isolated from and mainly characterized in rodent brain, where it constitutes a major component of the extracellular matrix (ECM) during early postnatal development (*reviewed in refs.* (1, 2)). Notably, neurocan was identified as a component of the soluble fraction of detergent-free brain homogenates (3), and neurocan has never been reported to be a part-time PG, a PG occurring, in part, unsubstituted with glycosaminoglycan chains.

The knowledge about the presence and possible function of neurocan in the human central nervous system (CNS) is scarce. Human neurocan was found to distribute in a fine reticular pattern in white matter and is associated with blood vessel walls in brain

Françoise Rédini (ed.), *Proteoglycans: Methods and Protocols*, Methods in Molecular Biology, vol. 836,
DOI 10.1007/978-1-61779-498-8_6,

and spinal cord (4, 5). Interestingly, after spinal cord injury neurocan has been observed associated with Schwann cell myelin on regenerating peripheral nerve fibers, thus appears unlikely to contribute to failed CNS axon regeneration in humans (5), while due to its association with glial scars it has been repeatedly implicated in axon regeneration failure in rodents (6–8). However, in brains of multiple sclerosis (MS) patients, increases in neurocan immunostaining were observed at edges of most active MS plaques, areas which are also characterized by high glial fibrillary acidic protein (GFAP) levels (4). In turn was a loss of ECM staining for neurocan apparent in centers of active MS lesion, where foamy macrophages showed dense intracytoplasmic granular staining for neurocan (4).

Cerebrospinal fluid (CFS) is the most easily accessible component of the CNS of living humans and has been used repeatedly for the analysis of ECM components ((9, 10), *also reviewed in ref.* (2)). However, a first attempt to detect neurocan directly in total human CSF by Western blot analysis failed. To elucidate the presence of neurocan in CSF and eventually find consistent changes in MS patients, two methods to enrich PGs from CFS and to analyze the neurocan component in the PG-fraction were evaluated. The presence of neurocan in human CSF could be successfully demonstrated by Western blot analysis of the anion exchange-enriched PG-fraction. In addition, the development of a 96-well plate PG capturing assay was initiated, which might, upon further refinement, be an interesting alternative option for the analysis of PGs in general in biological fluid and tissue samples.

2. Materials

2.1. Anion Exchange Enrichment of PGs

1. Samples of CSF: The described experiments were performed with five samples of CSF with normal (AI–AV) and five samples of CSF with high immunoglobulin content (BI–BV), with volumes of 4–7 mL per sample.
2. 10× TBPI (Tris-buffered protease inhibitors): 200 mM Tris[=Tris(hydroxymethyl)-aminomethan]–HCl, pH 8, 50 mM EDTA, 50 mM benzamidiniumchloride, 50 mM *N*-ethyl maleimide.
3. PMSF-solution: 1% PMSF (phenylmethylsulfonyl fluoride) in isopropanol.
4. 10 and 0.1% Triton X-100 (Sigma) solutions in water.

5a. DEAE-Sephacel-slurry: DEAE-Sephacel (Pharmacia) equilibrated in 1× TBPI, 150 mM NaCl and 0.1% Triton.

5b. DEAE-Sephacel-slurry-B: DEAE-Sephacel equilibrated in 1× TBPI, 150 mM NaCl, 0.1% Triton, and 20 mg/mL BSA (bovine serum albumine).

6a. Wash-Buffer: 1× TBPI, 250 mM NaCl, and 0.1% Triton.

6b. Wash-Buffer-B: Wash-Buffer with 20 mg/mL BSA.

7a. Elution-Buffer: 1× TBPI, 750 mM NaCl, and 0.1% Triton.

7b. Elution-Buffer-B: Elution-Buffer with 20 mg/mL BSA.

8. TCA: 55% (w/w) Trichlor-acetic acid/water.

9. Acetone: Acetone stored at −20°C.

2.2. Chondroitinase Treatment

Chondroitinase-Buffer: 100 mM Tris–HCl, pH 8, 30 mM NaAc.

2.3. SDS-PAGE (Sodium Dodecylsulfate: Polyacrylamide Gel Electrophoresis) and Western Blotting

1. 3× SDS-sample buffer: 6% SDS, 30% glycerol, 187.5 mM Tris–HCl, pH 6.8, Bromphenol blue.
2. Separating gel solution: 6% acrylamide/bisacrylamide (37.5:1), 0.1% SDS, 375 mM Tris–HCl, pH 8.8, 10% glycerol, 0.02% APS (ammonium-peroxodisulfate), 0.1% TEMED (tetramethylen-diamine).
3. Stacking gel solution: 4% acrylamide/bisacrylamide (37.5:1), 0.1% SDS, 375 mM Tris–HCl, pH 8.8, 0.02% APS, 0.1% TEMED.
4. Electrode buffer: 25 mM Tris, 192 mM glycine, pH 8.3, 0.1% SDS.
5. Blotting buffer: 25 mM Tris, 192 mM glycine, pH 8.3, 10% methanol.
6. TBST: 20 mM Tris–HCl, pH 7.6, 150 mM NaCl, 0.1% Tween 20 (polyoxyethylene-sorbitan-monolaurate; Roth).

2.4. 96-Well Plate PG Capturing Assay

1. PBS: 20 mM Na-phosphate, pH 7.2, 150 mM NaCl.
2. Polybrene solution: 20 mg polybrene (see Note 1) in 1 mL PBS.
3. PBST: 0.02% Tween-20 in PBS.
4. CSF-sample-solution (see Note 2): 100 µL CSF + 500 µL DMEM/F12 (Gibco) + 20 µL FCS (Gibco) + 60 µL 200 mM Na-phosphate, pH 7.2 + 15 µL 5% CHAPS (Sigma).
5. Anti-neurocan-antiserum-solution: NC-2 serum 1:1,000 in 1% BSA in PBST.
6. Alkaline phosphatase-(AP-) linked anti-rabbit antibodies solution: AP-linked antibodies (Sigma) 1: 3,500 in 1% BSA in PBST.
7. pNPP-substrate-solution: 3 mg/mL para-nitrophenylphosphate in 1 M diethylamine/HCl, pH 9.6, 1 mM MgCl.

3. Methods

Methods of enrichment of PGs by anion exchange resins have probably been reported hundreds of times, either by chromatography in a column or batch-wise, as described below. The following description, in particular the description for individual CSF samples, has been especially adapted for many small individual samples. However, a critical step in this protocol is the resuspension/solubilization of the acetone pellet in digestion buffer, which is likely to be not transferable to many other PGs, like membrane PGs, or PGs, which depend for solubility on an intact tertiary structure, like leucine-rich repeat PGs. Neurocan can due to its abundant substitution with O- and N-linked oligosaccharides, in addition to its chondroitin sulfate chains, be considered as a particularly soluble PG.

While chondroitinase ABC digestions and SDS-PAGE/Western blot procedures are also described elsewhere, likely in this volume as well, a solid phase PG capturing approach based on polybrene, although a known heparin antagonist, has to my knowledge not been described before. However, while first tests with AP-fusion PGs, which can be detected directly with pNPP-substrate-solution, were very promising, the below-described detection of the PG with primary and secondary antibodies can only be considered to be semiquantitative. While the results presented here discouraged from further analysis of human CSF samples and thus refinement of the method for this application, the method in general might be an interesting alternative option for other analytical questions regarding PGs in biological fluid and tissue samples of limited availability.

3.1. Anion Exchange Enrichment of PGs

3.1.1. Enrichment from Pooled (See Note 3) CSF Samples

1. 2.1 mL pooled CSF was supplemented with 220 μL of 10× TBPI and 22 μL PMSF-solution.
2. Samples with 1 mL of this mixture were supplemented with 10 μL of 10%Triton and 100 μL DEAE-Sephacel slurry in a 1.5-mL test tube and rotated slowly for 1 h at 6°C.
3. The samples were centrifuged for 2 min with 900 RCF and the supernatant was removed.
4. The DEAE-Sephacel-slurry was washed 2 times with 1.2 mL Wash-Buffer, by 5 min rotation, 2 min centrifugation with 900 RCF, and removal of the supernatant each time.
5. The DEAE-Sephacel-slurry was eluted with 1 mL Elution-Buffer by 5 min rotation, 5 min centrifugation with 10,000 RCF, and recovery of the supernatant.
6. The recovered Elution-Buffer was supplemented and mixed with 250 μL ice-cold TCA, left for 10 min on ice, and centrifuged with 10,000 RCF for 10 min at 6°C (see Note 4).

7. The supernatant was removed and the pellet resuspended by vortexing in 1 mL –20°C acetone, left for 10 min on ice, and centrifuged with 10,000 RCF for 10 min at 6°C.
8. The supernatant was removed and the pellet resuspended by vortexing in 200 μL 0.1%Triton (see Note 5), mixed with 50 μL ice-cold TCA, left for 10 min on ice, and centrifuged with 10,000 RCF for 10 min at 6°C.
9. The supernatant was removed and the pellet resuspended by vortexing in 1 mL –20°C acetone, left for 10 min on ice, and centrifuged with 10,000 RCF for 10 min at 6°C.
10. The washing step with –20°C acetone was repeated, the supernatant removed again, and the residual pellet air-dried.

3.1.2. Enrichment from Individual CSF Samples

1. 0.3 mL individual CSF was supplemented with 30 μL of 10× TBPI, 3 μL PMSF-solution, 3 μL of 10%Triton, and 30 μL DEAE-Sephacel-slurry-B in 0.65-mL test tube and rotated slowly for 1 h at 6°C.
2. The samples were centrifuged for 2 min with 900 RCF and the supernatant was removed.
3. The DEAE-Sephacel slurry was washed 2× with 0.5 mL Wash-Buffer-B, by 5 min rotation, 2 min centrifugation with 900 RCF, and removal of the supernatant each time.
4. The DEAE-Sephacel slurry was eluted with 0.3 mL Elution-Buffer-B, by 5 min rotation, 5 min centrifugation with 10,000 RCF, and recovery of the supernatant in 1.5-mL test tubes.
5. The recovered Elution-Buffer was diluted with 0.7 mL 0.1% Triton solution, supplemented and mixed with 250 μL ice-cold TCA, left for 10 min on ice, and centrifuged with 10,000 RCF for 10 min at 6°C.
6. The supernatant was removed and the pellet resuspended in 1 mL –20°C acetone, left for 10 min on ice, and centrifuged with 10,000 RCF for 10 min at 6°C.
7. The washing step with –20°C acetone was repeated, the supernatant removed again, and the residual pellet air-dried.

3.2. Chondroitinase Treatment

1. The air-dried pellet was resuspended/dissolved in 10 μL Chondroitinase-Buffer containing 1× TBPI by vortexing.
2. The 10 μL of Chondroitinase-Buffer used for the digestion of one individual sample contained 50 mU (1% of the content of one vial) protease-free Chondroitinase ABC (Seikagaku). For undigested control samples, the enzyme was omitted.
3. The samples were incubated for 1 h at 37°C, and finally supplemented with 5 μL 3× SDS-sample buffer.

3.3. SDS-PAGE and Western Blotting

1. SDS-PAGE was performed on the Bio-Rad mini gel system II with 0.75-mm spacers and 15-well combs with a 4% polyacrylamide containing stacking and 6% polyacrylamide containing separating gel. 4 mL of the Separating gel solution was poured between the glass plates and carefully overlaid with water (see Note 6).
2. After polymerization, water was replaced by Stacking gel solution and the surface sealed by pushing the comb deep enough to avoid direct contact of Stacking gel solution with air.
3. Electrophoresis was performed with 120 mL Electrode buffer in the upper and 280 mL Electrode buffer in the lower chamber in usually less than a hour with 200 V (constant).
4. Blotting on PVDF membrane (Amersham) was performed in methanol-containing Blotting buffer for 1 h at 100 V using the Bio-Rad mini blot system.
5. The membrane was blocked with 5% dry milk in TBST overnight, incubated 1 h with NC-2 anti-neurocan antiserum (1:5,000) in TBST, washed in TBST, incubated 1 h with HRP-linked donkey anti-rabbit antibody (Jackson) (1:20,000) in TBST, washed in TBST, and developed with the ECL plus detection system (Amersham) (see Figs. 1 and 2).

3.4. 96-Well Plate PG Capturing Assay

1. Wells (see Note 7) of a flat bottom 96-well plates (High binding, Corning) were coated with 50 μL of polybrene solution overnight at 6°C.
2. Wells were washed with PBST and blocked with 360 μL 1% BSA in PBST.

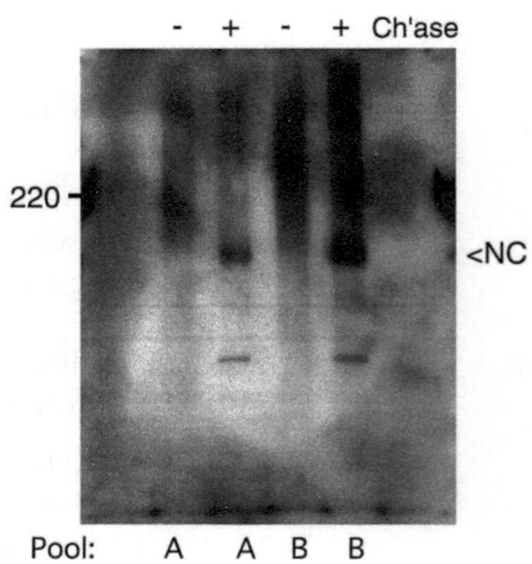

Fig. 1. Western blot analysis of pooled CSF samples (see Note 3) from patients with normal (**A**) and high (**B**) CSF immunoglobulin content. Note the appearance of a core protein band at a size expected for the C-terminal neurocan fragment neurocan-C (NC) only after chondroitinase digestion of the sample. The molecular mass of the visible prestained marker in kDa is indicated on the *left*.

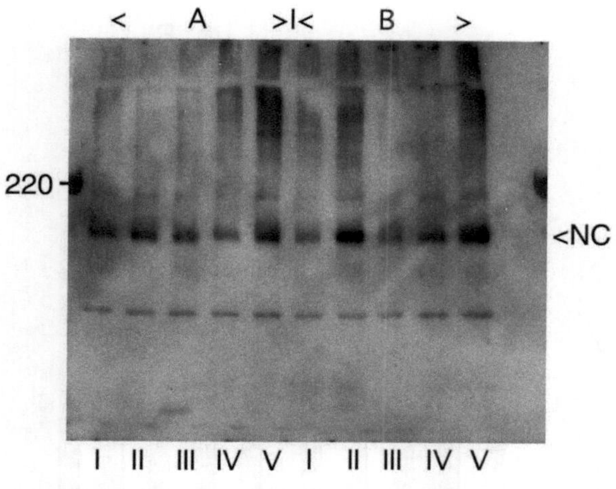

Fig. 2. Western blot analysis of five individual CSF samples from patients with normal (**A**) and high (**B**) CSF immunoglobulin content. All samples are chondroitinase-treated and reveal the C-terminal neurocan fragment neurocan-C (NC). The molecular mass of the visible prestained marker in kDa is indicated on the *left*. Note the high variability among the patients with high CSF immunoglobulin content, which indicates that neurocan may but is not always increased in CSF with high immunoglobulin content, which is indicative for MS.

3. Wells were supplemented in triplicates with 200 μL CSF-sample-solution each, and incubated overnight at 6°C.
4. Wells were washed with PBST (3×) and incubated 90 min at RT with 100 μL anti-neurocan antiserum-solution.
5. Wells were washed with PBST (3×) and incubated 90 min at RT with 100 μL AP-linked anti-rabbit antibodies solution.
6. Wells were washed with PBST (5×) and the color was developed with 200 μL pNPP-substrate-solution.
7. Absorptions at 415 nm were taken immediately after adding the pNPP-substrate-solution, and consecutively several times with intervals in the 5–10 min range. The differences to the absorption values taken immediately are calculated and presented in Fig. 3.

4. Notes

1. Polybrene (1,5-dimethyl-1,5-diazaundecanmethylene polymethobromide, hexadimethrine bromide; Sigma) is a positively charged polymer.
2. The presence of DMEM/F12 medium and FCS in the sample solution is due to the fact that the first studies to evaluate the possibility to capture PGs in polybrene-coated 96-well plates had been performed with conditioned medium from HEK 293 cells, which were secreting recombinant PGs. Initial experiments

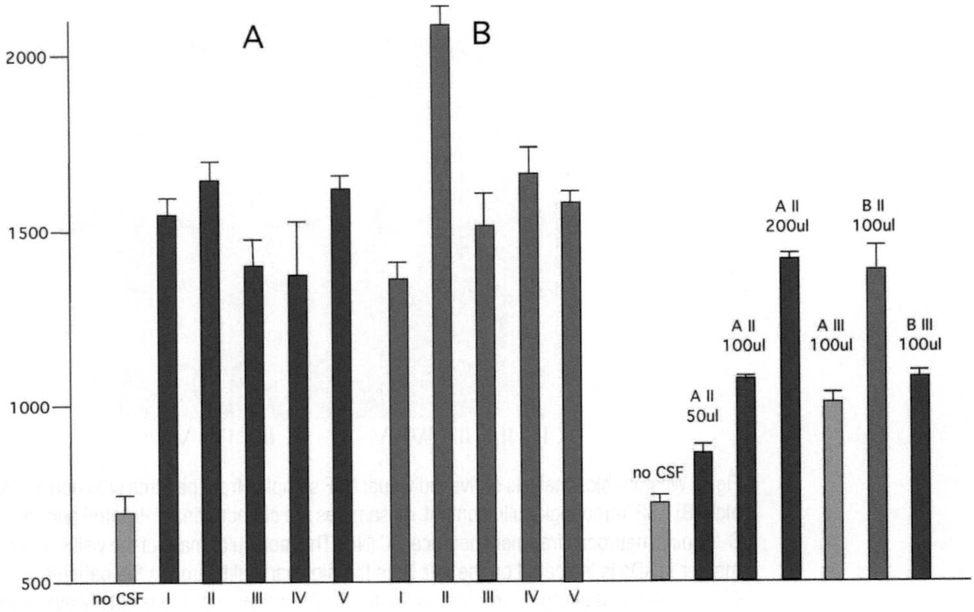

Fig. 3. 96-well plate PG capturing analysis of CSF samples. Presented are the results from two different, independent experiments. Shown are the average extinction and standard deviations of triplicates. To the *left* are the values for the same five individual CSF samples from patients with normal (**A**) and high (**B**) CSF immunoglobulin content, which had been analyzed by Western blot in Fig. 2. To the *right* are values from an independent experiment with four of these samples, where sample A II has been applied in three different concentrations.

were performed with alkaline phosphatase (AP)-fusion PGs, containing the N-terminal hyaluronan-binding domain and the first part of the CS-chain attachment domain of neurocan (unpublished), designed analogous to other neurocan-AP fusion proteins used for hyaluronan-binding studies (11). Further studies were performed with recombinant CS-chain substituted fragments of neurocan (D925) (12), which were detected by anti-neurocan antiserum and AP-linked secondary antibodies. In these studies, it was noticed that including a small amount of FCS in the sample was helpful for background reduction.

3. Due to the different total volumes of the supplied samples, for this proof of principle experiment different amounts and not all samples were used. (Pool A: 600 μL AI, 800 μL AII, and 700 μL AIII; Pool B: 400 μL BI, 550 μL BII, 550 μL BIII, 300 μL BIV, and 300 μL BV).
4. TCA precipitates not only protein, but also Triton X-100, which serves to generate a pellet even in solutions with a very small amount of protein, which would be difficult to precipitate alone. The following washing step with acetone removes the Triton X-100, leaving only a protein pellet.

5. This step was necessary because of the large size of the pellet, likely due to high salt concentration in the Elution-Buffer.
6. The overlay with water was possible because the separating gel solution contained 10% glycerol.
7. The 36 outer wells of the plate (row A and H, as well as row 1 and 12) were never used for samples. From the central 60 wells for the 3 triplicate samples, always the wells B, C, and D or the wells E, F, and G were used, thereby, if possible, avoiding rows 2 and 11 for the most important samples.

Acknowledgments

I would like to thank Dr. A. Grubb and J. Warenholt for help with the human CSF samples, and the Alfred Österlunds, the H and J Forssmans, the G and J Kocks, and the Crafoords foundations, the Swedish Research Council and Lunds Universities Medical Faculty for support.

References

1. Rauch, U., Feng, K., and Zhou, X. H. (2001) Neurocan: a brain chondroitin sulfate proteoglycan. *Cell. Mol. Life Sci.* **58**, 1842–1856.
2. Rauch, U. (2004) Extracellular matrix components associated with remodeling processes in brain. *Cell. Mol. Life Sci.* **61**, 2031–2045.
3. Rauch, U., Gao, P., Janetzko, A., Flaccus, A., Hilgenberg, L., Tekotte, H., et al. (1991) Isolation and characterization of developmentally regulated chondroitin sulfate and chondroitin/keratan sulfate proteoglycans of brain identified with monoclonal antibodies. *J. Biol. Chem.* **266**, 14785–14801.
4. Sobel, R. A. and Ahmed, A. S. (2001) White matter extracellular matrix chondroitin sulfate/dermatan sulfate proteoglycans in multiple sclerosis. *J. Neuropathol. Exp. Neurol.* **60**, 1198–1207.
5. Buss, A., Pech, K., Kakulas, B. A., Martin, D., Schoenen, J., Noth, J., et al. (2009) NG2 and phosphacan are present in the astroglial scar after human traumatic spinal cord injury. *BMC Neurol.* **9**, 32.
6. McKeon, R. J., Jurynec, M. J., and Buck, C. R. (1999) The chondroitin sulfate proteoglycans neurocan and phosphacan are expressed by reactive astrocytes in the chronic CNS glial scar. *J. Neurosci.* **19**, 10778–10788.
7. Asher, R. A., Morgenstern, D. A., Fidler, P. S., Adcock, K. H., Oohira, A., Braistead, J. E., et al. (2000) Neurocan is upregulated in injured brain and in cytokine-treated astrocytes. *J. Neurosci.* **20**, 2427–2438.
8. Huang, X., Kim, J. M., Kong, T. H., Park, S. R., Ha, Y., Kim, M. H., et al. (2009) GM-CSF inhibits glial scar formation and shows long-term protective effect after spinal cord injury. *J. Neurol. Sci.* **277**, 87–97.
9. Leppert, D., Ford, J., Stabler, G., Grygar, C., Lienert, C., Huber, S., et al. (1998) Matrix metalloproteinase-9 (gelatinase B) is selectively elevated in CSF during relapses and stable phases of multiple sclerosis. *Brain* **121 (Pt 12),** 2327–2334.
10. Saez-Valero, J., Costell, M., Sjogren, M., Andreasen, N., Blennow, K., and Luque, J. M. (2003) Altered levels of cerebrospinal fluid reelin in frontotemporal dementia and Alzheimer's disease. *J. Neurosci. Res.* **72**, 132–136.
11. Rauch, U., Hirakawa, S., Oohashi, T., Kappler, J., and Roos, G. (2004) Cartilage link protein interacts with neurocan, which shows hyaluronan binding characteristics different from CD44 and TSG-6. *Matrix Biol.* **22**, 629–639.
12. Feng, K., Arnold-Ammer, I., and Rauch, U. (2000) Neurocan is a heparin binding proteoglycan. *Biochem. Biophys. Res. Commun.* **272**, 449–455.

Part II

Glycosaminoglycans Analysis

Chapter 7

Glycosaminoglycan Chain Analysis and Characterization (Glycosylation/Epimerization)

Shuji Mizumoto and Kazuyuki Sugahara

Abstract

Glycosaminoglycans (GAGs) including chondroitin sulfate (CS), dermatan sulfate (DS), and heparan sulfate/heparin (HS/Hep) are linear polysaccharides and involved in the regulation of various biological events through interaction with functional proteins. GAGs are modified by sulfation at various positions of each saccharide residue and the epimerization of uronic acid residues during the chain's biosynthesis, resulting in enormous structural diversity. This structural diversity is the basis for the wide range of biological activities of GAGs. Thus, the structural analysis of GAGs is key to understanding their biological functions. This chapter describes detailed instructions for the extraction and structural analysis of GAGs from cultured cells and tissues using a combination of GAG-degrading enzymes and high-performance liquid chromatography.

Key words: Glycosaminoglycans, Chondroitin sulfate, Dermatan sulfate, Heparan sulfate, Heparin, Hyaluronan, Proteoglycan, Chondroitinase, Heparinase, Heparitinase

1. Introduction

Sulfated glycosaminoglycan (GAG) chains are covalently attached to various core proteins forming proteoglycans, which are ubiquitously distributed at cell surfaces and to extracellular matrices (1). Proteoglycans play roles through the GAG side chains in a variety of biological processes such as cell adhesion, proliferation, tissue morphogenesis, neurite outgrowth, infections by viruses and bacteria, and the regulation of various growth factors and cytokine effectors (2–4). Linear sulfated GAGs are largely classified based on structural units into chondroitin sulfate/dermatan sulfate (CS/DS) and heparan sulfate/heparin (HS/Hep), which are composed of the repeating disaccharide units [-4GlcUAβ1(IdoUAα1)-3GalNAcβ1-]n and [-4GlcUAβ1(IdoUAα1)-4GlcNAcα1-]n, respectively, where

Françoise Rédini (ed.), *Proteoglycans: Methods and Protocols*, Methods in Molecular Biology, vol. 836,
DOI 10.1007/978-1-61779-498-8_7, © Springer Science+Business Media, LLC 2012

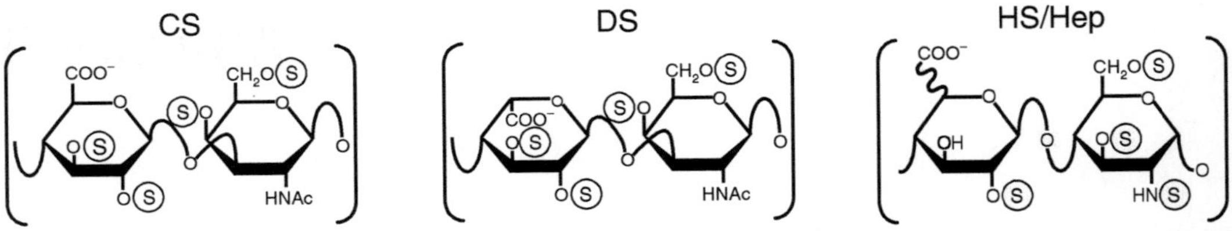

[-4GlcUAβ1-3GalNAcβ1-]

[-4IdoUAα1-3GalNAcβ1-]

[-4GlcUAβ1(IdoUAα1)-4GlcNα1-]

Symbol	Sequence
O unit	GlcUA-GalNAc
A unit	GlcUA-GalNAc(4S)
C unit	GlcUA-GalNAc(6S)
B unit	GlcUA(2S)-GalNAc(4S)
D unit	GlcUA(2S)-GalNAc(6S)
E unit	GlcUA-GalNAc(4S,6S)
K unit	GlcUA(3S)-GalNAc(4S)
L unit	GlcUA(3S)-GalNAc(6S)
M unit	GlcUA(3S)-GalNAc(4S,6S)
T unit	GlcUA(2S)-GalNAc(4S,6S)

Symbol	Sequence
iO unit	IdoUA-GalNAc
iA unit	IdoUA-GalNAc(4S)
iC unit	IdoUA-GalNAc(6S)
iB unit	IdoUA(2S)-GalNAc(4S)
iD unit	IdoUA(2S)-GalNAc(6S)
iE(H) unit	IdoUA-GalNAc(4S,6S)
iT unit	IdoUA(2S)-GalNAc(4S,6S)

Symbol	Sequence
HS-0S	HexUA-GlcNAc
HS-UA2S	HexUA(2S)-GlcNAc
HS-6S	HexUA-GlcNAc(6S)
HS-NS	HexUA-GlcN(NS)
HS-diS1	HexUA-GlcN(NS,6S)
HS-diS2	HexUA(2S)-GlcN(NS)
HS-diS3	HexUA(2S)-GlcNAc(6S)
HS-triS	HexUA(2S)-GlcN(NS,6S)

Fig. 1. Typical repeating disaccharide units in CS/DS and HS/Hep and their potential sulfation sites. CS and DS are constituted of uronic acid and GalNAc. DS is a stereoisomer of CS including IdoUA instead of or in addition to GlcUA. Both polysaccharides are frequently found as CS-DS hybrid chains in mammalian systems. HS and Hep consist of uronic acid and GlcN residues with varying proportions of IdoUA. These sugar residues can be esterified by sulfate at various positions as indicated by S . We propose that CS units be named traditionally and the corresponding DS units be indicated with "i," standing for IdoUA.

GlcUA, IdoUA, GalNAc, and GlcNAc represent D-glucuronic acid, L-iduronic acid, *N*-acetyl-D-galactosamine, and *N*-acetyl-D-glucosamine, respectively (see Fig. 1), which are covalently attached to the core proteins through the so-called common GAG-protein linkage tetrasaccharide, GlcUAβ1-3Galβ1-3Galβ1-4Xylβ1-*O*- (where Gal and Xyl stand for D-galactose and D-xylose, respectively) attached to the hydroxy group of an L-serine (Ser) residue of a core protein (5). During the biosynthesis of GAGs, various modifications such as sulfation and epimerization occur in both CS/DS and HS/Hep. The structure–function relationship in CS/DS and HS/Hep has revealed important roles for GAGs (3, 4). Thus, structural analysis is required to understand the functions of GAGs.

GAG-degrading enzymes, chondroitinase, heparitinase, and heparinase, are commercially available, specifically cleave their respective polysaccharides, and are powerful tools for structural analyses (6, 7). These enzymes were originally purified from bacteria and catalyze eliminative cleavage of *N*-acetyl-D-hexosaminidic linkages in CS/DS and HS/Hep, yielding mainly disaccharides with a 4,5-unsaturated uronic acid, 4-deoxy-α-L-*threo*-hex-4-enepyranosyluronic acid (ΔHexUA), on the nonreducing side, which can be detected and quantified by measuring absorbance at 232 nm. Although the identification of original uronic acids using bacterial GAG-degrading enzymes is difficult, it can be achieved by using a combination of distinct chondroitinases AC and B for CS/DS (8) or treatment with nitrous acid for HS/Hep (9).

Chondroitinase ABC (not a protease-free but a plain preparation) purified from *Proteus vulgaris* cleaves the *N*-acetyl-D-galactosaminidic linkage not only in the GalNAcβ1-4GlcUA units, but also in the GalNAcβ1-4IdoUA units of CS, DS, and CS-DS hybrid chains (10). There are two chondroitinases AC, AC-I (endolyase) and AC-II (exolyase) purified from *Flavobacterium heparinum* and *Arthrobacter aurescens*, respectively, which act on the *N*-acetyl-D-galactosaminidic linkage in the GalNAcβ1-4GlcUA units of CS and CS-DS hybrid chains (10, 11). Chondroitinase B (also known as dermatanase) purified from *F. heparinum* cleaves the *N*-acetyl-D-galactosaminidic linkage in the GalNAcβ1-4IdoUA units of DS and CS-DS hybrid chains (12). The following properties of each enzyme should be noted: (1) Chondroitinases ABC and AC digest hyaluronan as well, which is structurally similar to CS (8); (2) Although chondroitinase AC-I hardly cleaves the *N*-acetyl-D-galactosaminidic linkage in the GalNAcβ1-4GlcUA(2-*O*-sulfate) units, it is susceptible to chondroitinase AC-II under harsh conditions (8); (3) Chondroitinase B requires the 4-*O*-sulfated modification of GalNAc residues in DS to split the galactosaminidic bonds (8).

Heparitinase predominantly cleaves the *N*-acetyl-D-glucosaminidic linkage in the GlcNAcα1-4GlcUA units of HS/Hep chains (13). The enzyme preparation, which has been commercialized by Seikagaku Corp. and named "heparitinase," contains predominantly heparitinase-I, but also a small proportion of heparitinase-II (7). Heparinase mainly acts on the *N*-acetyl-D-glucosaminidic linkage in the GlcNAcα1-4IdoUA units of HS/Hep chains (13). The detailed substrate specificity of heparitinases and heparinase has been summarized in ref. (7). Although HS/Hep-degrading enzymes are frequently used, their nomenclature differs among manufactures. The nomenclature of heparitinases and heparinases is summarized in Subheading 2.1.

This chapter will describe how GAG chains are usually extracted from cells and tissues, and how to perform anion-exchange high-performance liquid chromatography (HPLC) for the structural analysis of CS/DS and HS/Hep chains.

2. Materials

2.1. GAG-Degrading Enzymes

1. Chondroitinase ABC (plain) from *P. vulgaris* (EC 4.2.2.20) (Seikagaku, Tokyo, Japan; Sigma, St. Louis, MO).
2. Chondroitinase ABC (protease-free) from *P. vulgaris* (EC 4.2.2.20) (Seikagaku).
3. Chondroitinase AC-I from *F. heparinum* (EC 4.2.2.5) (Seikagaku; Sigma; IBEX Technologies, Montreal, Canada).
4. Chondroitinase AC-II from *A. aurescens* (EC 4.2.2.5) (Seikagaku; Sigma).

5. Chondroitinase B from *F. heparinum* (EC 4.2.2.19) (Seikagaku; Sigma; IBEX Technologies).
6. Chondroitinase C from *F. heparinum* (EC 4.2.2.-) (Sigma).
7. Heparinase/Heparinase-I from *F. heparinum* (EC 4.2.2.7) (Seikagaku; Sigma; IBEX Technologies).
8. Heparitinase-II/Heparinase-II from *F. heparinum* (no EC number) (Seikagaku; Sigma; IBEX Technologies).
9. Heparitinase-I/Heparinase-III from *F. heparinum* (EC 4.2.2.8) (Seikagaku; Sigma; IBEX Technologies).
10. Heparitinase from *F. heparinum* (EC 4.2.2.8) (Seikagaku).
11. Hyaluronidase from *Streptomyces hyalurolytics* (EC 4.2.2.1) (Seikagaku; Sigma).

It should be noted that the GAG-degrading enzymes produced by IBEX Technologies are recombinant enzymes, and other GAG-degrading enzymes (see Note 1) are also available for the structural analysis of GAGs.

2.2. Buffers for Enzymatic Digestion

1. Chondroitinase ABC: 50 mM Tris–HCl, pH 8.0 containing 60 mM CH_3COONa.
2. Chondroitinase AC-I: 50 mM Tris–HCl, pH 7.3.
3. Chondroitinase AC-II: 50 mM CH_3COONa, pH 6.0.
4. Chondroitinase B: 50 mM Tris–HCl, pH 8.0 containing 2 mM $CaCl_2$.
5. Heparinase: 100 mM CH_3COONa, pH 7.0 containing 3 mM $Ca(CH_3COO)_2$.
6. Heparitinase: 20 mM CH_3COONa, pH 7.0 containing 2 mM $Ca(CH_3COO)_2$.

2.3. Extraction of GAGs, Labeling with a Fluorophore, and Anion-Exchange HPLC

1. Actinase E (Kaken Pharmaceutical Co., Tokyo, Japan).
2. BCA Protein Assay Kit (Pierce, Thermo Scientific, Rockford, IL).
3. Amicon Ultra-4 or -15 (MWCO, 3,000 or 10,000) centrifugal filter unit (Millipore, Billerica, MA).
4. PD-10 column (GE Healthcare Bio-Sciences, Uppsala, Sweden).
5. 0.025 M Sodium tetraborate-10 H_2O in sulfuric acid.
6. 0.125% Carbazole in ethanol.
7. 40 μg/mL Glucuronate standards.
8. 2-Aminobenzamide (2AB).
9. Sodium cyanoborohydride, $NaCNBH_3$.
10. Fifteen kinds of unsaturated standard disaccharides (ΔDi-HA, ΔDi-0S, ΔDi-4S, ΔDi-6S, ΔDi-UA2S, ΔDi-diS_B, ΔDi-diS_D, ΔDi-diS_E, ΔDi-triS, ΔDiHS-0S, ΔDiHS-6S, ΔDiHS-NS, ΔDiHS-diS_1, ΔDiHS-diS_2, and ΔDiHS-triS) derived from CS, DS, HS, Hep, and hyaluronan (Seikagaku).

11. An amine-bound silica PA-03 column (4.6×250 mm, YMC Co., Kyoto, Japan).
12. AG 50W-X2 resin, 200–400 mesh, hydrogen form (Bio-Rad, Hercules, CA).

3. Methods

3.1. Extraction and Purification of GAGs from Cell Cultures and Tissues

1. For adherent cultured cells, cells are washed with phosphate-buffered saline (PBS), collected using a cell scraper or by treatment with 2 mM EDTA-PBS, and centrifuged (see Note 2). For tissues, chop into chunks of 0.5–1 cm after washing with PBS.
2. To dehydrate and delipidate, cells and tissues are homogenized with cold acetone 3 times. After removal of the acetone by centrifugation, the samples are dried with P_2O_5 in a desiccator at 4°C overnight (see Note 3).
3. The acetone powder yielded is weighed, suspended in distilled water, and boiled for 10 min to inactivate endogenous glycosidases. To calibrate the amount of GAG, the total amount of protein is determined using a BCA protein assay kit.
4. To release GAGs from core proteins, two methodologies are used as described below.
   (a) To remove the core protein of proteoglycans, the samples are treated with a proteinase, yielding the GAG-peptides (14, 15). The reaction mixture of the samples contains 0.1 M borate buffer, pH 8.0/10 mM $CaCl_2$, and actinase E and is then incubated at 60°C for 1–2 days (see Note 4).
   (b) To liberate *O*-linked saccharides including GAG chains from the core proteins, the samples are treated with 1 M $NaBH_4$/0.5 M NaOH at room temperature for 2–24 h, yielding free GAGs (16, 17). The reaction is stopped by neutralization with 1 M acetic acid.
5. To remove insoluble materials, the reaction mixtures are centrifuged at 4°C for 5–10 min and then the supernatant fluid containing GAG-peptides or free GAG chains is recovered by treatment with a proteinase or alkali, respectively, into a centrifugal tube. To remove proteins, the soluble fraction is treated with 5% (w/v) trichloroacetic acid for 1 h on ice. Subsequently, the reaction mixture is centrifuged at 3,000 rpm for 30 min at 4°C, and the supernatant is recovered.
6. For the removal of trichloroacetic acid from the soluble fraction, an equal volume of ether is mixed with the sample and centrifuged at 2,000 rpm for 2 min at 4°C. The upper layer (ether fraction) is removed using a Pasteur pipette. This step is repeated 3 times.

After air-drying to completely remove the ether, the soluble fraction is neutralized with 0.5 M Na_2CO_3.

7. A GAG-containing fraction is recovered by precipitation with 80% ethanol by adding a one-fifth volume of 3 M CH_3COONa, pH 5.2, and a fourfold volume of ethanol. The sample is placed on ice for 1–2 h or overnight and centrifuged at 4°C at full-speed. The supernatant is discarded, and the precipitate is washed with cold 80% ethanol. After air-drying of the precipitate, the GAG fraction is dissolved in distilled water or PBS.
8. For desalting, the crude GAG-peptide or a GAG preparation is applied to an Amicon Ultra-4 or -15 (NMWL, 3,000) centrifugal filter unit using PBS or distilled water as a running solution, or to a PD-10 desalting column (see Note 5). To further purify GAGs, if necessary, anion-exchange chromatography is carried out using a Sep-Pak Plus Accell Plus QMA cartridge (Waters, Milford, MA) by stepwise elution using 50 mM phosphate buffers, pH 6.0 containing 0.15, 0.5, 1.0, or 2.0 M NaCl (17), and then each eluate is desalted with an Amicon Ultra-4 or a PD-10 column.
9. To confirm the extraction of GAGs, a GAG-peptide or a GAG preparation is subjected to the reaction of uronic acids with carbazole (18). Glucuronate standards of 4–40 μg/mL are prepared in a total volume of 60 μL in a glass tube for generating a calibration curve. Following the mixing of 300 μL of 0.025 M $Na_2B_4O_7/H_2SO_4$ with an aliquot of a GAG solution or glucuronate standards on ice, the samples are incubated at 100°C for 10 min and then cooled on ice. Subsequently, 12 μL of 0.125% carbazole/ethanol is added to each reaction tube, which is then incubated at 100°C for 10 min on ice. Aliquots (100–200 μL) of the samples are placed into a 96-well plate. The absorption at 530 nm is measured using a Microplate reader.

3.2. Treatment of GAGs with Chondroitinases or Heparinases

1. Chondroitinases. For the analysis of total disaccharides of both CS and DS, a crude GAG-peptide or a GAG preparation is digested using chondroitinase ABC for 30–120 min at 37°C in a total volume of 20 μL (see Note 6) of 50 mM Tris–HCl, pH 8.0 containing 60 mM CH_3COONa. For the analysis of disaccharides of CS (i.e., GlcUA-containing structure), a crude GAG-peptide or a GAG preparation is digested using a mixture of chondroitinases AC-I and AC-II for 30–120 min at 37°C in a total volume of 20 μL of 50 mM Tris–HCl, pH 7.3 (see Note 7). For the analysis of the disaccharides of DS (i.e., IdoUA-containing structure), a crude GAG-peptide or a GAG preparation is digested using chondroitinase B (see Note 8) for 30–120 min at 37°C in a total volume of 20 μL of 50 mM Tris–HCl, pH 8.0 containing 2 mM $CaCl_2$.

2. Heparinase and Heparitinase. For the analysis of total disaccharides of Hep and HS, a crude GAG-peptide or a GAG preparation is digested using a mixture of heparinase and heparitinase (Seikagaku) or a mixture of heparinases-I, -II, and -III (IBEX Technologies) for 30–120 min at 37°C in a total volume of 20 μL (see Note 6) of 100 mM CH_3COONa, pH 7.0 containing 3 mM $Ca(CH_3COO)_2$ for heparinase or 20 mM CH_3COONa, pH 7.0 containing 2 mM $Ca(CH_3COO)_2$ for heparitinase (see Note 9).
3. The incubation mixtures are boiled for 5 min to terminate the reaction and dried by a Speed-Vac centrifugal concentrator (Savant).

3.3. Derivatization of Disaccharides with a Fluorophore, 2-Aminobenzamide

1. The 2AB-derivatization reagent, 0.35 M 2AB/1.0 M $NaCNBH_4$/30% (v/v) acetic acid in dimethyl sulfoxide, is first prepared (19). Briefly, 350 μL of dimethyl sulfoxide is mixed with 150 μL of acetic acid (reagent A). Then, 9.5 mg of 2AB is dissolved in 200 μL of reagent A (reagent B). Subsequently, 6.2 mg of $NaCNBH_3$ is dissolved in 100 μL of reagent B (reagent C).
2. 5–10 μL of the 2AB reagent mix (reagent C) is added to the dried digests.
3. Incubate at 65°C for 2 h.
4. For removal of excess 2AB, the derivatized disaccharides (or oligosaccharides) are purified by extraction with chloroform (20) or by paper chromatography (19) as described below (see Note 10).
   (a) Extraction of excess 2AB with chloroform (20).

   500 μL of distilled water is added to the reaction tube, and the mixture is transfered to a glass centrifuge tube. An equal volume of chloroform (500 μL) is added and mixed. The mixture is centrifuged at room temperature at 1,000 rpm for 15 s. The chloroform phase is discarded using a Pasteur pipette or a long plastic pipette. This step is repeated 5–10 times.

   (b) Paper chromatography (19).

   Whatman 3MM paper is cut 3 × 10 cm pieces, washed with distilled water, and air-dried. The 2AB-derivatives are spotted about 1.5 cm from the edge, and the paper is immersed with a solvent, *n*-butanol/ethanol/distilled water (4:1:1), in a chromatographic chamber. After development, the paper is dried, and the original position and the UV-absorbing upper positions are cut out (see Note 11) under a UV light to visualize the 2AB-derivatives. Each piece of paper is placed into an Ultrafree-MC centrifugal filter (Millipore), and the 2AB-derivatives are eluted with distilled water (300 μL × 4 times), and the eluate is dried using a Speed-Vac centrifugator.

3.4. Anion-Exchange HPLC

Experimental conditions for anion-exchange HPLC (19).

Column: Amine-bound silica PA-03 (4.6×250 mm) (see Note 12)

Flow rate: 1 mL/min

Running buffers: 16 mM and 1 M NaH_2PO_4, which are filtered and degassed.

Elution conditions: A linear gradient of NaH_2PO_4. Namely, isocratic conditions with 16 mM of NaH_2PO_4 are employed for the first 10 min followed by a linear gradient from 16 to 530 or 800 mM NaH_2PO_4 over 60 min (see Note 13).

Detection: A fluorometric detector is used with excitation and emission wavelengths of 330 and 420 nm, respectively (see Note 14).

Representative chromatograms of the chondroitinase and heparitinase digests from cultured cells and mouse tissues are shown in Figs. 2–4.

3.5. Analysis of the Chain Length of GAG

1. A purified GAG-peptide or a proteoglycan preparation is dried (see Note 15).
2. 30 μL of 0.5 M LiOH is added and incubated at 4°C overnight (16).
3. 10 μL of 1.7 M acetate is added to terminate the reaction.
4. A blue pipette tip is filled with 0.3–0.5 mL of AG50W-X2 resine (H^+ form).
5. The reaction product is applied to the AG50W-X2 column.
6. Free GAG chains are eluted with 2–5 mL of distilled water.
7. The eluate is neutralized with 1 M NH_4HCO_3 and dried up by Speed-Vac centrifugation (see Note 16).
8. Reducing termini of the free GAG chains are derivatized with 2AB (see Subheading 3.3).
9. Excess 2AB is removed by paper chromatography (see Subheading 3.3).
10. The 2AB-labeled GAGs are further purified with an Amicon Ultra-4 or a PD-10 column, if necessary.
11. The 2AB-derivatives are subjected to gel filtration using Superdex 200 10/300 GL, HiPrep 16/60 Sephacryl S-400 HR, or HiPrep 16/60 Sephacryl S-500 HR (GE Healthcare Bio-Sciences), monitored by a fluorometric detector with excitation and emission wavelengths of 330 and 420 nm, respectively, using 0.2 M ammonium bicarbonate or ammonium acetate as an effluent at a flow rate of 0.4–1.0 mL/min (17, 21) (see Note 17).

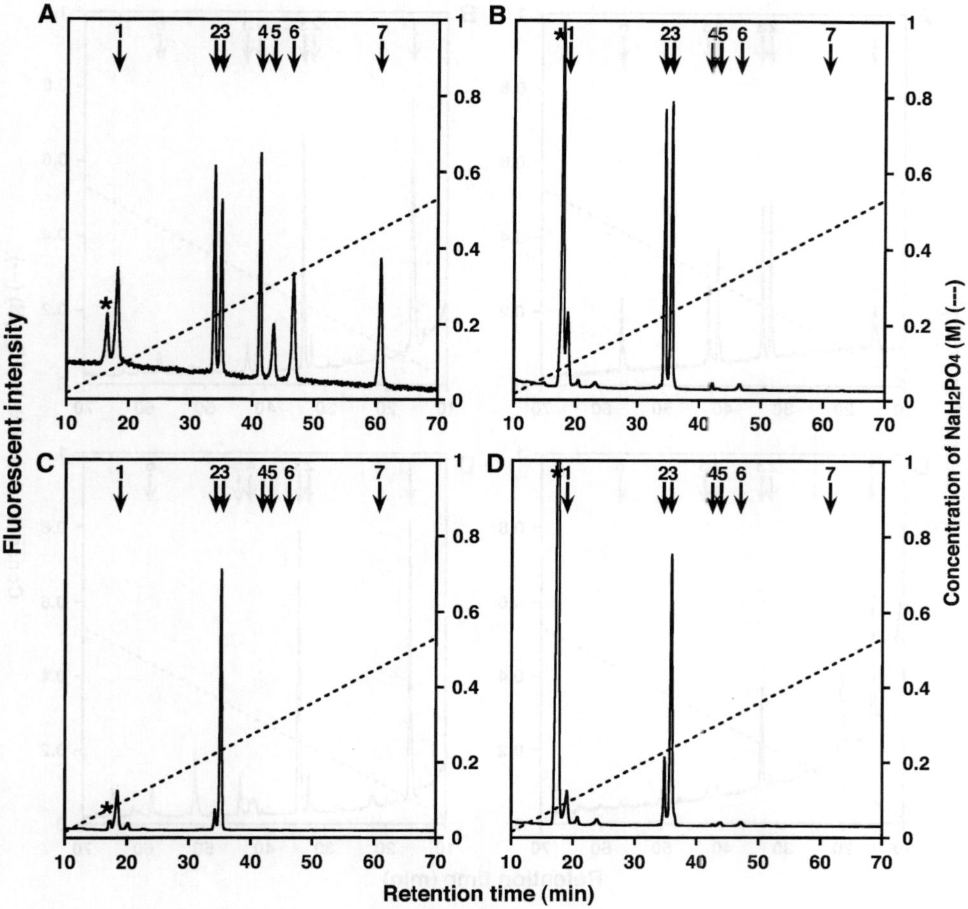

Fig. 2. Anion-exchange HPLC of CS/DS derived from various mouse tissues. (**A**) The authentic CS and DS standard disaccharides were separated by anion-exchange HPLC on an amine-bound silica PA-03 column using a linear gradient of NaH_2PO_4 as indicated by the *dashed line*. The elution positions of 2AB-labeled CS and DS disaccharide standards are indicated by *numbered arrows*. (1) ΔHexUA-GalNAc; (2) ΔHexUA-GalNAc(6S); (3) ΔHexUA-GalNAc(4S); (4) ΔHexUA(2S)-GalNAc(6S); (5) ΔHexUA(2S)-GalNAc(4S); (6) ΔHexUA-GalNAc(4S,6S); (7) ΔHexUA(2S)-GalNAc(4S,6S). The *asterisk* indicates the position of 2AB-labeled ΔHexUA-GlcNAc derived from hyaluronan, which is also cleaved by chondroitinases. (**B**–**D**) The GAGs preparations purified from mouse brain (**B**), cartilage (**C**), and skin (**D**) at postnatal day 0 were digested with a mixture of chondroitinases ABC and AC-II, and the digests were labeled with a fluorophore 2AB and analyzed by anion-exchange HPLC as described in Subheading 3.4.

3.6. Structural Analysis of a Common GAG-Protein Linkage Tetrasaccharide

1. For preparation of a 2AB-labeled GAG preparation, see Subheading 3.5.
2. The GAG-2AB preparation is treated with a mixture of chondroitinases ABC and AC-II (see Note 18) or a mixture of heparitinase and heparinase to yield the 2AB-labeled linkage region tetrasaccharides (or hexasaccharides) from CS/DS or HS/Hep, respectively.
3. The 2AB-derivatives are subjected to anion-exchange HPLC using on a PA-03 column (see Subheading 3.4). For identification of

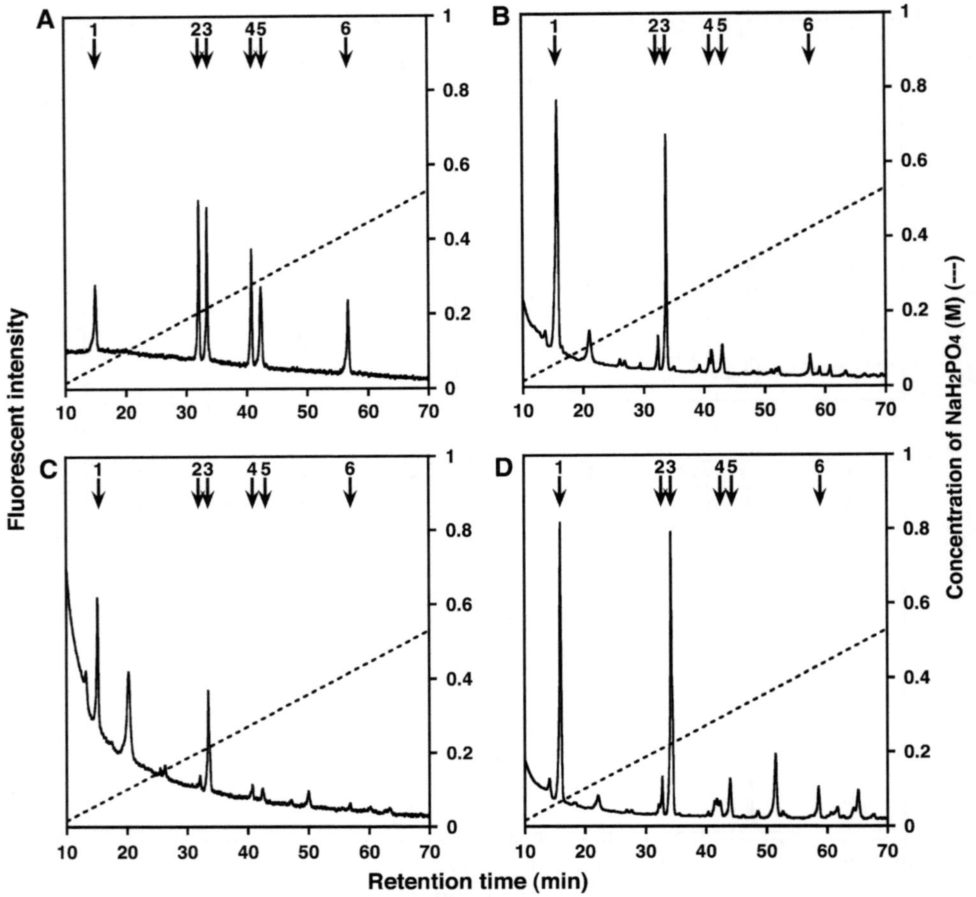

Fig. 3. Anion-exchange HPLC of HS derived from various mouse tissues. (**A**) The authentic HS standard disaccharides were separated by anion-exchange HPLC on an amine-bound silica PA-03 column using a linear gradient of NaH_2PO_4 as indicated by the *dashed line*. The elution positions of 2AB-labeled HS disaccharide standards are indicated by *numbered arrows*. (1) ΔHexUA-GlcNAc; (2) ΔHexUA-GlcNAc(6S); (3) ΔHexUA-GlcN(NS); (4) ΔHexUA-GlcN(NS,6S); (5) ΔHexUA(2S)-GlcN(NS); (6) ΔHexUA(2S)-GlcN(NS,6S). (**B**–**D**) The GAGs preparations purified from mouse brain (**B**), cartilage (**C**), and skin (**D**) at postnatal day 0 were digested with a mixture of heparinases-I and -III, and the digests were labeled with a fluorophore 2AB and analyzed by anion-exchange HPLC as described in Subheading 3.4.

the modifications by sulfation at the galactose residues and the phosphorylation at the xylose residue in the linkage region, the elution position of each peak from the chromatogram is compared to that of the structurally defined linkage tetrasaccharides (16, 22) (see Fig. 5).

3.7. Exosequencing of the CS-Oligosaccharides

To identify the functional domains (or sequences) in GAG chains, which interact with specific proteins, sequencing of the oligo- or polysaccharides is required after determination of the disaccharide composition (3, 19, 23–28). The exosequencing is depicted in Fig. 6.

1. Oligosaccharides are isolated using gel filtration chromatography on a Superdex Peptide 10/300 GL column

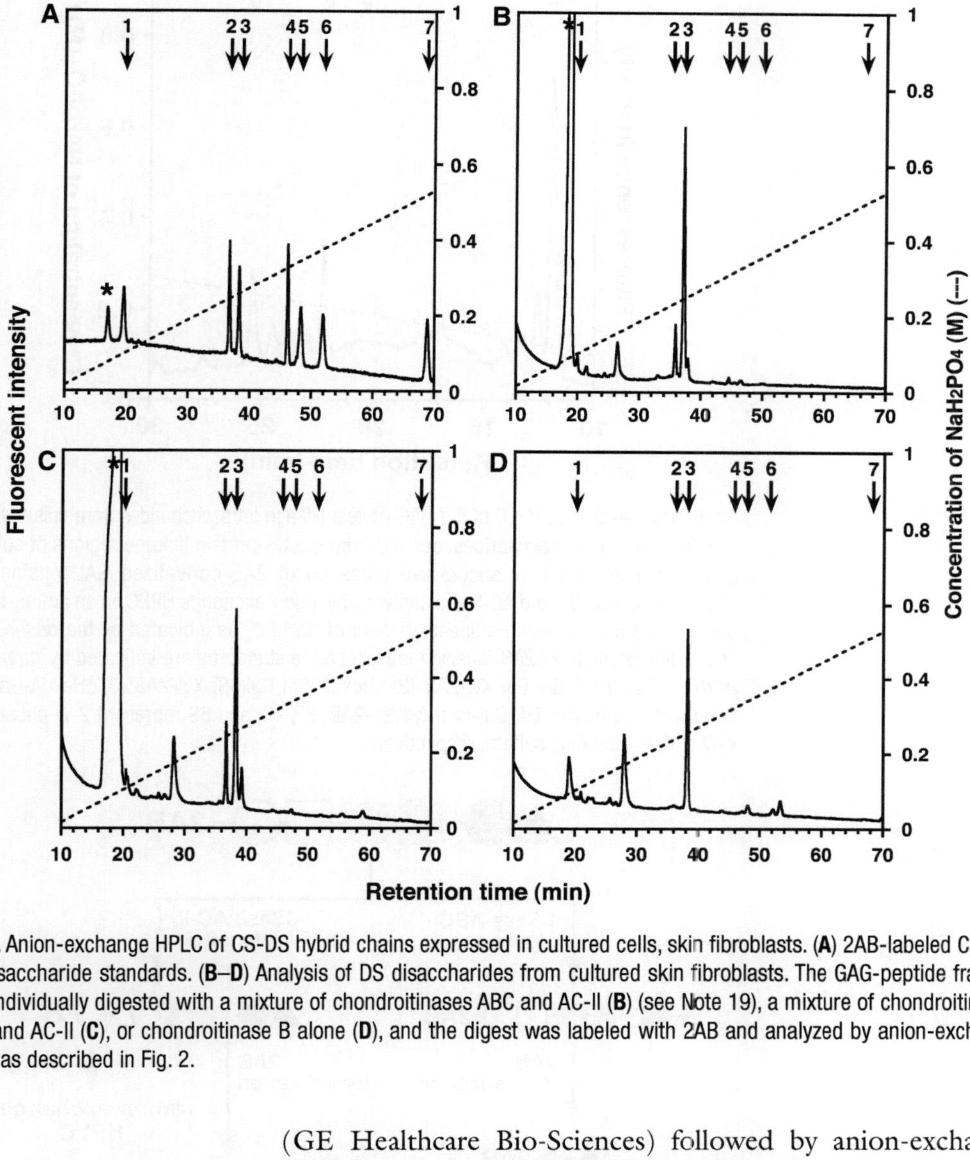

Fig. 4. Anion-exchange HPLC of CS-DS hybrid chains expressed in cultured cells, skin fibroblasts. (**A**) 2AB-labeled CS and DS disaccharide standards. (**B**–**D**) Analysis of DS disaccharides from cultured skin fibroblasts. The GAG-peptide fraction was individually digested with a mixture of chondroitinases ABC and AC-II (**B**) (see Note 19), a mixture of chondroitinases AC-I and AC-II (**C**), or chondroitinase B alone (**D**), and the digest was labeled with 2AB and analyzed by anion-exchange HPLC as described in Fig. 2.

(GE Healthcare Bio-Sciences) followed by anion-exchange HPLC on an amine-bound PA-03 column.

2. Isolated oligosaccharides are derivatized with 2AB (see Subheading 3.3).
3. Excess 2AB is removed by paper chromatography (see Subheading 3.3).
4. Each AB-derivatized oligosaccharide is treated with chondroitinase ABC and/or AC-II.
5. The enzyme digest is retreated with or without 2AB to relabel the newly released disaccharides, and the excess 2AB is removed by extraction with chloroform (see Subheading 3.3).
6. Each digest is subjected to anion-exchange HPLC using a PA-03 column (see Subheading 3.4).

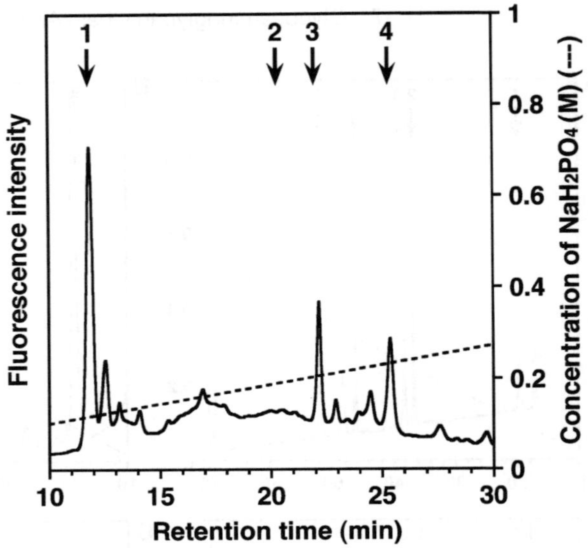

Fig. 5. Anion-exchange HPLC of the GAG-protein linkage tetrasaccharides from cultured cells. The 2AB-labeled tetrasaccharides, derived from the CS-protein linkage regions of cultured cells, were prepared by successive digestion of 2AB-derivatized GAG chains with chondroitinases ABC and AC-II and analyzed by anion-exchange HPLC on an amine-bound silica PA-03 column using a linear gradient of NaH_2PO_4 as indicated by the *dashed line*. The elution positions of 2AB-labeled tetrasaccharide standards are indicated by *numbered arrows*. (1) ΔHexUA-Gal-Gal-Xyl-2AB; (2) ΔHexUA-Gal-Gal(6S)-Xyl-2AB; (3) ΔHexUA-Gal(4S)-Gal-Xyl-2AB; (4) ΔHexUA-Gal-Gal-Xyl(2P)-2AB. 2P, 4S, and 6S represent 2-*O*-phosphate, 4-*O*-sulfate, and 6-*O*-sulfate, respectively.

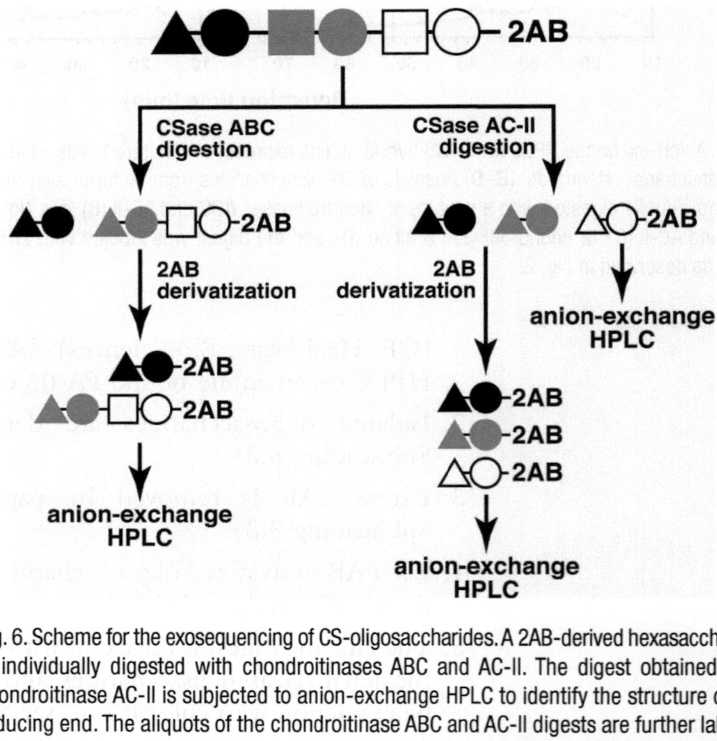

Fig. 6. Scheme for the exosequencing of CS-oligosaccharides. A 2AB-derived hexasaccharide is individually digested with chondroitinases ABC and AC-II. The digest obtained with chondroitinase AC-II is subjected to anion-exchange HPLC to identify the structure of the reducing end. The aliquots of the chondroitinase ABC and AC-II digests are further labeled with 2AB, and the 2AB-derivatives are subjected to anion-exchange HPLC to identify the structure of the disaccharides derived from the nonreducing end. The *triangles*, *squares*, and *circles* represent the ΔHexUA, GlcUA, and GalNAc, respectively. CSase stands for chondroitinase.

4. Notes

1. Other GAG-degrading enzymes, Chondro-4-sulfatase and Chondro-6-sulfatase from *P. vulgaris* (EC 3.1.6.9 and 3.1.6.10) (Seikagaku), are utilized for desulfation of ΔHexUA-GalNAc(4S) and ΔHexUA-GalNAc(6S), respectively (10). $\Delta^{4,5}$ Glycuronate-2-sulfatase, which is not commercially available, hydrolyzes the sulfate ester bond of the 4,5-unsaturated uronic acid 2-*O*-sulfate structure at the nonreducing end of di- or oligosaccharides (29). HS-$\Delta^{4,5}$-Glycuronidase, which is not available commercially either, acts on disaccharides containing the 4,5-unsaturated uronic acid obtained from HS/Hep (30). This enzyme is not active if an unsaturated uronic acid is 2-*O*-sulfated. Hyaluronidase from sheep testis (EC 3.2.1.35) (Sigma) is a hydrolase and can be also used for depolymerization of CS chains (24). In addition, hyaluronidase SD from *Streptococcus dysgalactiae* (EC 4.2.2.) (Seikagaku) catalyzes the eliminative cleavage of *N*-acetyl-D-hexosamine linkages of hyaluronan and chondroitin, yielding disaccharides with an unsaturated hexuronic acid at the nonreducing ends (31). This enzyme does not act on CS variants, DS, HS, heparin, or keratan sulfate. On the other hand, hyaluronidase from *Streptomyces hyalurolyticus* (EC 4.2.2.1) (Seikagaku and Sigma) is an eliminase and quite specific for hyaluronan. Various types of CS (A, C, D, and E), chondroiitn, DS, heparin, and keratan sulfate are not susceptible to this enzyme (32). Keratanase from *Pseudomonas* sp. (EC 3.2.1.103) (Seikagaku) hydrolyzes beta-galactosidic linkages in keratan sulfate, in which nonsulfated galactosyl residues participate (33). Keratanase II from *Bacillus* sp. (EC 3.2.1) (Seikagaku) catalyzes the endohydrolytic cleavage of glucosaminidic linkages in keratan sulfate (34). Endo-β-galactosidase from *Escherichia freundii* (EC 3.2.1.103) (Seikagaku) hydrolyzes specifically the beta-galactosidic linkages of nonsulfated galactosyl residues in keratan sulfate and various oligosaccharides in glucoconjugates such as glycoproteins, milk oligosaccharides, and glycosphingolipids (35).
2. The GAGs or proteoglycans can also be extracted from the serum-free conditioned medium of a cell culture and concentrated by ethanol precipitation or using an Amicon Ultra centrifugal filter unit.
3. To shorten the procedure, detergents can be utilized to homogenize cells or tissues.
4. It is necessary to preactivate the actinase E by incubation at 60°C for 30 min. For exhaustive digestion of proteins, additional actinase E (half the amount added initially) should be added once a day during the reaction for 2–3 days.

5. For removal of salts, a PD-10 desalting column is also available, which is eluted using 50 mM pyridine-acetic acid buffer, pH 5.0 (15). Subsequently, the eluates containing GAGs are repeatedly dried using a Speed-Vac concentrator.
6. The total volume of an enzyme mixture should be less than 20 μL to avoid the interference of 2AB-derivatization by salts.
7. For comparison of the disaccharide composition after digestion with chondroitinases ABC and AC, GAG samples are individually digested with these enzymes using the same buffer, 50 mM Tris–HCl pH 8.0, containing 60 mM sodium acetate to equalize the efficacy of 2AB-labeling.
8. It should be noted that a trace amount of 4-sulfatase is often present in the commercial chondroitinase B preparation purified from *F. heparinum* (36). Thus, a recombinant chondroitinase B preparation (IBEX Technologies) free of 4-sulfatase has to be utilized to determine the disaccharide composition of DS.
9. It is difficult to determine the type of uronic acids (GlcUA or IdoUA) by treatment with heparinase and heparitinase (7). However, when HS and Hep are treated with nitrous acid, they are specifically cleaved into fragments without a change in the type of uronic acids (7). Furthermore, this methodology can identify the rare 3-*O*-sulfated glucosamine-containing structure, which is resistant to heparinase and heparitinase, in HS/Hep.
10. For removal of the excess 2AB, the reaction mixture is spotted onto filter paper, which is washed with 1 mL of acetonitrile 5 times by centrifugation. Subsequently, the paper is placed into an Ultrafree-MC centrifugal filter, eluted with distilled water (300 μL×4 times), and dried by a Speed-Vac centrifuge (19).
11. This is because nonsulfated disaccharides migrate to the middle of the paper.
12. YMC-Pack PA-G (Cat# PG12S05-2546WT) can be used instead of YMC-Pack PA-03. Before being loaded onto the column, the sample is filtered by an Ultrafree MC to remove insoluble materials.
13. A linear gradient of 16–800 mM NaH_2PO_4 is used for elution of each disaccharide when using a new column, which retains trisulfated disaccharides strongly.
14. Alternatively, unsaturated hexuronic acid produced by the treatment with bacterial chondroitinases and heparitinases is detected by measuring absorbance at 232 nm using a UV detector (19).
15. For the extraction and purification of proteoglycans, refer to the methodologies in ref. (37).
16. If necessary, free GAGs can be further purified by gel filtration or hydrophobic chromatography using a Superdex Peptide or a Sep-Pak C_{18} cartridge (Waters), respectively.

17. Before gel filtration chromatography, purified GAGs are treated with heparitinase and hyaluronidase from *S. hyalurolyticus* to depolymerize and remove HS and hyaluronan, respectively, to determine the chain length of CS/DS. For analysis of the chain length of HS chains, 2AB-labeled GAGs are treated with a mixture of chondroitinases ABC and AC-II. After digestion, the reaction mixture is applied to a Microcon YM-3, an Amicon Ultra-4 (MWCO, 3,000), or PD-10 desalting column to remove the di- or oligosaccharides.

    Molecular mass was determined using Superdex 200, Sephacryl S-400, or Sephacryl S-500 columns calibrated with a series of known molecular mass markers, including dextran and dextran sulfate preparations. Dextrans and dextran sulfates are monitored by the orcinol (38) and dimethyl-methylene blue (39) methods for neutral and sulfate sugars, respectively, after fractionation using 0.2 M ammonium acetate as an eluate.

18. Chondroitinases ABC and AC-II yield the linkage region hexasaccharide ΔHexUA-GalNAc-GlcUA-Gal-Gal-Xyl-2AB and tetrasaccharide ΔHexUA-Gal-Gal-Xyl-2AB, respectively.

19. Although chondroitinase ABC can depolymerize not only CS/DS but also hyaluronan, which is abundantly expressed in skin fibroblasts, the action on hyaluronan is at a significantly reduced rate (8). On the other hand, chondroitinase AC-II exhibits comparable activities toward hyaluronan and CS (8).

Acknowledgments

This work was supported in part by Grants-in-Aid for Young Scientists (B) 23790066 (to S.M.) from Japan Society for the Promotion of Science, Japan, and Future Drug Discovery and Medical Care Innovation Program (to K.S.) from the Ministry of Education, Culture, Sports, Science, and Technology of Japan (MEXT).

References

1. Iozzo, R. V. (1998) Matrix proteoglycans: From molecular design to cellular function. *Annu. Rev. Biochem.* **67**, 609–652.
2. Sugahara, K., Mikami, T., Uyama, T., Mizuguchi, S., Nomura, K., and Kitagawa, H. (2003) Recent advances in the structural biology of chondroitin sulfate and dermatan sulfate. *Curr. Opin. Struct. Biol.* **13**, 612–620.
3. Sugahara, K and Mikami, T. (2007) Chondroitin/dermatan sulfate in the central nervous system. *Curr. Opin. Struct. Biol.* **17**, 536–545.
4. Bishop, J. R., Schuksz, M., and Esko, J. D. (2007) Heparan sulphate proteoglycans fine-tune mammalian physiology. *Nature* **446**, 1030–1037.
5. Sugahara, K. and Kitagawa, H. (2000) Recent advances in the study of the biosynthesis and functions of sulfated glycosaminoglycans. *Curr. Opin. Struct. Biol.* **10**, 518–527.
6. Petit, E., Delattre, C., Papy-Garcia, D., and Michaud, P. (2006) Chondroitin sulfate lyases: Applications in analysis and glycobiology, in

Advanced in Pharmacology, vol, 53 "Chondroitin sulfate: structure, role and pharmacological activity" (Volpi, N., ed.), Academic Press, London, UK, pp. 337–356.

7. Yamada, S. and Sugahara, K. (1998) Structure of oligosaccharides isolated from heparan sulfate/heparin and substrate specificities of the degrading enzymes of bacterial origin. *Trends in Glycosci. and Glycotechnol.* **10**, 95–123.
8. Yoshida, K., Arai, M., Kohno, Y., Maeyama, K., Miyazono, H., Kikuchi, H., Morikawa, K., Tawada, A., and Suzuki, S. (1993) Activity of bacterial eliminases towards dermatan sulphates and dermatan sulphate proteoglycan, in *Dermatan sulphate proteoglycans: chemistry, biology, chemical pathology* (Scott, J. E., ed.), Portland Press, London, pp. 55–80.
9. Conrad, H. E. (2001) Degradation of heparan sulfate by nitrous acid, in *Methods in Molecular Biology, vol. 171 "Proteoglycan protocols"* (Iozzo, R. V., ed.), Humana Press, Totowa, New Jersey, pp. 347–351.
10. Yamagata, T., Saito, H., Habuchi, O., and Suzuki, S. (1968) Purification and properties of bacterial chondroitinases and chondrosulfatases. *J. Biol. Chem.* **243**, 1523–1535.
11. Hiyama, K. and Okada, S. (1975) Crystallization and some properties of chondroitinase from *Arthrobacter aurescens. J. Biol. Chem.* **250**, 1824–1828.
12. Michelacci, Y. M. and Dietrich, C. P. (1974) Isolation and partial characterization of an induced chondroitinase B from *Flavobacterium heparinum. Biochem. Biophys. Res. Commun.* **56**, 973–980.
13. Ototani, N., Kikuchi, M., and Yosizawa, Z. (1981) Purification of heparinase and heparitinase by affinity chromatography on glycosaminoglycan-bound AH-sepharose 4B. *Carbohydr. Res.* **88**, 291–303.
14. Uyama, T., Ishida, M., Izumikawa, T., Trybala, E., Tufaro, F., Bergstrom, T., et al. (2006) Chondroitin 4-*O*-sulfotransferase-1 regulates "E" disaccharide expression of chondroitin sulfate required for herpes simplex virus infectivity. *J. Biol. Chem.* **281**, 38668–38674.
15. Mizumoto, S., Mikami, T., Yasunaga, D., Kobayashi, N., Yamauchi, H., Miyake, A., et al. (2009) Chondroitin 4-*O*-sulfotransferase-1 is required for somitic muscle development and motor axon guidance in zebrafish. *Biochem. J.* **419**, 387–399.
16. Sakaguchi, H., Watanabe, M., Ueoka, C., Sugiyama, E., Taketomi, T., Yamada, S., and Sugahara, K. (2001) Isolation of oligosaccharides from the chondroitin/dermatan sulfate-protein linkage region as reducing sugar chains and preparation of the analytical probes by fluorescent labeling with 2- aminobenzamide. *J. Biochem.* **129**, 107–118
17. Hashiguchi, T., Mizumoto, S., Yamada, S., and Sugahara, K. (2010) Analysis of the structure and neuritogenic activity of chondroitin sulfate/dermatan sulfate hybrid chains from porcine fetal membranes. *Glycoconjugate J.* **27**, 49–60.
18. Bitter, T. and Muir, H. (1962) A modified uronic acid carbazole reaction. *Anal. Biochem.* **4**, 330–334.
19. Kinoshita, A. and Sugahara, K. (1999) Microanalysis of glycosaminoglycan -derived oligosaccharides labeled with the fluorophore 2- aminobenzamide by high-performance liquid chromatography: Application to disaccharide composition analysis and exo-sequencing of oligosaccharides. *Anal. Biochem.* **269**, 367–378.
20. Kawashima, H., Atarashi, K., Hirose, M., Hirose, J., Yamada, S., Sugahara, K., and Miyasaka, M. (2002) Oversulfated chondroitin/dermatan sulfates containing GlcAβ1/IdoAα1-3GalNAc(4,6-*O*-disulfate) interact with L- and P-selectin and chemokines. *J. Biol. Chem.* **277**, 12921–12930.
21. Nakagawa, H., Hama, Y., Sumi, T., Li, S. C., Maskos, K., Kalayanamitra, K., et al. (2007) Occurrence of a non-sulfated chondroitin proteoglycan in the dried saliva of *Collocalia* swiftlets (edible bird's nest). *Glycobiology* **17**, 157–164.
22. Yamada, S., Okada, Y., Ueno, M., Iwata, S., Deepa, S.S., Nishimura, S., et al. (2002) Determination of the glycosaminoglycan-protein linkage region oligosaccharide structures of proteoglycans from *Drosophila melanogaster* and *Caenorhabditis elegans. J. Biol. Chem.* **277**, 31877–31886.
23. Bao, X., Muramatsu, T., and Sugahara, K. (2005) Demonstration of the pleiotrophin-binding oligosaccharide sequences isolated from chondroitin sulfate/dermatan sulfate hybrid chains of embryonic pig brains. *J. Biol. Chem.* **280**, 35318–35328.
24. Deepa, S. S., Kalayanamitra, K., Ito, Y., Kongtawelert, P., Fukui, S., Yamada, S., et al. (2007) Novel sulfated octa- and decasaccharides from squid cartilage chondroitin sulfate-E: Sequencing and their application for determination of the epitope structure of monoclonal antibody MO-225. *Biochemistry* **46**, 2453–2465.
25. Deepa, S. S., Yamada, S., Fukui, S., and Sugahara, K. (2007) Structural determination of novel sulfated octasaccharides isolated from chondroitin sulfate of shark cartilage and their application for characterizing monoclonal antibody epitopes. *Glycobiology* **17**, 631–645.
26. Pothacharoen, P., Kalayanamitra, K., Deepa, S. S., Fukui, S., Hattori, T., Fukushima, N., et al. (2007)

Two related but distinct chondroitin sulfate mimetope octasaccharide sequences recognized by monoclonal antibody WF6. *J. Biol. Chem.* **282**, 35232–35246.

27. Li, F., Nandini, C. D., Hattori, T., Bao, X., Murayama, D., Nakamura, T., et al. (2010) Structure of pleiotrophin- and hepatocyte growth factor-binding sulfated hexasaccharide determined by biochemical and computational approaches. *J. Biol. Chem.* **285**, 27673–27685.
28. Malavaki, C., Mizumoto, S., Karamanos, N., and Sugahara, K. (2008) Recent advances in the structural study of functional chondroitin sulfate and dermatan sulfate in health and disease. *Connect. Tissue Res.* **49**, 133–139.
29. McLean M. W., Bruce, J. S., Long, W. F., and Williamson, F. B. (1984) *Flavobacterium heparinum* 2-*O*-sulphatase for 2-*O*-sulphato -$\Delta_{4,5}$-glycuronate-terminated oligosaccharides from heparin. *Eur. J. Biochem.* **145**, 607–615.
30. Warnick, C. T. and Linker, A. (1972) Purification of an unusual α-glycuronidase from Flavobacteria. *Biochemistry* **11**, 568–572.
31. Hamai, A., Morikawa, K., Horie, K., and Tokuyasu, K. (1989) Purification and characterization of hyaluronidase from *Streptococcus dysgalactiae*. *Agric. Biol. Chem.* **53**, 2163–2168.
32. Ohya, T. and Kaneko, Y. (1970) Novel hyaluronidase from Streptomyces. *Biochim. Biophys. Acta* **198**, 607–609.
33. Nakazawa, K., Suzuki, N., and Suzuki, S. (1975) Sequential degradation of keratan sulfate by bacterial enzymes and purification of a sulfatase in the enzymatic system. *J. Biol. Chem.* **250**, 905–911.
34. Hashimoto, N., Morikawa, K., Kikuchi, H., Yoshida. K., and Tokuyasu, K. (1988) Purification and characterization of kerdtanase. *Seikagaku* **60**, 935.
35. Nakagawa, H., Yamada, T., Chien, J. L., Gardas, A., Kitamikado, M., Li, S. C., and Li, Y. T. (1980) Isolation and characterization of an endo-β-galactosidase from a new strain of *Escherichia freundii*. *J. Biol. Chem.* **255**, 5955–5959.
36. Nandini, C.D., Mikami, T., Ohta, M., Itoh, N., Akiyama-Nambu, F., and Sugahara, K. (2004) Structural and functional characterization of oversulfated chondroitin sulfate/dermatan sulfate hybrid chains from the notochord of hagfish: Neuritogenic activity and binding activities toward growth factors and neurotrophic factors. *J. Biol. Chem.* **279**, 50799–50809.
37. Yanagishita, M. (2001) Isolation of proteoglycans from cell cultures and tissues, in *Methods in Molecular Biology, vol. 171 "Proteoglycan protocols"* (Iozzo, R. V., ed.), Humana Press, Totowa, New Jersey, pp. 3–10.
38. Brückner, J. (1955) Estimation of monosaccharides by the orcinol-sulphuric acid reaction. *Biochem. J.* **60**, 200–205.
39. Chandrasekhar, S., Esterman, M. A., and Hoffman, H. A. (1987) Microdetermination of proteoglycans and glycosaminoglycans in the presence of guanidine hydrochloride. *Anal. Biochem.* **161**, 103–108.

Chapter 8

Characterization of Glycosaminoglycans by Tandem Vibrational Microspectroscopy and Multivariate Data Analysis

Nathalie Mainreck, Stéphane Brézillon, Ganesh D. Sockalingum, François-Xavier Maquart, Michel Manfait, and Yanusz Wegrowski

Abstract

Vibrational spectroscopies (VS), INFRARED SPECTROSCOPY and RAMAN SPECTROSCOPY, are well-established techniques for exploring the chemical composition of samples. VS are based on the molecular vibrations and give a spectral signature also called "molecular fingerprint" characteristic of the studied material. Recent advances in these techniques have rendered them faster, more sensitive, and easier to use. This chapter describes their application to characterize the main glycosaminoglycans—without any sample destruction or degradation. Nowadays, the use of multivariate statistical analysis for analyzing spectral data allows to extract rapidly the discriminant spectral information from large data sets. The combination of VS and this type of data analysis is also discussed in this chapter.

Key words: Glycosaminoglycans, Raman spectroscopy, Infrared spectroscopy, Multivariate statistical analysis, Hierarchical cluster analysis, Principal components analysis

1. Introduction

1.1. Basics of Vibrational Spectroscopy

1.1.1. Infrared and Raman Spectroscopies

Infrared (IR) and Raman spectroscopies are vibrational spectroscopic techniques based on the interaction between light and matter that allow to probe the chemical content of all forms of matter *via* its vibrational energy. The information obtained represents a "molecular fingerprint" characteristic of the state, molecular structure, and environment of the sample (spectral signature). These techniques are currently used to characterize simple molecules, complex biological macromolecules (lipids, proteins, nucleic acids, polysaccharides), and even highly complex systems like cells and tissues (1).

Françoise Rédini (ed.), *Proteoglycans: Methods and Protocols*, Methods in Molecular Biology, vol. 836,
DOI 10.1007/978-1-61779-498-8_8, © Springer Science+Business Media, LLC 2012

In infrared spectroscopy, the sample interacts with an incident light from a polychromatic source comprising a set of wavelengths that span over the midinfrared range (2.5–25 μm). The midinfrared source carries the energy that exactly corresponds to the energy differences between vibrational energy levels of the molecule. As a consequence, a molecule excited by a midinfrared source may absorb this energy and in this case will start vibrating. This is the case when the molecule is infrared-active, i.e., it undergoes a change in its dipole moment during this interaction. Infrared spectroscopy analyses the energetic difference between the incident light and light transmitted or reflected by the sample. The infrared spectrum therefore displays the absorbed energy at each wavelength. This spectrum can be interpreted qualitatively *via* the position of the band, which is directly related to the frequency of vibration and therefore to the chemical information, and quantitatively *via* its intensity, which can be associated to its concentration.

Raman spectroscopy is a scattering process that is based on the interaction of the sample with a monochromatic source, generally a laser. During this interaction, the molecule gains energy and reaches a "polarized" or "virtual" state which is not energetically stable and consequently the molecule comes down to the initial energy level by emission of photons. For a large majority of cases, the molecule will come back to its initial energy level, i.e., the lowest vibrational energy level of the ground state. This gives rise to elastically scattered photons with a wavelength that has the same energy as that of the incident laser. This event is the most probable and therefore predominates and is termed as "Rayleigh scattering." Since one gets back the same frequency as the incident excitation, Rayleigh light does not carry any chemical signature. A second possibility is that the scattered light is inelastic in nature, i.e., the transition to the ground state gives rise to scattered photons that have either gained energy (anti-Stokes Raman) or given away some energy (Stokes Raman). In both cases, the difference between the initial and final states gives the vibrational energy of the system and also corresponds to energy levels measured in IR spectroscopy (see Fig. 1). However, it must be noted that, while in IR spectroscopy a measurement of the absolute frequency is made, in Raman spectroscopy the measured frequency is relative since a difference is calculated. This frequency difference is also called "Raman shift." The spectral information obtained with the anti-Stokes and Stokes scattering is identical, but since in the former the initial state of the molecules is not the lowest energy level, the anti-Stokes is less intense because the number of events involved is less. Therefore, Raman spectrometers are generally designed to capture mainly the Raman Stokes signal, while the Rayleigh and anti-Stokes scattering are rejected. The Raman spectrum therefore displays the Stokes scattering intensity with respect to the Raman shift. In the same way as with IR, the position of the peaks is related to the chemical nature of the sample. Its intensity can be used in a semiquantitative manner.

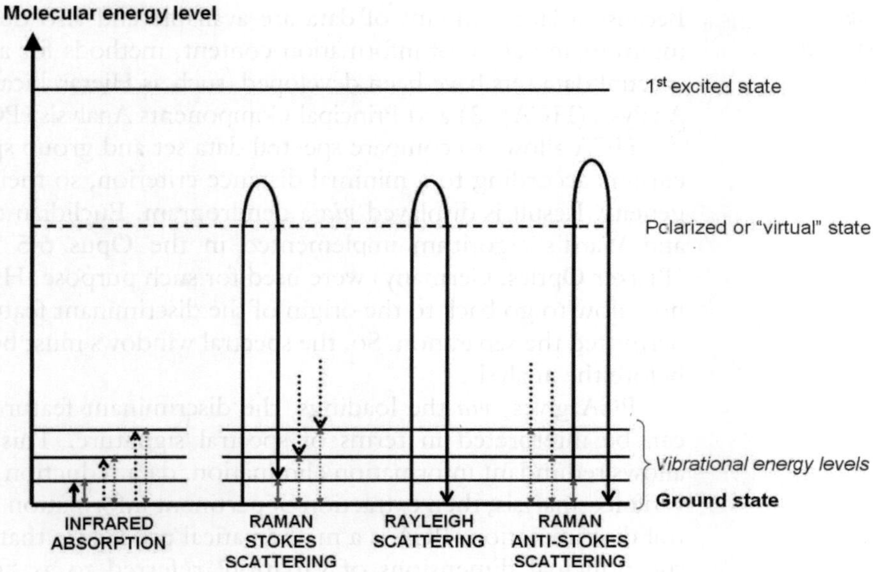

Fig. 1. Energy diagram showing transitions involved during infrared absorption and Raman scattering processes.

On the other hand, a molecule is said to be Raman-active if it undergoes a change in its polarizability during the interaction. Although the origins of the physical phenomena are not the same (absorption for IR and scattering for Raman), both techniques inform on the chemical nature of the sample. They are also complementary because certain vibrational modes that are active in IR may not be active in Raman and vice-versa. So, combining both techniques can give the full vibrational signature. These signatures consist of several vibrational modes of molecular linkages in a molecular group: symmetrical or asymmetrical stretching, scissoring, rocking, wagging, and twisting. Depending on the applications, lasers for Raman spectroscopy can range from the UV to the near-infrared. UV and visible lasers carry more energy and therefore can cause sample heating. Also, the "polarized" state can coincide with the first excited electronic level giving rise to fluorescence emission which will mask the Raman signal. Sample degradation and fluorescence can be circumvented by increasing the laser wavelength to the near-infrared (less energetic photons).

1.1.2. New Technology Contribution

Infrared and Raman spectroscopies have both benefited from last decade's methodology (e.g., Fourier Transform spectroscopy, microscope coupling, laser and multichannel detector development, holographic filters use) (2). This, associated with informatics progress and signal treatment development, has greatly increased the sensitivity of these methods, and the spectral, spatial, or temporal resolution. It is now possible to rapidly extract useful information from large data sets thanks to multivariate statistical analyses (MVA). Two of these analyses are detailed in the paragraph below.

1.1.3. Multivariate Statistical Analyses Contribution

Because a large amount of data are available and also due to the multivariate nature of information content, methods for analyzing spectral data sets have been developed, such as Hierarchical Cluster Analysis (HCA) (3) and Principal Components Analysis (PCA) (4).

HCA allows to compare spectral data set and group spectra in clusters according to a minimal distance criterion, so their homogeneity. Result is displayed *via* a dendrogram. Euclidian distances and Ward's algorithm implemented in the Opus 6.5 software (Bruker Optics, Germany) were used for such purpose. HCA does not allow to go back to the origin of the discriminant features that permitted the separation. So, the spectral windows must be chosen before the analysis.

PCA gives, *via* the loadings, the discriminant features which can be interpreted in terms of spectral signature. This analysis allows redundant information elimination, data reduction, spectral variance analysis, then extraction of pertinent information for spectral discrimination. PCA is a mathematical procedure that extracts the principal dimensions of variation, referred to as "principal components," in a set of correlated variables. Principal Components (PCs) are uncorrelated. PCA is therefore a method for visualizing high-dimensionality data by transforming them into a very low-dimensionality space (usually 1, 2, or 3D). The scores of the PCs are plotted for this visualization and each spectrum, which originally contains a large number of variables, is represented by only one point in this graph. The loadings of the PCs can also be compared to retrieve back the pertinent chemical information.

1.2. Glycosaminoglycans Analysis Application

The first infrared and Raman spectra of glycosaminoglycans (GAGs) were published more than 30 years ago (5, 6). Later studies analyzed the differences of the spectra in solid state and in aqueous solutions (7), or with different hydration rates (8, 9) or from different species (chondroitin sulfate from different sources of cartilage: crocodile, shark, chicken, ray) (10). Several peaks were identified by spectral comparison (11), especially those associated with sulfate vibrations (6, 12), and different conformations of iduronate (13). Prepared GAGs (14) or GAGs present in natural proteoglycan forms (15) were also characterized. The evolution of molecular changes of GAGs with aging was also studied (16, 17). More recently, the association of chemometrics with sensitive spectroscopic micro-techniques showed a good classification of some types of GAGs (15–18). However, these studies concerned only some particular types of GAGs or digested molecules. We recently showed the possibility to use these techniques and to obtain good differentiation and classification with all main types and some subtypes of glycosaminoglycans (19).

2. Materials

Infrared and Raman devices and optical substrates associated with each methodology are shown in Fig. 2.

2.1. Infrared Microspectroscopy

Infrared spectra were acquired with the Spectrum Spotlight 300 microspectrometer (Perkin Elmer, Courtaboeuf, France) and using the Spotlight software (Perkin Elmer). This setup (see Fig. 2A1) results from the coupling of the Fourier Transform infrared spectrometer Spectrum One (Perkin Elmer) with an infrared microscope that is equipped with both a single element MCT detector ($100 \times 100\ \mu m^2$) and a multielement detector (MCT is composed of 16 elements). The latter is more appropriate for imaging purposes. In our experiments, the measurements were carried out in a point mode using a $100 \times 100\,\mu m^2$ aperture.

2.2. Raman Microspectroscopy

Raman spectra were acquired with the Labram microspectrometer (Horiba Jobin-Yvon SAS, Lille, France) and using the LabSpec software version 4 (Horiba Jobin-Yvon SAS, Lille, France). The incident beam was a near-infrared laser excitation at 785 nm (Spectra Physics, Les Ulis, France) focused on the sample by a

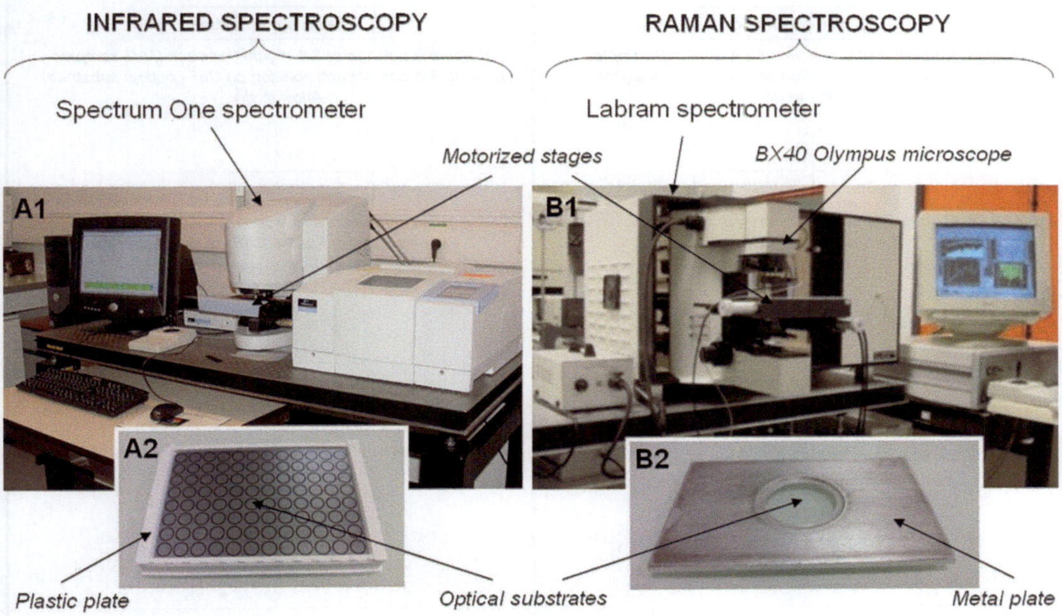

Fig. 2. (**A1**) Infrared microspectrometer Spotlight 300 (Perkin Elmer). (**A2**) Silicon 96-well optical substrate (mounted in plastic plate) for sampling in infrared microspectroscopy. This will be placed on motorized stage of spectrum one spectrometer for spectral acquisition. (**B1**) Raman microspectrometer Labram (Horiba Jobin Yvon). (**B2**) CaF_2 optical substrate (mounted in metal plate) for sampling in Raman microspectroscopy. This will be placed on motorized stage of Labram spectrometer for spectral acquisition.

BX40 Olympus microscope and a ×100 long working distance objective (see Fig. 2B1). Stokes Raman scattering was selected by a Notch filter and directed by mirrors to a 950-line diffraction grating, and then to a CCD detector.

3. Methods

An operational scheme of both infrared and Raman methodologies is displayed in Fig. 3. Four steps are distinguished: sample preparation, spectral acquisition, preprocessing of spectra, and processing by MVA.

3.1. Infrared Microspectroscopy

3.1.1. Sample Preparation

For each type of glycosaminoglycans:

1. Prepare a solution of 0.1 mg/mL (see Note 1).
2. Deposit 3 drops of 30 μL each onto 3 wells of a silicon 96-well optical substrate (Bruker Optics, Ettlingen, Germany) (see Note 2 and Fig. 2A2).
3. Dry at open air or in a dessicator under mild vacuum in presence of silicagel.

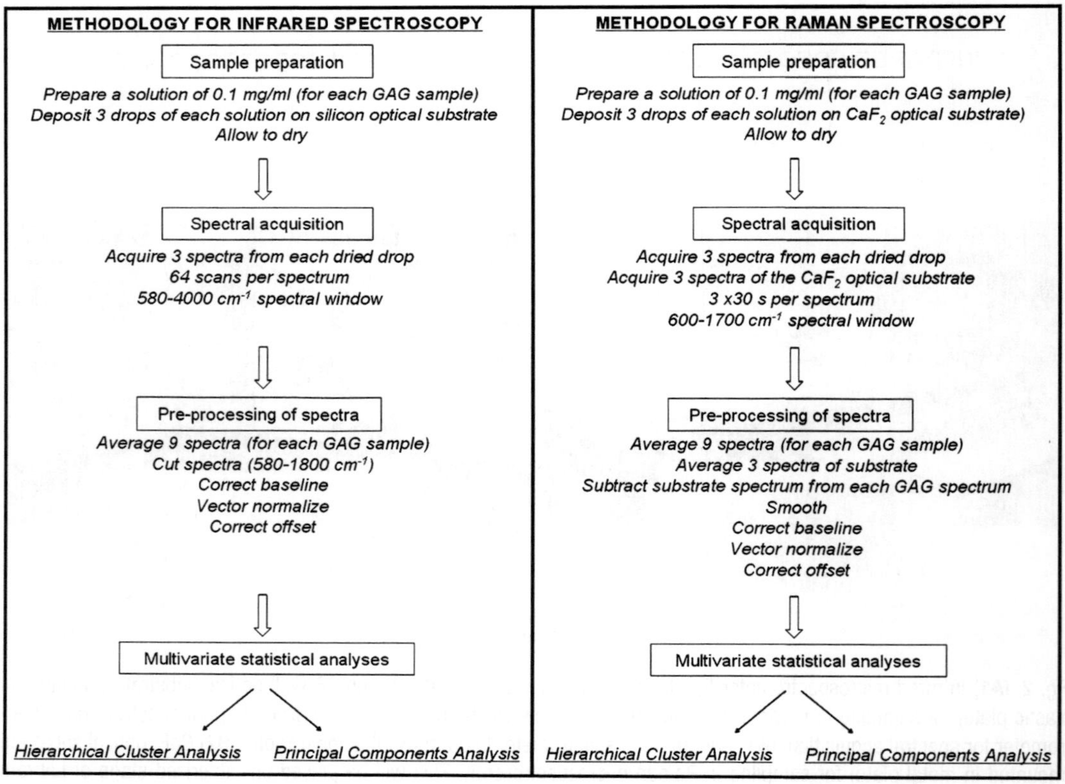

Fig. 3. Operational schemes for infrared and Raman spectroscopy.

3.1.2. Spectral Acquisition

1. Define acquisition parameters in the Spotlight software (see Note 3): point spectra (see Note 4), transmission mode, $100 \times 100\ \mu m^2$ aperture, 580–4,000 cm^{-1} spectral window.
2. Acquire background spectrum from plain optical substrate to subtract its signal from the sample signal. Choose a point of the optical substrate further from the sample deposit. Choose 128 coadded scans per spectrum on the software. The subtraction is automatically performed by the Spotlight Software.
3. Acquire spectra of each GAG sample. Inside each of the three drops, randomly choose three distinct points (see Note 4) and record a spectrum for each point. Choose on the software 64 coadded scans per spectrum.

3.1.3. Preprocessing of Spectra

Different procedures are used to preprocess spectra. This can be done with the Opus 6.5 software (see Fig. 4). For infrared spectra, three steps are involved:

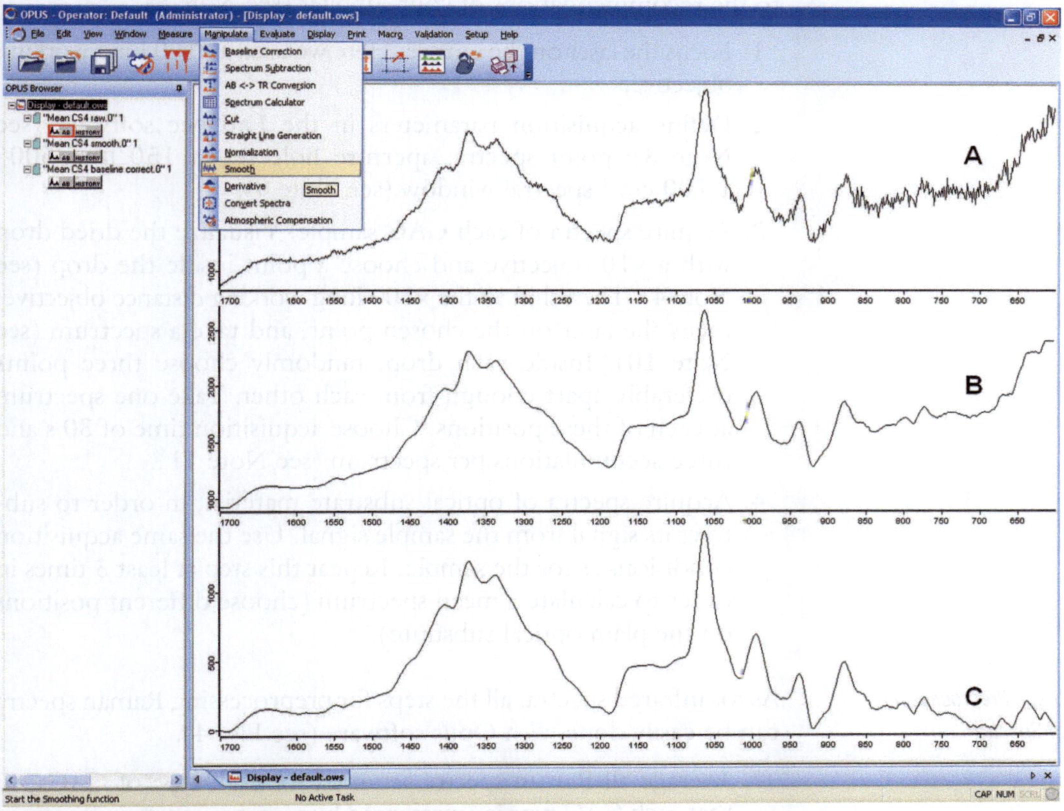

Fig. 4. Preprocessing steps with Opus software (Bruker optics). Screenshot displays mean spectrum of chondroitin-4-sulfate (C4S). Preprocessing procedures are available in the "manipulate" menu. This menu does not include all the possible preprocessing procedures by default: it has to be customized within the "view" menu. (**A**) Mean of raw spectra of C4S. (**B**) Resulting spectrum after smoothing. (**C**) Resulting spectrum after smoothing and baseline correction.

1. Average all spectra acquired for each type of GAGs, so that each type is represented by a mean spectrum (average of 9 spectra—3 from each of the 3 dried drops—each spectrum is the mean of 64 scans) (see Note 5).
2. Cut the spectra in the range from 580 to 1,800 cm^{-1} corresponding to the fingerprint spectral region of biological molecules.
3. Correct for baseline, vector normalize, and correct for offset. Save the resulting spectra.

3.2. Methods for Raman Microspectroscopy

3.2.1. Sample Preparation

For each type of glycosaminoglycans (see Note 6):

1. Prepare a solution of 0.1 mg/mL (see Note 1).
2. Deposit 3 drops of 10 μL each onto a calcium fluoride optical substrate (Crystran Ltd., Dorset, UK) (see Note 7 and Fig. 2B2).
3. Dry at open air or in a dessicator under mild vacuum in presence of silicagel.

3.2.2. Spectral Acquisition

Before use, Raman microspectrometer needs callibration. Please refer to the recommendations of your supplier (see Note 8).

1. Focus the laser on the sample. Here we used a ×100 long working objective.
2. Define acquisition parameters in the LabSpec software (see Note 3): point spectra, aperture hole set at 150 μm, 600–1,700 cm^{-1} spectral window (see Note 9).
3. Acquire spectra of each GAG sample. Visualize the dried drop with a ×10 objective and choose a point inside the drop (see Note 4). Then shift to the ×100 long working distance objective, focus the laser on the chosen point, and take a spectrum (see Note 10). Inside each drop, randomly choose three points preferably apart enough from each other. Take one spectrum at each of these positions. Choose acquisition time of 30 s and three accumulations per spectrum (see Note 11).
4. Acquire spectra of optical substrate material, in order to subtract its signal from the sample signal. Use the same acquisition conditions as for the sample. Repeat this step at least 3 times in order to calculate a mean spectrum (choose different positions on the plain optical substrate).

3.2.3. Preprocessing of Spectra

As for infrared spectra, all the steps for preprocessing Raman spectra can be easily done with Opus software (see Fig. 4).

1. Average all Raman spectra acquired for each type of GAGs, so that each GAG type is represented by a unique spectrum (average of nine spectra—three from each of the three dried drops—each spectrum is the mean of three accumulations) (see Note 5).
2. Average the three spectra acquired from the optical substrate for each type of GAGs (see Note 5).

3. Subtract the optical substrate mean spectrum from the sample mean spectrum. In this way, only the sample signal remains in the spectrum. (Please, note that this step is automatically performed in infrared spectroscopy during the spectral acquisition).
4. Smooth spectra using a 25 points Savitzky-Golay procedure (20). This step is necessary to reduce noise in Raman spectra which is more important than in infrared spectra.
5. Correct for baseline, vector normalize, and correct for offset. Save the resulting spectra.

3.3. Multivariate Statistical Analysis

3.3.1. Hierarchical Cluster Analysis

The HCA algorithm is implemented in the Opus software (see Fig. 5). Select "Cluster Analysis" in the "Evaluate" menu: a new window opens.

1. Open all the preprocessed spectra to include in the HCA.
2. Choose the standard clustering method.
3. Make the distance matrix for displaying the resulting dendrogram.

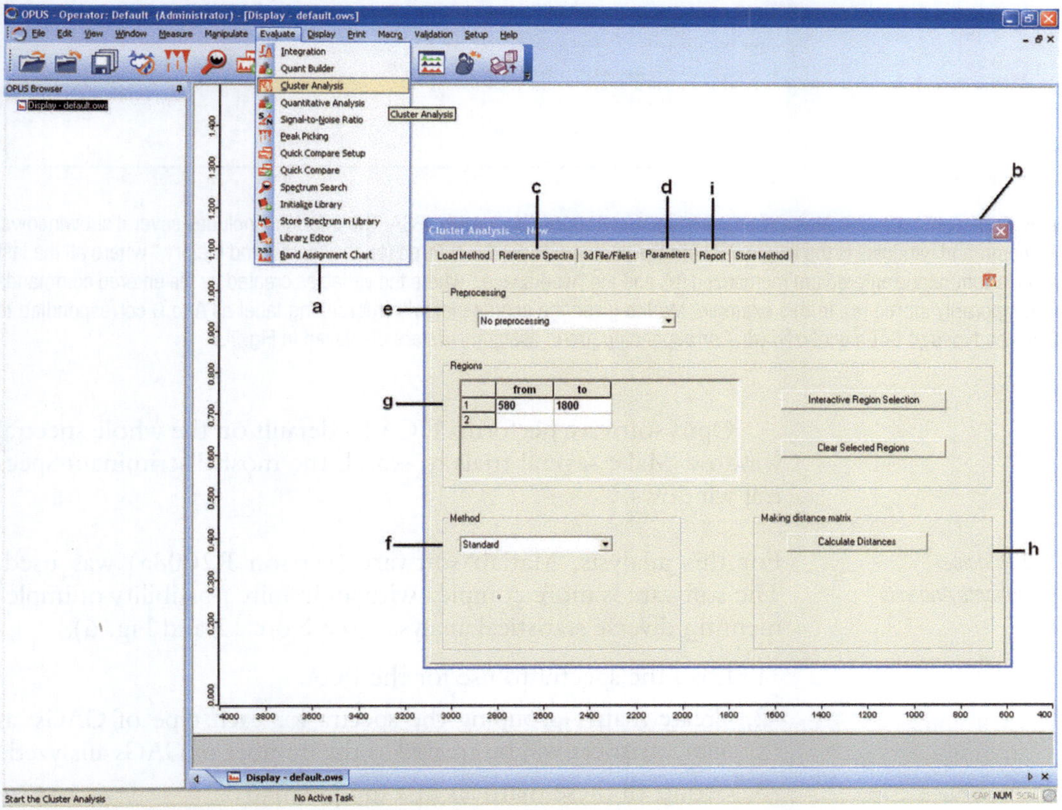

Fig. 5. Screenshot displays Opus software performing Hierarchical cluster analysis via the "evaluate" menu (**a**). A new window opens (**b**). Preprocessed spectra are loaded in the "reference spectra" tab (**c**). Cluster analysis parameters are selected in the "parameters" tab (**d**): no preprocessing (**e**), standard method (**f**). The exact limits of the window can be entered directly or selected manually with the "interactive selection" button (**g**). One can also combine several windows. The "calculate distances" button computes the distance matrix (**h**). The resulting dendrogram is displayed in the "report" tab (**i**).

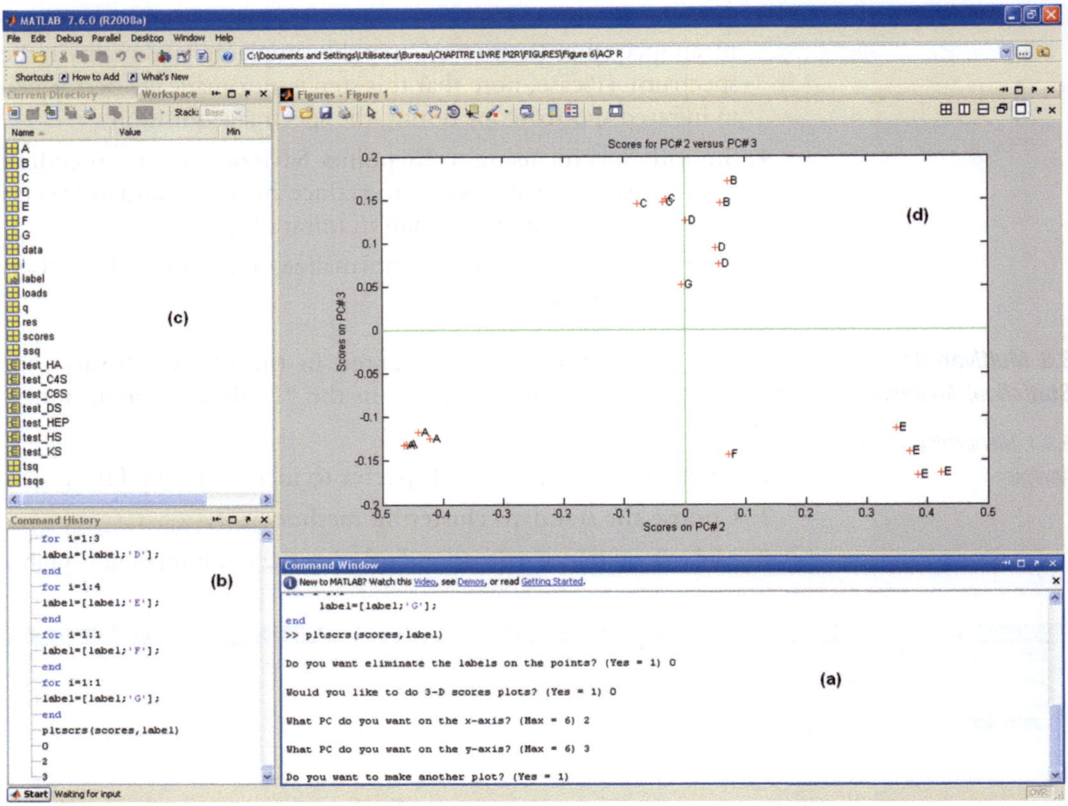

Fig. 6. Screenshot displays Matlab software (version R2008a) performing PCA. The interface includes several subwindows: the "command window" is the window where the commands will be entered (**a**), the "command history," where all the last previous commands entered are memorized (**b**), and the "workspace," where the variables created by the entered commands are temporarily stored (**c**). In this example, Matlab gives the scores plot with regrouping label as A to G corresponding to the seven types of GAGs analyzed (**d**). Corresponding group assignments are displayed in Fig. 7.

Opus software performs HCA by default on the whole spectral window. Make several trials to search the most discriminant spectral windows.

3.3.2. Principal Components Analysis

For this analysis, Matlab software (version R2008a) was used. The software is more complex with an infinite possibility of implementing diverse statistical analyses (see Note 12 and Fig. 6).

1. Load the spectra to use for the PCA.
2. Create matrix grouping the spectra for each type of GAGs: as many matrices will be created as the number of GAGs analyzed.
3. Group all these matrices in a unique matrix.
4. Perform the PCA using this final matrix and the "pca" Matlab function (see Note 13).
5. Choose the principal components to use for displaying the results (2 or 3, for a 2D or 3D plot, respectively).

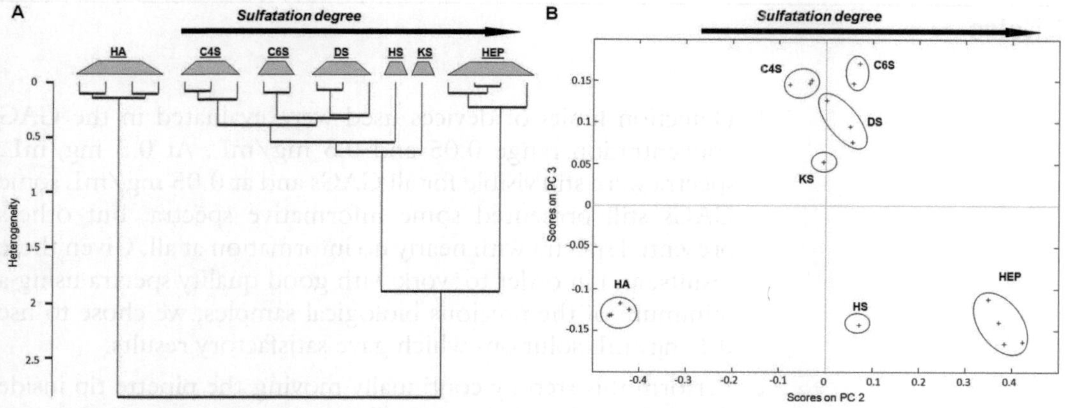

Fig. 7. Hierarchical cluster analysis of infrared spectra using the 900–1,800 cm^{-1} spectral window (**A**) and principal components analysis of Raman spectra (**B**) of main types of GAGs: hyaluronic acid (HA), dermatan sulfate (DS), chondroitin-4-sulfate (C4S), chondroitin-6-sulfate (C6S), keratan-sulfate (KS), heparan-sulfate (HS), and heparin (HEP). All types of GAGs are well distinguished. Sulfatation degree of the molecules is clearly identified.

3.4. Standard GAG Characterization by Combination of Spectroscopic and MVA Methods

This protocol was used to analyze the major types of GAGs: hyaluronic acid (HA), dermatan sulfate (DS), chondroitin-4-sulfate (C4S), chondroitin-6-sulfate (C6S), keratan-sulfate (KS), heparan-sulfate (HS), and heparin (HN). The details of GAG sources were already described (16). Hierarchical Cluster Analysis of infrared spectra and also Raman spectra gave good discrimination of all molecular types. The increasing sulfatation degree appears in the resulting dendrogram. Results were confirmed by Principal Components Analysis (see Fig. 7).

4. Conclusions

This protocol describes a direct and robust method based on the combination of spectroscopy and multivariate statistical analyses, to differentiate and classify complex macromolecules like gylcosaminoglycans. As already mentioned, infrared and Raman spectroscopies are complementary techniques, but in case of glycosaminoglycans, both carry enough molecular information that allows to differentiate molecular types and subtypes. The method is quantitative and can be used to rapidly perform quality tests and to assess GAG composition in more complex samples like tissues. It also gives information about dynamics of macromolecular changes. In the case of Raman microspectroscopy, which is a nondestructive technique, the method is adapted for in vivo approaches. In fact, numerous handled Raman devices with probe are currently in development, and it is now possible to shift from an in vitro study to an in vivo one. This allows the characterization of GAGs in living organism to obtain the insights of these macromolecules dynamics in normal and pathological processes (e.g., wound healing or aging).

5. Notes

1. Detection limits of devices used were evaluated in the GAG concentration range 0.05 and 0.5 mg/mL. At 0.5 mg/mL, spectra were still visible for all GAGs and at 0.05 mg/mL some GAGs still presented some informative spectra, but others presented spectra with nearly no information at all. Given these results, and in order to work with good quality spectra using a minimum of the precious biological samples, we chose to use 0.1 mg/mL solutions which gave satisfactory results.
2. Perform this step by continually moving the pipette tip inside the well forming circles to facilitate a homogeneous deposit (or use the robotic deposit device). This optical substrate allows to deposit numerous samples (up to 96). Another version of the optical substrate with 384 wells is also available for larger number of sample. The deposited sample volume must be adapted to the diameter of the well (e.g., 3–5 μL).
3. The Spotlight and LabSpec softwares are commercialized by Perkin Elmer and Horiba Jobin-Yvon, respectively. The use of spectrometers from other suppliers than those mentioned here implies the use of different softwares.
4. Drying of sugar solutions on the optical substrates is a process that gives an inhomogeneous sample deposit. Because of this, scanning the whole dried drop with the image mode is not applicable. As a consequence, points of interest are chosen manually at the periphery of the dried drop, where GAG molecules are more concentrated, so as to obtain a good signal-to-noise ratio.
5. Averaging reduces spectral noise and gives a representative spectrum of the studied GAGs. Care must be taken to remove any saturated or outlier spectra before averaging.
6. Raman microspectroscopy enables spectral acquisition of biochemical standards in the powdered form, without any preparation. However, we chose to work with aqueous solutions which were dried on the optical substrates, in order to perform both Raman and infrared measurements on the same sample preparation.
7. Drops of 10 μL gave satisfactory results in Raman microspectroscopy. Decreasing the volume may compromise spectral quality.
8. Before use, the Raman microspectrometer needs to be calibrated so that day-to-day measurements can be compared.
9. Spectral window 600–1,700 cm^{-1} was chosen because it represents the fingerprint spectral region for biological molecules.

In the case of GAG molecules, this spectral window contains the most relevant information.

10. In our case, the ×10 objective was used for the visualization of samples and to choose the region of interest in the dried drops. Spectra were acquired with the ×100 long working distance objective in order to collect a maximum amount of signal. Shifting from one objective to the other implies that both objectives have to be properly aligned. It is important to verify (and correct if necessary) that after objective shifting, the laser position on the sample is exactly the same.
11. Spectral quality depends on the chosen acquisition time and accumulation number (successive measurements averaged by LabSpec software). Increasing accumulation number leads to a decrease in the spectral noise (random phenomenon responsible of nonsmoothed spectra). Increasing acquisition time leads to an improvement of the Raman signal-to-noise ratio.
12. Contrary to other softwares mentioned above, the Matlab interface is not really intuitive for nonspecialists. However, the benefits of this kind of programming software are its speed for analyzing large data sets, and a wide possibility of statistical data analyses.
13. The "pca" function used is not a default function, but was included in the PLS_Toolbox supplied by Eigenvector Research Incorporated. This toolbox is a suite of essential and advanced chemometric multivariate analysis tools for use within the Matlab computational environment.

Acknowledgments

The authors' study received the financial support of the Ligue Nationale contre le Cancer (Comité de la Marne), the Fonds européen de développement régional (FEDER), and the Région Champagne-Ardenne (CPER 2007–2013).

References

1. Parker, F.S. (1983) Applications of infrared, Raman, and resonance Raman spectroscopy in biochemistry. *Springer*, Heidelberg, Germany.
2. Carrabba, M.M., Spencer, K.M., Rich, C., and Rauh, D. (1990) The utilization of a holographic Bragg diffraction filter for Rayleigh line rejection in Raman spectroscopy. *Appl Spectrosc.* **44**, 1558–1561.
3. Ward, J.H. (1963) Hierarchical grouping to optimize an objective function. *J Am Stat Assoc.* **58**, 236–244.
4. Pearson, K. (1901) On lines and planes of closest fit to systems of points in space. *Philos Mag,* **2**, 559–572.
5. Orr, S.F.D. (1954) Infra-red spectroscopic studies of some polysaccharides. *Biochim Biophys Acta.* **14**, 173–181.
6. Bansil, R., Yannas, I., and Stanley, H. (1978) Raman spectroscopy: a structural probe of glycosaminoglycans. *Biochim Biophys Acta.* **541**, 535–542.

7. Gilli, R., Kacuráková, M., Mathlouthi, M., Navarini, L., and Paoletti, S. (1994) FTIR studies of sodium hyaluronate and its oligomers in the amorphous solid phase and in aqueous solution. *Carbohydr Res.* **263**, 315–326.

8. Servaty, R., Schiller, J., Binder, H., and Arnold, K. (2001) Hydration of polymeric components of cartilage--an infrared spectroscopic study on hyaluronic acid and chondroitin sulfate. *Int J Biol Macromol.* **28**, 121–127.

9. Haxaire, K., Maréchal, Y., Milas, M., and Rinaudo, M. (2003) Hydration of polysaccharide hyaluronan observed by IR spectrometry. I. Preliminary experiments and peak assignments. *Biopolymers.* **72**, 10–20.

10. Garnjanagoonchorn, W., Wongekalak, L., and Engkagul, A. (2007) Determination of chondroitin sulfate from different sources of cartilage. Chemical Engineering and Processing: *Process Intensification.* **46**, 465–471.

11. Bychkov, S., Bogatov, V., and Kuz'mina, S. (1981) Comparative study of the IR-spectra of glycosaminoglycans and their monomers. *Biull Eksp Biol Med.* **91**, 442–445.

12. Longas, M., and Breitweiser, K. (1991) Sulfate composition of glycosaminoglycans determined by infrared spectroscopy. *Anal Biochem.* **192**, 193–196.

13. Grant, D., Long, W., Moffat, C., and Williamson, F. (1991) Infrared spectroscopy of heparins suggests that the region 750-950 cm-1 is sensitive to changes in iduronate residue ring conformation. *Biochem J.* **275**, 193–197.

14. Grant, D., Long, W., Moffat, C., and Williamson, F. (1990) Infrared spectroscopy as a method for investigating the conformations of iduronate saccharide residues in glycosaminoglycans. *Biochem Soc Trans.* **18**, 1277–1279.

15. Ellis, R., Green, E., and Winlove, C. (2009) Structural analysis of glycosaminoglycans and proteoglycans by means of Raman microspectrometry. *Connect Tissue Res.* **50**, 29–36.

16. Longas, M., Russell, C., and He, X. (1986) Chemical alterations of hyaluronic acid and dermatan sulfate detected in aging human skin by infrared spectroscopy. *Biochim Biophys Acta.* **884**, 265–269.

17. Longas, M., Russell, C., and He, X. (1987) Evidence for structural changes in dermatan sulfate and hyaluronic acid with aging. *Carbohydr Res.* **159**, 127–136.

18. Foot, M., and Mulholland, M. (2005) Classification of chondroitin sulfate A, chondroitin sulfate C, glucosamine hydrochloride and glucosamine 6 sulfate using chemometric techniques. *J Pharm Biomed Anal.* **38**, 397–407.

19. Mainreck, N., Brezillon, S., Sockalingum, G. D., Maquart, F. X., Manfait, M., and Wegrowski, Y. (2011) Rapid characterization of glycosaminoglycans using a combined approach by infrared and Raman microspectroscopies. *J Pharm Sc.* **100**, 441–450.

20. Savitzky, A. and Golay, M.J.E. (1964) Smoothing and differentiation of data by simplified least squares procedures. *Anal Chem.* **36**, 1627–1639.

Chapter 9

Glycosaminoglycans: Oligosaccharide Analysis by Liquid Chromatography, Capillary Electrophoresis, and Specific Labeling

Derek J. Langeslay, Christopher J. Jones, Szabolcs Beni, and Cynthia K. Larive

Abstract

Glycosaminoglycans (GAGs) are a class of biopolymers that include chondrotin sulfate, dermatan sulfate, keratan sulfate, hyaluronic acid, heparin, and heparan sulfate. The GAGs are linear polysaccharides that are microheterogeneous in composition and polydisperse in size. Because they have the most complex structures, this article is aimed at describing a step-by-step procedure for processing and analyzing heparin and heparan sulfate-derived oligosaccharides, although the basic protocols and procedures apply equally well to other members of the GAG family. The methods described in this manuscript include the preparation of oligosaccharides through enzymatic depolymerization, size fractionation by preparative scale size-exclusion chromatography (SEC), and disaccharide isomer analysis by reverse-phase ion-pair high-performance liquid chromatography (RPIP-HPLC) and capillary electrophoresis (CE).

Key words: Heparin, Heparan sulfate, Heparinase, Glycosaminoglycan, RPIP, Depolymerization, Isomer, Oligosaccharide, Anion-exchange, Disaccharide, Compositional analysis

1. Introduction

Heparin and heparan sulfate are of considerable interest to the scientific community. They are comprised of disaccharide building blocks of alternating iduronic acid (IdoA) or glucuronic acid (GlcA) and glucosamine (GlcN) residues. A key difference between heparin and heparan sulfate is the ratio of sulfonated disaccharides incorporated into the polymer. Heparin is more highly sulfonated, while heparan sulfate contains more *N*-acetylglucosamine residues. Heparan sulfate also contains a greater proportion of GlcA residues. The complex nature of glycosaminoglycans (GAGs) leads to complica-

Françoise Rédini (ed.), *Proteoglycans: Methods and Protocols*, Methods in Molecular Biology, vol. 836, DOI 10.1007/978-1-61779-498-8_9, © Springer Science+Business Media, LLC 2012

tions in their analysis. As chain lengths increase, sequence information becomes more difficult to extract. To compensate for this complexity, GAGs are often analyzed after depolymerization to smaller oligosaccharides, e.g., disaccharides, which can be more easily structurally characterized. Various methods for GAG depolymerization have been developed. The most common methods include enzymatic, base catalyzed beta-elimination, oxidation, and reductive deamination. Each of these methods has its own benefits and drawbacks which should be considered when selecting a depolymerization method. This chapter is aimed at providing information and references for all these methods, but the discussion will focus primarily on the enzymatic treatment of heparin and heparan sulfate.

1.1. Depolymerization

Often, GAG depolymerization is performed in preparation for biological studies. In this case, structural changes resulting from the depolymerization reaction must be considered. Enzymatic depolymerization introduces an unsaturated bond at the nonreducing end of the oligosaccharide (1). The geometry of the double bond forces the uronic acid ring to adopt a planar conformation causing loss of the chiral center that distinguishes IdoA from GlcA. Beta-elimination reactions have a similar effect on the nonreducing end (2), while requiring additional steps to esterify the carboxylate groups (3).The β-elimination is base catalyzed and the high pH can result in the formation of a bicyclic acetal between the positions C-1 and C-6 of the reducing-end monosaccharide (4). Oxidative depolymerization involving hydrogen peroxide and Cu(II) or Fe(II) results in the destruction of unsulfonated uronic acid residues (5) and residual oxidized uronic acid fragments at the reducing end (6). Reductive deamination using nitrous acid causes a ring contraction at the reducing end resulting in the conversion of terminal *N*-sulfonated D-glucosamines to anhydro-D-mannose residues (7). This process also introduces a reactive carbonyl at the reducing end which can be used for conjugation or can be reduced to a hydroxyl moiety. The glycosidic bonds of GAGs can also be hydrolyzed at low pH; unfortunately, low pH conditions also cause desulfonation, which eliminates much of the information regarding the oligosaccharide's identity.

1.2. Labeling Reactions

There are many methods for labeling GAG-derived oligosaccharides. Typical approaches involve conjugating an amine to the carboxylate groups of an uronic acid residue. This is a multistep process that usually uses a carbodiimide reagent. This has been shown to be effective for disaccharide analysis (8), but because each disaccharide contains a carboxylate group, larger oligosaccharides can be multiply labeled, complicating quantitative analysis. Another common technique for labeling depolymerized oligosaccharides is reductive amination, which conjugates an amine to the anomeric carbon at the reducing end of the oligosaccharide. Although this method has the advantage of introducing a single label to each oligosaccharide, it requires extended reaction times under harsh reducing conditions and often using hazardous chemicals (9). Nitrous acid cleavage produces a reactive carbonyl group suitable for introduction of a label. A common approach is reduction of the carbonyl using tritiated sodium borohydride to introduce a sensitive radiolabel tag (10). Because a radioactive tag is not optimal for all applications, an alternative labeling scheme reacts the carbonyl group with a hydrazine derivative to form a Schiff base (11). This is a fast and relatively simple process, but the linkage is susceptible to cleavage by hydrolysis which can counteract its benefits. Common tags include chromophores, fluorophores, and biotin containing compounds.

An alternative specific labeling scheme is the introduction of an unsaturated bond by enzymatic depolymerization. Specific enzymes have been isolated that permit digestion of each member of the GAG family. Often isolated from *Flavobacterium heparinum*, the heparinase enzymes are extremely specific to heparin and heparan sulfate (1). Their specificity is subclassified into heparinase I, II, and III isoforms, as illustrated in Fig. 1. Heparinase I cleaves the polymer chain between *N*-sulfonated GlcN and 2-*O*-sulfonated IdoA residues. This motif is especially common in porcine mucosal heparin making heparinase I ideal for digesting pharmaceutical heparins. Heparinase I can also tolerate 3-*O*-sulfonation of GlcNS residues, making the Antithrombin III binding site a suitable substrate. Heparinase III cleaves specifically at sites between *N*-acetylated or *N*-sulfonated GlcN and 2-*O*-unsubstituted IdoA, a common non-sulfonated motif found in heparan sulfate, making heparinase III

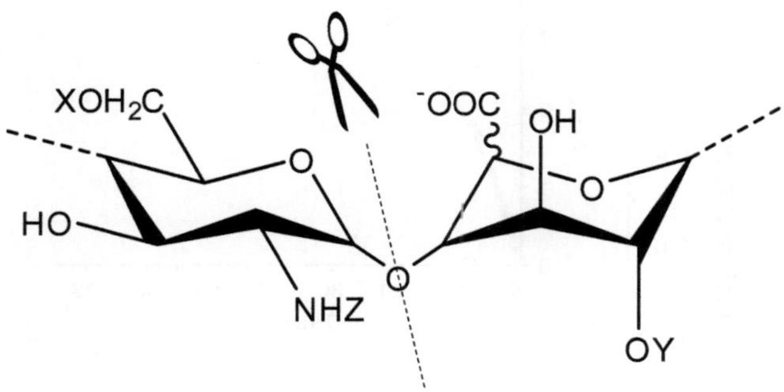

Heparinase I[a]	Heparinase II[b]	Heparinase III[b]
X = SO_3^- or H	X= SO_3^- or H	X = SO_3^- or H
Y = SO_3^-	Y= SO_3^- or H	Y = H
Z = SO_3^-	Z = SO_3^- or Ac	Z = SO_3^- or Ac
[a]Hexuronic acid must be IdoA		
[b]Hexuronic acid can be IdoA or GlcA		

Fig. 1. Cleavage specificity of the heparinase enzymes. Enzyme specificity is based on the substitution at locations X, Y, and Z.

ideal for heparan sulfate depolymerization. Heparinase II is less specific to substitution patterns, cleaving between GlcN residues that can be either *N*-sulfonated or *N*-acetylated and IdoA residues that can be 2-*O*-sulfonated or unsubstituted (1). The enzymatic depolymerization reaction can take several days; a kinetic profile for a heparinase I digestion of porcine intestinal mucosa heparin is shown in Fig. 2. An advantage of the relatively slow reaction rate is that it allows the degree of depolymerization to be fine-tuned. Not every case calls for exhaustive depolymerization and this reaction can be terminated prior to complete digestion to yield larger oligosaccharides which may be more relevant for structural characterization or protein-binding studies.

1.3. Size-Exclusion Chromatography (SEC)

If an exhaustive depolymerization is carried out, the GAG is reduced essentially to its component disaccharides and the digest solution can be analyzed directly by high-performance liquid chromatography (HPLC) or capillary electrophoresis (CE). However, if the depolymerization targets larger oligosaccharides, the first step in their separation is usually a fractionation based on size. As shown in Fig. 2, the SEC separation resolves the depolymerized oligosaccharides into peaks of uniformly sized even-numbered saccharide units. Collection of individual SEC peaks gives size-uniform fractions

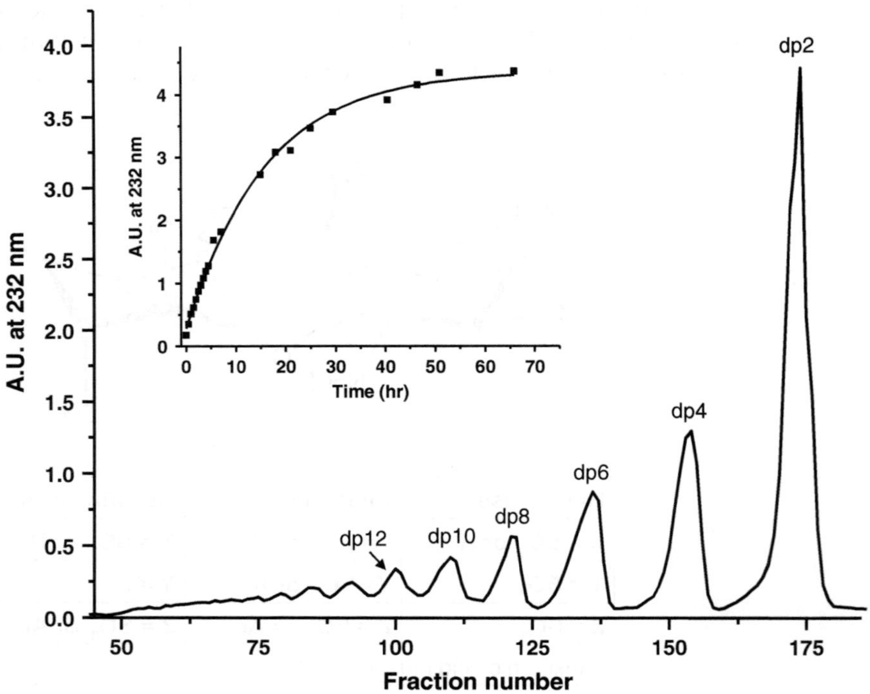

Fig. 2. Size fractionation by SEC of 1 g of porcine mucosal heparin digested with heparinase I. The separation was carried out on a 3 × 200-cm column as described in this method. The inset shows the extent of depolymerization as a function of time. The plateau in absorbance indicates that the reaction has ceased.

which can be further separated into their components in a subsequent HPLC or CE step. A high concentration of a volatile salt such as ammonium bicarbonate is generally used as the SEC eluent to prevent electrostatic interaction between the highly charged oligosaccharide and the stationary phase. The choice of volatile salts facilitates their removal by sequential lyophilization.

1.4. Reverse-Phase Ion-Pair High-Performance Liquid Chromatography (RPIP-HPLC)

Reverse-phase ion-pair HPLC (RPIP-HPLC) is a popular method for resolution of disaccharides obtained by exhaustive digestion or larger oligosaccharide fractions isolated by preparative scale SEC, although strong anion-exchange chromatography (SAX) can also be used. In RPIP-HPLC, a lipophilic ion of opposite charge to the analyte, known as an ion-pairing reagent (IPR), is used as a mobile phase modifier. The IPR aids in the retention and resolution of the charged analyte through electrostatic interactions creating a relatively neutral, hydrophobic ion-pair that can interact with the hydrophobic stationary phase of the reverse-phase HPLC column (12–14). Studies have demonstrated the utility of this method for the separation of many types of GAG oligosaccharides, including its use in determining the full disaccharide compositional analysis of heparin and HS (15). Combined with the specific labeling obtained through enzymatic depolymerization, this separation method allows the detection of individual isomeric oligomers using UV absorption at 232 nm (16). For analysis of GAG samples that are available in only trace quantities, specific labeling using fluorescent tags provides increased detection sensitivity due to the exceptionally low background of this method (17, 18).

While both UV absorbance and fluorescence detection are simple straight-forward methods that take advantage of specific labels, it is often necessary to identify an unknown peak using a detection method such as mass spectrometry (MS) that provides additional structural information. Using volatile IPR's several studies have demonstrated the amenability of RPIP-HPLC to MS detection (19–23). The use of smaller-sized column packing (2 μm) to achieve higher resolution in the RPIP-HPLC separation of heparin and HS has proven to be effective (24). Korir et al. showed that the full resolution of all 11 commercially available heparin and heparan sulfate disaccharides, including isomers, was possible through RPIP-ultra performance liquid chromatography (UPLC) using a commercially available UPLC in under 5 min with short equilibration times between runs (25). Recent work has improved upon this separation while also using the heparin disaccharides to probe the mechanism of RPIP (26). The speed and high-resolution capabilities of RPIP-UPLC make it an excellent choice for the full compositional analysis of heparin and heparan sulfate. Figure 3 shows the results obtained for the RPIP-UPLC separation of commercially available disaccharide standards and a heparin API digest sample (Table 1).

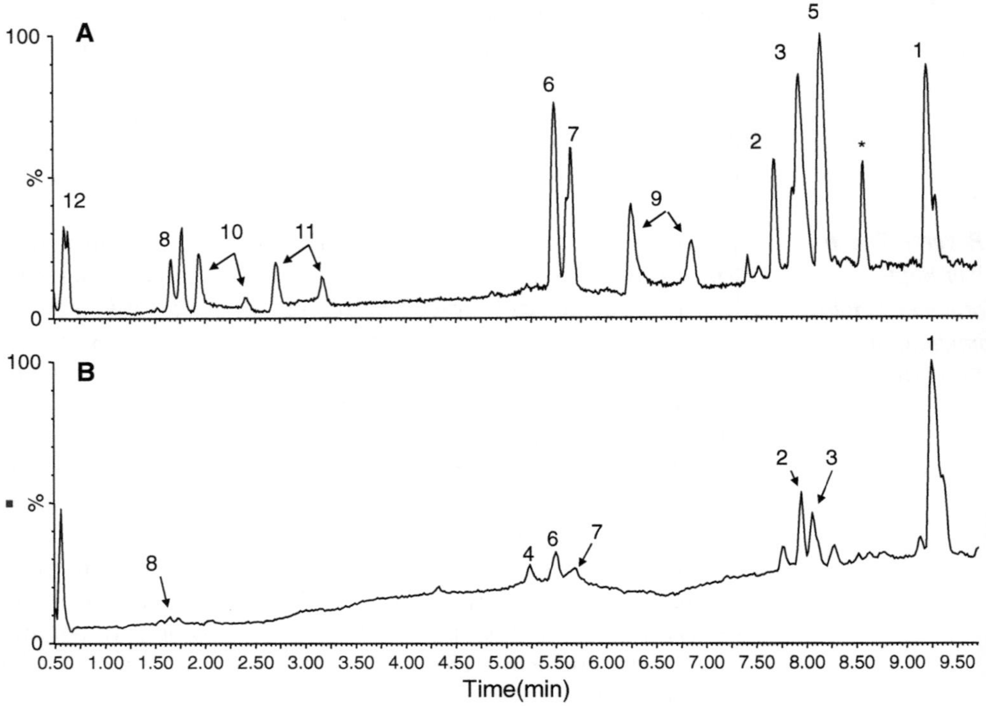

Fig. 3. Profile of heparin and heparan sulfate-derived disaccharides. Disaccharide designations are given in Table 1. (**a**) Chromatogram of disaccharide standards generated by the RPIP-UPLC-MS method described herein. The peak marked with an asterisk is the internal standard $\Delta^{4,5}$UA2S-GlcNCOEt6S. (**b**) Total ion chromatogram showing the compositional analysis of a heparin API sample treated to exhaustive digestion with a cocktail containing all three heparinase enzymes.

1.5. Capillary Electrophoresis (CE)

Among electrophoretic methods, CE provides solutions for both separation and physicochemical characterization of GAG disaccharides obtained by exhaustive depolymerization or of SEC fractions containing larger oligosaccharides. The advantages of CE over other chromatographic approaches are its high number of theoretical plates, rapid analysis time, and low sample consumption. This technique is commonly used with UV or fluorescence detection and therefore is well suited for analysis of oligosaccharides tagged with a specific label. It is also amenable to hyphenation with MS. Because of the negative charge of GAG oligosaccharides, CE separations are typically conducted in the reversed polarity configuration in which the sample is introduced at the negative end of the capillary and the analytes migrate in the opposite direction of the electro-osmotic flow (EOF), as shown in Fig. 4. At lower pH values, the EOF is naturally low, and as phosphate buffer is well known to coat the capillary surface, good reproducibility of migration times can be obtained (27). The pH selected for reversed polarity CE separations of heparin-derived disaccharides is typically

Table 1
Disaccharide structural identity, charge-to-mass ratio of disaccharide molecular ion, and electrophoretic mobility as reported by Eldridge et al. (28)

Number	Structure	m/z $[M-H]^-$	μ (10–5 cm^2/sV)
1	$\Delta^{4,5}$ UA2S-GlcNS6S	576	−47.1 ± 0.00
2	$\Delta^{4,5}$ UA-GlcNS6S	496	−37.5 ± 0.23
3	$\Delta^{4,5}$ UA2S-GlcNS	496	−38.5 ± 0.03
4	$\Delta^{4,5}$ UA-GlcNS	416	NA
5	$\Delta^{4,5}$ UA2S-GlcNAc6S	538	−37.2 ± 0.17
6	$\Delta^{4,5}$ UA-GlcNAc6S	458	−25.1 ± 0.02
7	$\Delta^{4,5}$ UA2S-GlcNAc	458	−26.0 ± 0.17
8	$\Delta^{4,5}$ UA-GlcNAc	378	−12.1 ± 0.16
9	$\Delta^{4,5}$ UA2S-GlcN6S	496	−27.7 ± 0.09
10	$\Delta^{4,5}$ UA-GlcN6S	416	−13.4 ± 0.19
11	$\Delta^{4,5}$ UA2S-GlcN	416	−14.0 ± 0.14
12	$\Delta^{4,5}$ UA-GlcN	336	2.02 ± 0.08

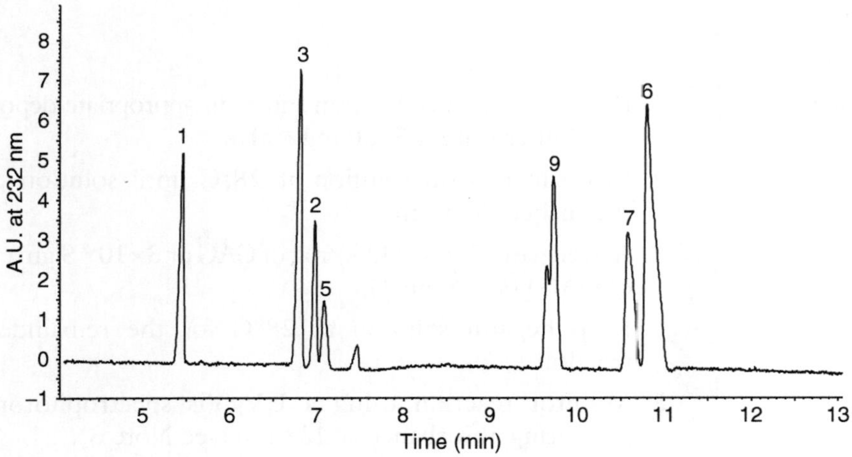

Fig. 4. CE electropherogram of the heparin and heparan sulfate-derived disaccharide standards.

around 3.5–4.0, near the pK_a values of the uronic acid carboxylate groups. The differential effective charge resulting from the subtle effects of structure on carboxylic acid pK_a values allows the resolution of structurally related disaccharides such as configurational isomers which would otherwise have very similar migration times (28).

2. Materials

1. Heparinase I depolymerization buffer: 100 mM Tris (12.11 g/L), 2.5 mM calcium acetate (0.395 g/L), pH 6.8.
2. Heparinase II depolymerization buffer: 20 mM Tris (2.42 g/L), calcium acetate (0.395 g/L), 50 mM sodium chloride (1.17 g/L), pH 7.3.
3. Heparinase III depolymerization buffer: 20 mM Tris (2.42 g/L), 2.5 mM calcium acetate (0.395 g/L), 20 mM sodium chloride (1.17 g/L), pH 7.5.
4. SEC buffer: 500 mM ammonium bicarbonate (39.53 g/L), 0.02% (0.2 g/L) sodium azide, pH 7.8 (see Note 1).
5. Reverse-phase ion-pair HPLC buffer A: 95% HPLC grade water, 5% Optima grade acetonitrile, 20 mM tributylamine (3.7 g/L), 2.5 mM ammonium acetate (0.19 g/L).
6. Reverse-phase ion-pair HPLC buffer B: 50% HPLC grade water, 50% Optima grade acetonitrile, 20 mM tributylamine (3.7 g/L), 2.5 mM ammonium acetate (0.19 g/L).
7. CE buffer: 50 mM mono sodium phosphate (5.9 g/L) pH 3.5.

3. Methods

3.1. Enzymatic Depolymerization

1. Dissolve heparin or heparan sulfate in appropriate depolymerization buffer (target 5–20 mg/mL).
2. Incubate heparin solution at 28°C until solution comes to thermal equilibrium.
3. Add enzyme (5×10^{-5} I.U./mg of GAG or 3×10^{-2} Sigma Unit/mg of GAG) (see Note 2).
4. Keep heparin solution at 28°C for the remainder of the reaction.
5. Monitor reaction using a UV/Vis spectrophotometer by measuring absorbance at 232 nm (see Note 3).
6. Lyophilize depolymerized heparin solution and store at −20°C until ready to separate.

3.2. Size-Exclusion Chromatography

1. Hydrate stationary phase by adding dry gel to 3 volumes of degassed SEC buffer in an oversized beaker and let sit for 4 h. Don't stir the solution; stirring can damage the gel and create more fines (see Notes 4–6).
2. Define gel by gently swirling the hydrated gel solution until the particles become suspended; let the solution sit covered until ~95% of the gel has settled and then remove as much of

the supernatant as possible without disturbing the settled gel bed. Add 1 volume of buffer back to the beaker and repeat this procedure 4 times for best results. At the end of the last define step, do not add buffer back to the beaker. There should be just enough liquid to cover the top of the gel bed.

3. Pack column by first filling it 20% full with SEC buffer. Gently swirl the beaker containing the gel until it all becomes suspended. Once suspended, quickly and in one fluid motion pour the gel into the column. Use of a funnel is recommended.
4. Let the gel settle until it forms a gel bed of 5–10 cm and start the column flow to pack the column. Flow buffer through column until the gel bed height stops decreasing, this signifies that the column is sufficiently packed.
5. Equilibrate column by passing 2–3 column volumes of SEC buffer at the desired separation flow rate. A linear flow rate of 0.01–0.02 cm/min is recommended (see Note 7).
6. Stop feeding the column buffer and allow the column to drain until there is approximately 0.5 mm of SEC buffer above the gel bed.
7. Add sample to the top of the column and allow it to load onto the column until there is a minimal amount of sample above the gel bed. The sample volume shouldn't be more than about 1% of total column volume. Dilute the sample with SEC buffer if it is excessively viscous (see Note 8).
8. Restart flow to column.
9. Collect column eluent in 2–4 mL fractions.
10. Assay fractions using UV/Vis spectrophotometer at a wavelength of 232 nm and plot absorption vs. fraction number (see Note 9).

3.3. Compositional Analysis by RPIP-HPLC

1. Compositional analysis should be performed on GAG samples subjected to exhaustive digestion to reduce all of the material to the constituent disaccharides. Analysis of SEC fractions containing larger components, such as tetra- or hexasaccharides, may require some modifications to the disaccharide separation method recommended here (see Note 10). Reconstitute the lyophilized material in an appropriate volume of HPLC grade water to reach an overall disaccharide concentration of 0.2–0.4 mM (see Note 11). For MS detection, the disaccharide $\Delta^{4,5}$UA2S-GlcNCOEt6S or a similar surrogate should be added at a concentration of 0.2 mM as an internal standard for quantitation.
2. Solvents for RPIP-HPLC consist of a binary solvent system for gradient elution. Solvent A should consist of 5% acetonitrile in water, while solvent B should contain 50% acetonitrile in water. Each buffer should also contain 20 mM tributylamine (TrBA) and 2.5 mM ammonium acetate titrated to pH 6.5 with glacial acetic acid (see Note 12).

3. The specifics of the separation described herein pertain to the UPLC separation. This method can be easily adapted with minor modification to a standard HPLC system (see Note 13). A Waters Corporation 2.1 × 100 mm Acquity™ UPLC BEH C18 column with 1.7 μm particles is recommended for all RPIP-UPLC chromatographic separations. Additionally, a guard column packed with the same 1.7 μm C18 particles should be utilized prior to the analytical column.
4. Prior to starting a chromatographic separation, the column should be equilibrated with the running buffer at a mobile phase composition of 50/50 solvent A and B for 30 min at a flow rate of 0.5 mL/min.
5. For full disaccharide compositional analysis, the following gradient elution method should be used: Start with a 1-min. isocratic step of 100% solvent A after which the fraction of solvent B should be increased to 4.8% over the next 1.5 min. The fraction of solvent B is then increased to 40% over the next 2.5 min and maintained at 40% for 1 min before it is increased to 56% over a 1-min period. Increase the fraction of solvent B over the next 3 min to 96%. Finally, the fraction of solvent B is increased to 100% over the next 2 min. A 5-min equilibration should be utilized prior to the next injection. The column temperature for the experiment should be maintained at 40°C and the flow rate should be kept at a constant 0.5 mL/min. Typical sample injection volumes are between 10 and 20 μL.
6. For UV detection, the separation should be monitored at a wavelength of 232 nm corresponding to the absorbance maximum of the double bond of the disaccharides.
7. For detection by MS (see Note 14), negative mode ESI-MS should be used. Ionization conditions should be optimized to minimize in-source collision-induced dissociation as the sulfate groups are easily cleaved in the gas phase under harder source conditions.
8. For absolute identification of each chromatographic peak, a sample of heparin-derived disaccharide standards should be run both before and after the chromatographic separation of the disaccharide sample from the digest. Standard concentrations should be around 0.2 mM for each disaccharide.

3.4. Compositional Analysis by CE

1. Compositional analysis should be performed for GAG samples subjected to exhaustive digestion to reduce all of the material to the constituent disaccharides. Alternatively, it is also possible to analyze the pooled and lyophilized SEC fractions corresponding to the disaccharide peak in the SEC chromatogram. Analysis of SEC fractions containing larger components, such as tetra- or hexasaccharides, may require some modifications to

the disaccharide separation method recommended here (see Note 15). Reconstitute the lyophilized material in an appropriate volume of 50 mM phosphate buffer pH 3.50 to reach an optimal concentration range of 0.02–1.2 mM for each component (see Note 11). As a reference marker, benzene sulfonic acid (BSFA) should be added at a concentration of 0.5 mM to account for fluctuations in EOF (see Note 16). Filter the sample and all solutions through a 0.2-μm membrane filter prior to analysis.

2. For the separation, a fused silica capillary with 75 μm i.d. and a minimum length of 50 cm is recommended (see Note 17). Absorbance should be monitored at 232 nm and the capillary should be maintained at a constant temperature of 25°C.
3. Between each separation, condition the capillary by rinsing with 0.1 M NaOH for 1 min and HPLC grade water for 5 min, followed by a rinse with 50 mM buffer at pH 3.5 for 5 min, all at 20 psi pressure.
4. A typical sample injection is 7 s long at 0.5 psi.
5. Separate the disaccharides using separation voltage of 23.3 kV with a 35-min separation run time (see Note 18).

4. Notes

1. Ammonium bicarbonate is chosen because it can be removed by successive lyophilization steps and eliminates an additional desalting step.
2. For frozen enzyme solutions, be sure to thaw over ice as thawing at room temperature can greatly reduce enzyme activity. Liquid enzyme samples come in set volumes and freeze/thaw steps may reduce activity. If performing a depolymerization on a small amount of material, it may be better to use enzyme sold in lyophilized form. Also note, 1 I.U. of heparinase = ~600 Sigma units.
3. Once the absorbance stops increasing, the reaction is complete. The reaction may take up to 4 days. Also note that allowing this reaction to progress until the enzyme is deactivated may produce a polydisperse mixture of oligosaccharides of various sizes. For exhaustive digestion for compositional analysis, it is typically necessary to use a cocktail containing all three heparinase enzymes. Naturally, increasing the concentration of the enzyme will also help to ensure complete digestion.
4. This material is hazardous! Use a respiratory mask when handling dry polyacrylamide gel.

5. For Bio-Gel P10 fine gel, 1 g dry weight is approximately 7.5 mL of gel.
6. BioRad Bio-Gel P10 Fine is a recommended stationary phase. This gel works very well in longer (>1 m) columns and can facilitate separation up to octadecasaccharides at an appropriate flow rate. For smaller columns (<1 m), the Biorad Bio-Gel P6 is a good choice. It will not resolve as many peaks but works very well for smaller oligosaccharides like hexa-, tetra-, and disaccharides.
7. The recommended flow rate is significantly lower than that recommended by the manufacturer, but is required for peak resolution. If the column is already packed, usually passing 1.5–2 column volumes through the column is sufficient.
8. Be careful not to disrupt the gel bed surface.
9. To save time, assay every 2–3 fractions.
10. To separate larger oligosaccharides, the starting solvent B percentage should be changed to 25%. A linear gradient to 100% solvent B can then be used over 12 min to elute the oligosaccharides. The length and slope of this solvent ramp can be optimized to increase resolution of desired oligosaccharides.
11. Lower concentrations can be used, but less abundant disaccharides may be difficult to detect. Concentration can be determined from the UV absorbance at 232 nm using the molar extinction coefficient of 5,500 M^{-1} cm^{-1} (29).
12. Buffers should be prepared initially as a 40-mM TrBA and 5.0 mM ammonium acetate or ammonium formate buffer in water and titrated to the desired pH with glacial acetic acid. The buffers can then be diluted to their final concentration with the appropriate amounts of water and acetonitrile.
13. To transfer methods to a standard reverse-phase HPLC column, the time for gradient changes recommended for UPLC columns should be lengthened to allow adequate retention and resolution of disaccharides. These changes will depend highly on the type and length of column used. To achieve similar results to the UPLC separation, a longer HPLC column is recommended to provide a similar number of theoretical plates to a 100-mm UPLC column. Also a higher flow rate is recommended to prevent diffusional broadening.
14. MS will yield higher sensitivity as well as provide structural information that UV alone cannot.
15. To separate larger oligosaccharides, the pH of the separation buffer should be further optimized in order to separate slightly different structures. Typically, a buffer in the range of pH 3.5–4.3 can be used to take advantage of subtle differences in carboxylate basicities. The separation voltage should also be

optimized to achieve satisfactory migration differences between sulfonated isomers.

16. A lower EOF results in a shorter separation.

17. The reported method is compatible with a Beckman Coulter ProteomeLab PA800 CE instrument. Although the method can be adapted for use with instruments from other manufacturers, some modification will likely be necessary.

18. The separation voltage is capillary length-dependent, ensuring the same voltage drop along the capillary results in a comparable separation.

Acknowledgments

The authors gratefully acknowledge financial support from the National Science Foundation grant CHE 0848976. Sz. B. gratefully acknowledges support from KTIA-OTKA MB08A/80066.

References

1. Desai, U. R., Wang, H. M., and Linhardt, R. J. (1993) Specificity studies on the heparin lyases from flavobacterium-heparinum. *Biochemistry* *32*, 8140–8145.
2. Linhardt, R. J., Loganathan, D., Alhakim, A., Wang, H. M., Walenga, J. M., Hoppensteadt, D., and Fareed, J. (1990) Oligossacharide mapping of low-molecular-weight heparins - Structure and activity differences. *J. Med. Chem.* *33*, 1639–1645.
3. Mardiguian, J. (April 3, 1984) U.S. Patent 4440926.
4. Mascellani, G., Guerrini, M., Torri, G., Liverani, L., Spelta, F., and Bianchini, P. (2007) Characterization of di- and monosulfated, unsaturated heparin disaccharides with terminal N-sulfated 1,6-anhydro-beta-D-glucosamine or N-sulfated 1,6-anhydro-beta-D-mannosamine residues. *Carbohyd. Res.* *342*, 835–842.
5. Vismara, E., Pierini, M., Mascellani, G., Liverani, L., Lima, M., Guerrini, M., and Torri, G. (2010) Low-molecular-weight heparin from Cu^{2+} and Fe^{2+} Fenton type depolymerisation processes. *Thromb. Haemostasis* *103*, 613–622.
6. Vismara, E., Pierini, M., Guglieri, S., Liverani, L., Mascellani, G., and Torri, G. (2007) Structural modification induced in heparin by a Fenton-type depolymerization process. *Semin. Thromb. Hemost.* *33*, 466–477.
7. Shively, J. E. and Conrad, H. E. (1976) Formation of anhydrosugars in chemical depolymerization of heparin. *Biochemistry* *15*, 3932–3942.
8. Rassi, Z. E., Postlewait, J., Mechref, Y., and Ostrander, G. K. (1997) Capillary electrophoresis of carboxylated carbohydrates: III. Selective precolumn derivatization of glycosaminoglycan disaccharides with 7-aminonaphthalene-1,3-disulfonic acid fluorescing tag for ultrasensitive laser-induced fluorescence detection. *Anal. Biochem.* *244*, 283–290.
9. Roger, O., Colliec-Jouault, S., Ratiskol, J., Sinquin, C., Guezennec, J., Fischer, A. M., and Chevolot, L. (2002) Polysaccharide labeling: impact on structural and biological properties. *Carbohyd. Polym.* *50*, 273–278.
10. Bienkowski, M. J. and Conrad, H. E. (1985) Structural charecterization of the oligosaccharides formed by depolymerization of heparin with nitrous-acid. *J. Biol. Chem.* *260*, 356–365.
11. Kariya, Y., Herrmann, J., Suzuki, K., Isomura, T., and Ishihara, M. (1998) Disaccharide analysis of heparin and heparan sulfate using deaminative cleavage with nitrous acid and subsequent labeling with paranitrophenyl hydrazine. *J. Biochem.* *123*, 240–246.
12. Cecchi, T. (2008) Ion Pairing Chromatography. *Crit. Rev. Anal. Chem.* *38*, 161–213.
13. Cecchi, T. (2009) Ion-Pair Chromatography and Related Techniques. Taylor & Francis Group.
14. Cecchi, T., Pucciarelli, F., and Passamonti, P. (2001) Extended thermodynamic approach to ion interaction chromatography. *Anal. Chem.* *73*, 2632–2639.

15. Karamanos, N. K., Vanky, P., Tzanakakis, G. N., Tsegenidis, T., and Hjerpe, A. (1997) Ion-pair high-performance liquid chromatography for determining disaccharide composition in heparin and heparan sulphate. *J. Chromatogr. A* ***765***, 169–179.
16. Jones, C. J., Membreno, N., and Larive, C. K. (2010) Insights into the mechanism of separation of heparin and heparan sulfate disaccharides by reverse-phase ion-pair chromatography. *J. Chromatogr. A* ***1217***, 479–488.
17. Imanari, T., Toida, T., Koshiishi, I., and Toyoda, H. (1996) High-performance liquid chromatographic analysis of glycosaminoglycan-derived oligosaccharides. *J. Chromatogr. A* ***720***, 275–293.
18. Sinnis, P., Coppi, A., Toida, T., Toyoda, H., Kinoshita-Toyoda, A., Xie, J., et al. (2007) Mosquito heparan sulfate and its potential role in malaria infection and transmission. *J. Biol. Chem.* ***282***, 25376–25384.
19. Doneanu, C. E., Chen, W., and Gebler, J. C. (2009) Analysis of oligosaccharides derived from heparin by ion-pair reversed-phase chromatography/mass spectrometry. *Anal. Chem.* ***81***, 3485–3499.
20. Henriksen, J., Roepstorff, P., and Ringborg, L. H. (2006) Ion-pairing reversed-phased chromatography/mass spectrometry of heparin. *Carbohydr. Res.* ***341***, 382–387.
21. Kuberan, B., Lech, M., Zhang, L., Wu, Z. L., Beeler, D. L., and Rosenberg, R. D. (2002) Analysis of heparan sulfate oligosaccharides with ion pair-reverse phase capillary high performance liquid chromatography-microelectrospray ionization time-of-flight mass spectrometry. *J. Am. Chem. Soc.* ***124***, 8707–8718.
22. Thanawiroon, C., Rice, K. G., Toida, T., and Linhardt, R. J. (2004) Liquid chromatography/mass spectrometry sequencing approach for highly sulfated heparin-derived oligosaccharides. *J. Biol. Chem.* ***279***, 2608–2615.
23. Zhang, Z., Xie, J., Liu, H., Liu, J., and Linhardt, R. J. (2009) Quantification of heparan sulfate disaccharides using ion-pairing reversed-phase microflow high-performance liquid chromatography with electrospray ionization trap mass spectrometry. *Anal. Chem.* ***81***, 4349–4355.
24. Toyoda, H., Yamamoto, H., Ogino, N., Toida, T., and Imanari, T. (1999) Rapid and sensitive analysis of disaccharide composition in heparin and heparan sulfate by reversed-phase ion-pair chromatography on a 2 μm porous silica gel column. *J. Chromatogr. A* ***830***, 197–201.
25. Korir, A. K., Limitiaco, J. F. K., Gutierrez, S. M., and Larive, C. K. (2008) Ultraperformance ion-pair liquid chromatography coupled to electrospray time-of-flight mass spectrometry for compositional profiling and quantification of heparin and heparan sulfate. *Anal. Chem.* ***80***, 1297–1306.
26. Jones, C. J., Membreno, N., and Larive, C. K. (2010) Insights into the mechanism of separation of heparin and heparan sulfate disaccharides by reverse-phase ion-pair chromatography, *J. Chromatogr. A* ***1217***, 479–488.
27. McCormick, R. M. (1988) Capillary zone electrophoretic separation of peptides and proteins using low pH buffers in modified silica capillaries. *Anal. Chem.* ***60***, 2322–2328.
28. Eldridge, S. L., Higgins, L. A., Dickey, B. J., and Larive, C. K. (2009) Insights into the capillary electrophoresis separation of heparin disaccharides from nuclear magnetic resonance, pK(a), and electrophoretic mobility measurements. *Anal. Chem.* ***81***, 7406–7415.
29. Linker, A. and Hovingh, P. (1972) Isolation and characterization of oligosaccharides obtained from heparin by action of heparinase. *Biochemistry* ***11***, 563–568.

Chapter 10

Brain Chondroitin/Dermatan Sulfate, from Cerebral Tissue to Fine Structure: Extraction, Preparation, and Fully Automated Chip-Electrospray Mass Spectrometric Analysis

Alina D. Zamfir, Corina Flangea, Alina Serb, Eugen Sisu, Leon Zagrean, Andreas Rizzi, and Daniela G. Seidler

Abstract

Chondroitin sulfate (CS) and dermatan sulfate (DS) glycosaminoglycans (GAGs) are covalently linked to proteins, building up a wide range of proteoglycans, with a prevalent expression in the extracellular matrix (ECM). In mammalian tissues, these GAG species are often found as hybrid CS/DS chains. Their structural diversity during chain elongation is produced by variability of sulfation in the repeating disaccharide units. In central nervous system, a large proportion of the ECM is composed of proteoglycans; therefore, CS/DS play a significant role in the functional diversity of neurons, brain development, and some brain diseases. A requirement for collecting consistent data on brain proteoglycan glycosylation is the development of adequate protocols for CS/DS extraction and detailed compositional and structure analysis. This chapter will present a strategy, which combines biochemical tools for brain CS/DS extraction, purification, and fractionation, with a modern analytical platform based on chip-nanoelectrospray multistage mass spectrometry (MS) able to provide information on the essential structural elements such as epimerization, chain length, sulfate content, and sulfation sites.

Key words: Glycosaminoglycan, Chondroitin sulfate, Dermatan sulfate, Cerebral tissue, Chip-nanoelectrospray, Multistage mass spectrometry

1. Introduction

Proteoglycans, the major components of the extracellular matrix (ECM) and basement membranes (1), are also present in the intracellular region of all tissues, including brain. They are composed of a core protein that bears covalently bound sulfated glycosaminoglycan (GAG) chains. In central nervous tissues, the majority of

Françoise Rédini (ed.), *Proteoglycans: Methods and Protocols*, Methods in Molecular Biology, vol. 836,
DOI 10.1007/978-1-61779-498-8_10,

the proteoglycans contain either chondroitin sulfate (CS) or heparan sulfate GAGs, while a minor proportion contain dermatan sulfate (DS) (2).

CS consists of repeating disaccharide units of D-glucuronic acid (GlcA) β (1–3) and *N*-acetylgalactosamine (GalNAc) joined together by β (1–4) glycosidic linkages. DS is a variant of CS in which D-GlcA is epimerized to L-iduronic acid (IdoA). In mammalian tissues, these chains are often found as hybrid CS/DS structures. Most of CS/DS species have a regular sulfation pattern with one sulfate group per disaccharide repeat, usually located at GalNAc; however, oversulfated and undersulfated glycoforms as well as structures having the sulfate group located at hexuronic acids were reported (3–7). According to previously acquired information (8, 9), sulfation may occur at C-2 of GlcA/IdoA and/or C-4 and/or C-6 of GalNAc in various combinations, thereby producing characteristic motifs leading to a high structural diversity of CS/DS species.

CS/DS chains of the mammalian brains are characterized by a high degree of microheterogeneity. The irregular sulfation pattern, especially the oversulfation of hybrid CS/DS structures, was found implicated in brain development and correlated with the functional diversity of neurons (10).

Recent studies have shown that mass spectrometry (MS) based on either electrospray ionization (ESI) or matrix-assisted laser desorption/ionization (MALDI) alone or in conjunction with hyphenated techniques such as capillary electrophoresis (CE) and high-performance liquid chromatography (HPLC) is one of the most efficient analytical methods for GAG analysis in general and CS/DS in particular (1, 4, 11–15). In recent years, due to the technical advancements in MS instrumentation, in particular of the systems able to rapidly perform multistage MS (MS^n) by either collision-induced dissociation (CID) (4, 7, 16, 17) or electron-detachment dissociation (EDD) (18) and the development of reliable analysis protocols, the determination of CS/DS structure in terms of chain length, epimerization, sulfation content, and pattern became feasible.

To increase the experiment throughput, sensitivity, and spray stability necessary for efficient screening and sequencing and to reduce the *in-source* loss of the labile sulfate groups often reported (2, 4, 5, 16, 19–22) as a major drawback in ESI MS of GAGs, we have introduced in CS/DS analysis a novel method based on robotized sample delivery by fully automated chip-based nanoelectrospray (nanoESI) (16) in conjunction with ion trap mass spectrometry.

In this chapter, our optimized methodology for extraction, preparation, and chip-based mass spectrometric analysis of brain CS/DS structures is described. The method requires the combination of biochemical tools for CS/DS extraction, purification, digestion, and fractionation, with a modern analytical system based on fully automated chip-nanoESI (NanoMate robot) coupled to an ion trap mass spectrometer. By following this strategy rigorously, accurate information on structural brain CS/DS characteristics such as epimerization, chain length, sulfate content, and sulfation sites can be obtained.

2. Materials

2.1. Brain Tissue Sampling and Preparation

1. Extraction buffer: 4 M guanidinium chloride, 50 mM sodium acetate, pH 6.0, stored at room temperature.
2. Fresh brain tissue (see Note 1).
3. Laboratory desktop Eppendorf 5804R centrifuge (Eppendorf AG, Hamburg, Germany).
4. Weighing digital balance, Sartorius, ED 153 (Sartorius AG, Göttingen, Germany).
5. Rotator/shaker. Example: Standard rotator SB2 from VWR (International GmbH, Vienna, Austria).

2.2. Extraction of CS/DS from Cerebral Tissue

1. DEAE-Tris-Acryl M ion exchange solution (Pall Life Sciences, Dreieich, Germany).
2. Start buffer: 20 mM Tris/HCl pH 7.4 containing 150 mM NaCl, 1% Triton X100 stored at 4°C.
3. Wash buffer 1: 20 mM Tris/HCl pH 7.4 containing 300 mM NaCl, 1% Triton X100, stored at 4°C.
4. Wash buffer 2: 20 mM Tris/HCl pH 7.4 containing 300 mM NaCl, stored at 4°C.
5. Elution buffer: 20 mM Tris/HCl pH 7.4 containing 1 M NaCl, stored at 4°C.
6. Cleaning buffer: 1 M NaCl pH 7.4, 1% Triton X100, store at 4°C.
7. Dialysis membrane 14,000 MWCO (Pierce, Rockford, IL, USA).
8. Aquacide I powder (Calbiochem, Bad Soden, Germany).
9. Sodium borohydride solution 1 M in water, freshly prepared (see Note 2).
10. Sodium hydroxide solution 1 M.

2.3. Purification of CS/DS Chains

1. 96% Ethanol (EtOH), sodium nitrite solution 1 M in water, hydrochloric acid solution 1 M, acetic acid glacial HPLC grade.
2. Carbazole solution 0.125% in absolute EtOH (see Note 3).
3. Ammonium sulfamate solution 4 M in water, sodium tetraborate decahydrated in H_2SO_4 solution 0.025 M.

4. Heparine standard (Sigma-Aldrich, St. Louis, MO, USA).
5. Digital SpeedVac system SPD111V (Thermo Electron, Asheville, NC, USA), coupled to a vacuum pump PC 2002 Vario with CVC 2000 Controller from Vaccubrand GmbH (Wertheim, Germany).
6. Laboratory desktop Eppendorf 5804R centrifuge (Eppendorf AG, Hamburg, Germany).
7. Ultracentrifuge Optima TLX (Beckman Coulter, Krefeld, Germany).
8. UV VIS spectrophotometer 180–800/nm (example: from Thermo Scientific).

2.4. Depolymerization

1. Chondroitin AC I lyase (Seikagaku Kogyo, Tokyo, Japan) 50 mU/assay in water.
2. Chondroitin B lyase (Seikagaku Kogyo, Tokyo, Japan) 5 mU/assay in water.
3. Chondroitinase digestion buffer: 50 mM Tris/HCl, pH 8.0, containing 60 mM sodium acetate, 60 mM NaCl, 0.01% bovine serum albumin, and 3 mM NaN_3 (see Note 4).
4. Thermomixer Eppendorf Comfort (Eppendorf AG, Hamburg, Germany).

2.5. Fractionation

1. $(NH_4)HCO_3$ solution 0.5 M in bidistilled water.
2. Superdex Peptide HR10/30 column (Amersham-Pharmacia, Freiburg, Germany) and HPLC equipped with diode array detection system (example: from Bio-Tek, Kontron).

2.6. Preparation for Chip-nanoESI MS

1. Analytical grade methanol (MeOH) from Merck (Darmstadt, Germany) to be used without further purification.
2. SpeedVac Concentrator. Example: SPD111V-230 from Thermo Electron Corporation, (Asheville, NC, USA), coupled to a vacuum pump PC 2002 Vario with CVC 2000 Controller from Vaccubrand (Wertheim, Germany).
3. Laboratory desktop Eppendorf 5804R centrifuge (Eppendorf AG, Hamburg, Germany).

2.7. Chip-ESI MS Analysis

2.7.1. Reagents

1. Acetonitrile (ACN) from Merck (Darmstadt, Germany) to be used without further purification.
2. Calibration standard G2421A electrospray "tuning mix" (Agilent Technologies, Santa Rosa, CA, USA).
3. Nitrogen and helium for ion trap mass spectrometer and NanoMate robot, purity ≥99.999% vol.

2.7.2. Fully Automated Chip-nanoESI Infusion

1. NanoMate™ 400 or NanoMate™ TriVersa 400 robot (Advion BioSciences, Ithaca, NY, USA) with commercial bracket (Burker Daltonics, Bremen, Germany) for NanoMate robot coupling to Bruker ion trap mass spectrometer.
2. Silicon nanoESI Chip™ with 400 nozzles (Advion BioSciences, Ithaca, NY, USA).
3. Microtiter sample plate with 96-wells, V-bottom, 200 μL, thermal cycler-compatible, polypropylene or glass-coated (Beckman Coulter Inc. or Eppendorf AG).
4. Rack of 96 disposable conductive pipette tips (Advion BioSciences, Ithaca, NY, USA).
5. ChipSoft 8.1.0 software operating under Windows system (Advion BioSciences, Ithaca, NY, USA) to control and manipulate the NanoMate robot.

2.7.3. Mass Spectrometry

1. Bruker ion trap mass spectrometer equipped with ESI source. Example: high capacity ion trap ultra (HCT Ultra, PTM discovery) mass spectrometer (Bruker Daltonics, Bremen, Germany).
2. Compass™ 1.2 Software (Bruker Daltonics, Bremen, Germany) for HCT Ultra mass spectrometer.

3. Methods

The strategy for brain CS/DS preparation and analysis by chip-nanoESI mass spectrometry includes the following mandatory steps to be carried out in the succession indicated by the workflow depicted in Fig. 1: (1) extraction of intact proteoglycans from brain tissue; (2) releasing of CS/DS chains from the core protein by β-elimination; (3) digestion of CS/DS chains resulting in the cleavage of the glycosidic linkage between GalNAc and D-GlcA with chondroitin AC lyase, and in parallel between GalNAc and L-IdoA with chondroitin B lyase (23); (4) size fractionation by gel filtration, separately for AC- and B-lyase sequences; (5) identification of species by chip-nanoESI MS screening; (6) detailed analysis of structural motifs by sequencing in CID MS^n experiments. Optionally, quantification of extracted brain CS/DS may be performed.

3.1. Brain Tissue Sampling and Preparation

1. Collect fresh brain tissue and excise with a scalpel a piece of 1.5 g.
2. Cut the excised tissue in small portions and introduce them in a Falcon tube.
3. Add 5 mL extraction buffer and place the tube in the rotator/shaker for gentle mixing overnight at 4°C.

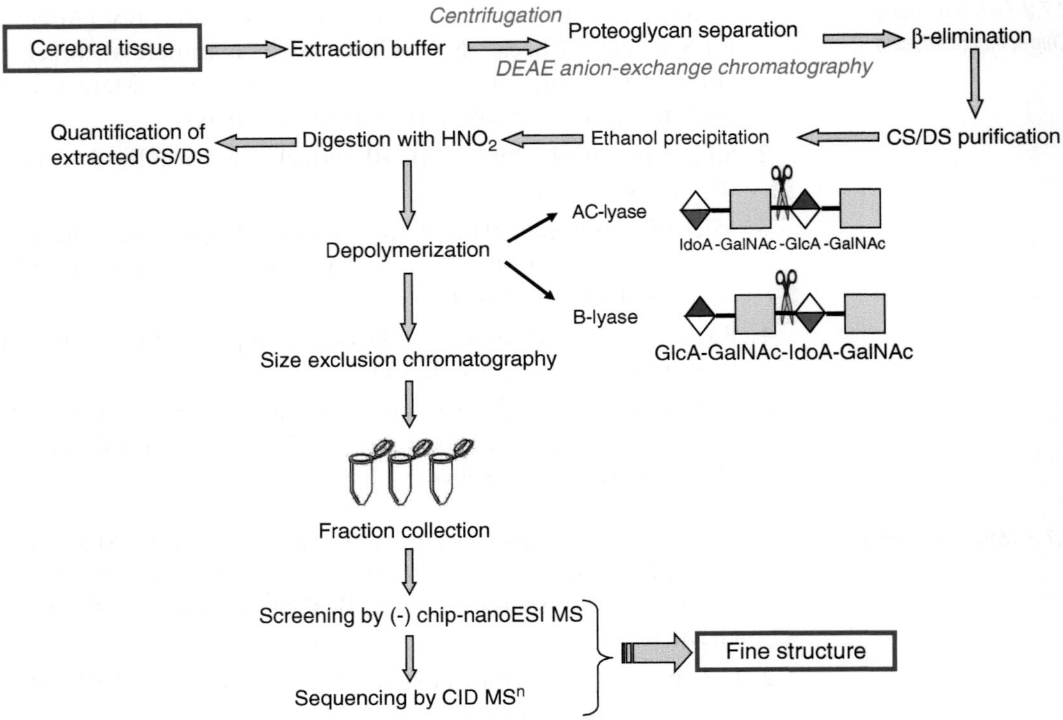

Fig. 1. Flow chart for extraction, preparation, and chip-electrospray mass spectrometric analysis of brain CS/DS.

3.2. Extraction of CS/DS from Cerebral Tissue

3.2.1. Proteoglycans Separation on DEAE Anion Exchange Chromatography

1. Centrifuge the viscous solution in the Falcon tube for 15 min, 5,000 rpm at 4°C; collect the supernatant.
2. Add the start buffer to complete 40 mL; centrifuge once more in the same conditions and collect again the supernatant.
3. Prepare the DEAE-Tris-Acryl M column with a 3 mL bedvolume and equilibrate with 3× bedvolume start buffer.
4. Apply the supernatant; wash with 3× bedvolume start buffer; wash with 5× bedvolume wash buffer, and 1× with wash buffer 2 to remove the Triton.
5. Apply 3× bedvolume elution buffer and collect the sample.
6. Regenerate the column by rinsing it with 3× bedvolume cleaning buffer followed by 3× bedvolume start buffer. Store it at 4°C.

3.2.2. Sample Concentration and Dialysis

1. Introduce the collected sample in a dialysis membrane, MWCO 14,000, shape it as a sack, and place it in Aquacide I powder.
2. Incubate for about 5 h at 4°C until approximately 1 mL sample remains in the membrane.
3. Shorten the dialysis sack to fit the corresponding new volume and submit the sample to dialysis overnight at 4°C against deionized water.

3.2.3. β-Elimination

1. Collect the sample from dialysis sack and measure the volume.
2. Introduce in a tube exhibiting a hole in the lid the following solutions: (a) sample solution; (b) the solution of 1 M $NaBH_4$; (c) the solution of 1 M NaOH v/v/v (see Notes 5 and 6).
3. Incubate the final solution overnight at 37°C.

3.3. Purification of CS/DS Chains

3.3.1. DEAE Anion Exchange Chromatography

1. Add acetic acid glacial drop by drop until pH 7.0 is reached. To reach pH 7.0, the estimated volume of acetic acid glacial is 150 μL. Dilute the solution 1:10 in start buffer.
2. Apply this solution to DEAE-Tris-Acryl M prepared in a Pasteur pipette with a bedvolume of 1 mL and washed with 9× bedvolume start buffer followed by 15× bedvolume wash buffer.
3. Elute with 4× bedvolume elution buffer and collect the sample.

3.3.2. Ethanol Precipitation

1. Dilute 1:4 the eluted sample with absolute EtOH.
2. Store it overnight at −20°C.
3. Centrifuge for 45 min at 20,000 × *g* at 4°C.
4. Extract three fourth of supernatant and dry the sediment in the hood under ventilation.

3.3.3. Digestion with HNO_2

Digestion with HNO_2 is a compulsory step, which helps eliminating the heparin and heparan sulfate that might be present in the sample.

1. Mix 1 M $NaNO_2$ and 1 M HCl in a glass tube on ice; cool down to 5°C.
2. Dissolve the dry pellet in 100 μL water; add 400 μL mixture $NaNO_2$/HCl and cool down to the room temperature (see Note 7).
3. Reduce the volume to 200 μL in vacuum.
4. Ethanol precipitation: dilute 1:4 with absolute EtOH and store it overnight at −20°C.
5. Centrifuge for 45 min at 20,000 × *g* at 4°C; extract the supernatant and leave the pellet which contains CS/DS to dry in the hood, at room temperature.
6. Dissolve the pellet in 100 μL deionized water, extract 10 μL for Carbazole reaction, and dry it again in vacuum.

3.3.4. Quantification of Extracted CS/DS: Carbazole Reaction

1. Dilute 10 μL sample 1:10 in water; prepare successive dilution: 1:20, 1:40, and 1:80.
2. Plot the standard curve: heparin solutions 4, 8, 16, and 32 μg/100 μL. For 4–40 μg/mL or 10–100 μmol GlcA, a direct linear correlation must be found.
3. On ice, add in a small tube 100 μL diluted sample or standard, 10 μL 4 M ammonium sulfamate, and 500 μL 0.025 M sodium tetraborate decahydrated in H_2SO_4.

4. Incubate 10 μL for 10 min at 100°C; cool down to room temperature.
5. Add 20 μL 0.125% Carbazole solution, incubate for 15 min at 100°C, and cool down to room temperature.
6. Read the absorbance of the sample at $\lambda = 530$ nm against water.

3.4. Depolymerization

Because of high molecular weight, the structural analysis of brain CS/DS by ESI or chip-nanoESI MS almost always requires a detachment of the chain(s) from proteoglycan core protein followed by their digestion/depolymerization. For discrimination between D-GlcA and L-IdoA, chondroitin AC lyase and chondroitin B lyase are specifically used. Chondroitin B lyase cleaves the linkage between GalNAc and L-IdoA, while chondroitin AC lyase cleaves the linkage between GalNAc and D-GlcA irrespective of the sulfation content and distribution along the chain or within the monomer (4, 24).

For MS profiling of brain disaccharide repeats differing in the number of sulfates (example in Fig. 2) and determination of oversulfation sites within monomer ring by CID MS^n, total digestion with both chondroitin AC and B lyase is recommended (7). For the assessment of sulfation status along the CS/DS chain, partial digestion that yields longer chains, however still detectable by ESI MS, is to be applied.

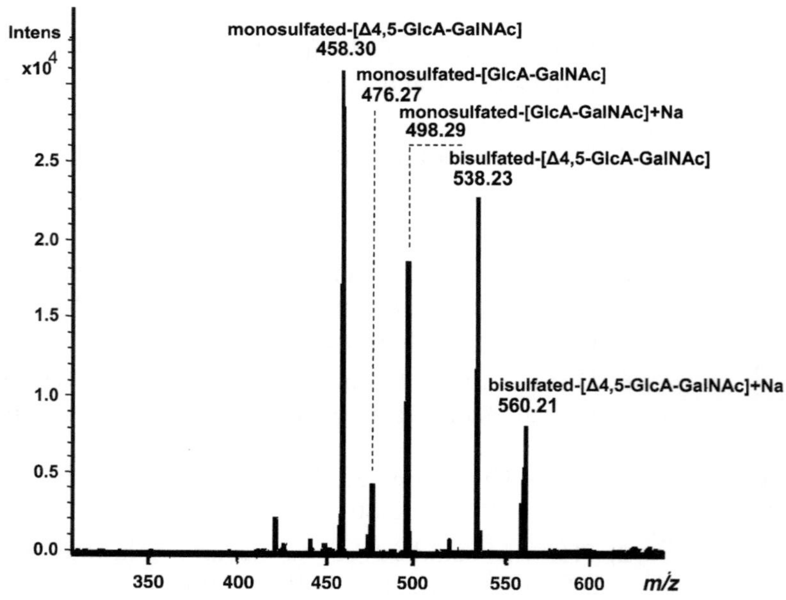

Fig. 2. Fully automated chip-nanoelectrospray (nanoESI) high capacity ion trap (HCT) mass spectrometry (MS) of brain chondroitin sulfate (CS) disaccharide obtained after depolymerization with chondroitin AC I lyase and separation on GFC-Superdex peptide column. Solvent: MeOH; ESI MS parameters: Chip-nanoESI: 1.4 kV; capillary exit: −50 V. Back nitrogen pressure 0.30 psi. Nitrogen nebullizer on MS at 50 psi (reproduced from ref. (7) with permission from Springer).

1. Dissolve the pellet in 20 μL water and vortex it (see Note 8).
2. Split the sample in two aliquots: 10 μL for digestion with chondroitin AC lyase and 10 μL for digestion with chondroitin B lyase.

3.4.1. Chondroitin B Lyase

1. On ice, add 200 μL chondroitinase digestion buffer and 5 μL 5 mU/assay chondroitin B lyase (see Note 9).
2. Incubate for 2 h at 37°C; stop the reaction by introducing the tubes in ice. The enzyme will be inactivated by increasing the temperature at 75°C for 20 min (see Note 10).

3.4.2. Chondroitin AC Lyase

1. On ice, add 200 μL chondroitinase digestion buffer and 5 μL 50 mU/assay chondroitin AC lyase (see Note 9).
2. Incubate 2 h at 37°C; stop the reaction by introducing the tubes in ice. The enzyme will be inactivated by increasing the temperature at 75°C for 20 min (see Note 10).

3.5. Fractionation

1. Add to digested sample 400 μL 0.5 M NH_4HCO_3 and centrifuge at 30,000 × *g* for 45 min at 4°C.
2. Collect the supernatant and apply 2 times 200 μL of it on the Superdex Peptide HR10/30 column, equilibrated and eluted in 150 μL fractions with 0.5 M NH_4HCO_3 at a flow rate of 0.5 mL/min and continuous UV detection at 232 nm.
3. Collect the oligosaccharide fractions (deca-, octa-, hexa-, tetra-, and disaccharide).

3.6. Preparation for Chip-nanoESI MS

1. Dry the fractions in vacuum (SpeedVac).
2. Prepare the stock solution by dissolving the dried sample in about 50 μL MeOH to be stored at −20°C until mass spectrometric analysis is performed.
3. Working aliquots can be obtained by successive dilutions in MeOH in (1:100) range.
4. Centrifuge for 20 min and collect the supernatant to be submitted to chip-nanoESI MS and CID MS^n (see Note 11).

3.7. Analysis of Brain CS/DS by Fully Automated Chip-nanoESI MS and CID MS^n

NanoESI MS is an attractive technique used for the analysis of CS/DS, due to the wealth of structural information that can be obtained (25). To improve the ESI process and significantly increase the experiment throughput, sensitivity, and reproducibility, fully automated chip-based nanoESI MS was implemented for CS/DS analysis (16). In brain CS/DS oligosaccharide analysis, under suitable instrumental and solution parameters, chip-nanoESI MS and CID MS^n in the negative ion mode are able to provide: (a) high ionization yield; (b) detection of long and short chains in mixtures; (c) identification of sulfation degree; (d) determination of over-, regularly-, and undersulfated regions; (e) generation of diagnostic sequence ions in tandem MS experiments; (f) multistage MS of sulfated fragment ions for determination of sulfate group location along the chain and within a certain monomer ring.

3.7.1. Brain CS/DS Screening by Chip-nano ESI MS

1. Couple the ion trap mass spectrometer and NanoMate robot via Bruker commercial interface following the coupling principle described before (26).
2. Turn on the gas supply, the mass spectrometer, and the NanoMate robot.
3. Tune the ion trap mass spectrometer for proper detection of CS/DS negative ions by setting on the Compass™ software the operating parameters as follows: "negative ion mode"; HCT capillary exit: 50 V; source block temperature: 200°C; nebulizer nitrogen pressure: 50 psi; nitrogen flow rate: 0.5 L/min; detection in a range of (100–2,000) *m/z*, scan speed: 8,000 *m/z* per s.
4. Tune the NanoMate robot for generating negative ions by setting on the ChipSoft program the following values of the spray parameters: "negative ion mode"; pipette tip voltage: –1.4 kV; nitrogen back pressure: 0.30 psi (see Note 12); sample aspiration volume: 5 μL; air aspiration volume: 2 μL (see Note 13).
5. Insert the silicon 400 ESI Chip in the slot of the NanoMate chip holder with the frame notches oriented down and the chip frame tab oriented up while the serial number is facing the NanoMate sample and tip stage.
6. Select in the ChipSoft program the ESI Chip with the corresponding serial number.
7. Fix the rack of 96 disposable conductive pipette tips in their dedicated NanoMate holder.
8. Load 10 μL aliquot of the working sample solution (see Note 14) into a well of the microtiter plate and select in the ChipSoft program the corresponding well position.
9. Slide the microtiter plate into the sample plate holder of the NanoMate robot.
10. Set on the ChipSoft the following options: (1) aspirate 5 μL sample volume followed by 2 μL of air into the pipette tip; (2) return to the initial well the unused sample.
11. Select "deliver sample" on the ChipSoft software of the NanoMate robot and "start acquisition" on the Compass software of the ion trap MS.
12. After initiation of the electrospray, optimize the signal intensity and stability by fine adjustment of the chip position with respect to the mass spectrometer inlet. Use the back–forward, right–left options in the "spray optimization" sector of the ChipSoft software (see Notes 15–18).

13. Acquire the spectrum until a fair *signal-to-noise* ratio is obtained. Recommended acquisition time is above 2 min.
14. Switch to DataAnalysis portal of Compass™ software, import the total ion chromatogram (TIC), and generate the spectrum by combining it over all TIC scans.
15. Load in a new well a tune mix solution (1:100) diluted in ACN. Generate the calibration spectrum by repeating the steps 9–14. Apply the calibration file to the sample spectrum. On a Bruker HCT mass spectrometer, by proper calibration a mass accuracy of at least 30 ppm can be achieved for brain CS/DS disaccharides.
16. For multiple samples in high throughput regime, load the aliquots in different wells, insert the sample plate into the robot holder, and infuse them one after the other by changing successively the number of the well followed by choosing "deliver sample" in the ChipSoft.

3.7.2. Sequencing by CID MS^n

CID MS^n gives the possibility to sequence in several stages the precursor ion and its derived fragment ions and obtain, in a single experiment, straightforward information on the sulfate group location along the CS/DS chain (see Fig. 3) and/or within a certain monomer ring. Besides significant gain in sensitivity, infusion by chip-nanoESI reduces to minimum the *in-source* loss of labile sulfate groups and ensures a steady spray over long analysis time. Both features are beneficial to MS^n experiments since multiple fragmentations require long-lasting signals of high and constant intensity associated to intact precursor ions, which did not undergo *in-source* decay.

1. Identify in the MS run the ion, which according to the calculated mass corresponds to a structure of interest for further detailed investigation.
2. Select "manual mode" of CID MS^2 fragmentation.
3. Leave the MS settings unaltered and set the isolation window at 2u (see Note 19) and the fragmentation amplitude at 0.40 V (see Note 20).
4. Acquire the MS^2 signal for 5–10 min while increasing gradually the fragmentation amplitude up to 0.80 V (see Note 21) with a ramp from 30 to 200% within 40 ms per single spectrum and a fragmentation cutoff default of 27% of the precursor ion *m/z*.
5. Stop the acquisition, switch to DataAnalysis portal, generate, and calibrate the spectrum as described at Subheading 3.7.1.
6. For CID MS^3, identify in the MS^2 the ion of interest for additional fragmentation (see an example in Fig. 3), leave unchanged the selection and settings for MS and MS^2, select CID MS^3, and proceed further as indicated at steps 3–5.

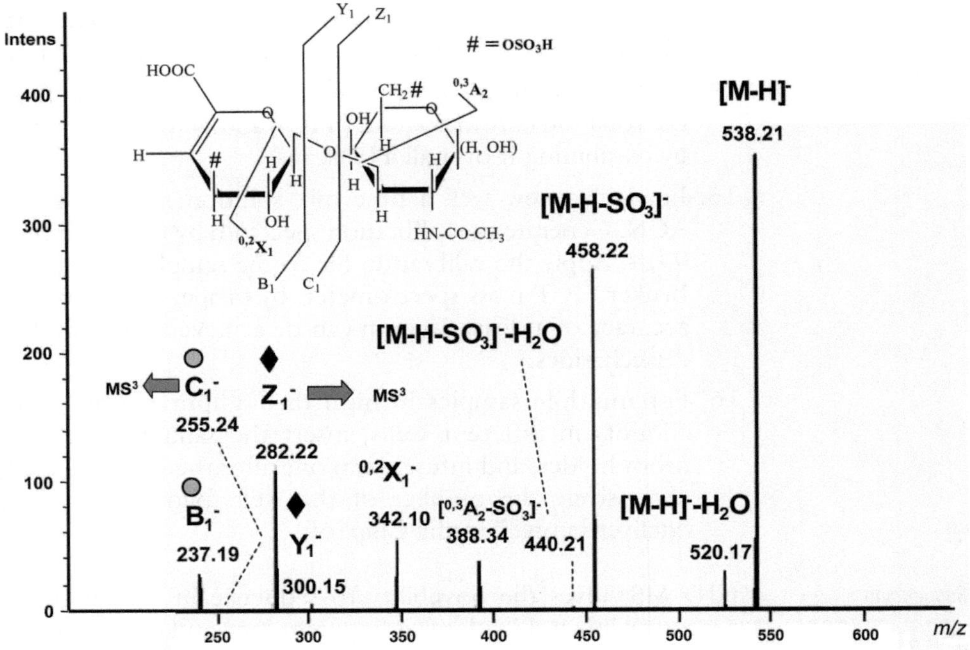

Fig. 3. Fully automated chip-nanoESI HCT collision-induced dissociation (CID) MS^2 of the singly deprotonated ions at *m/z* 538.21 assigned according to mass calculation to bisulfated 4,5-Δ-(GlcA-GalNAc). Diagnostic ions for D-glucuronic acid (GlcA) β (1–3) and *N*-acetylgalactosamine (GalNAc) monosulfation are marked by *filled diamond*; diagnostic ions for GlcA monosulfation are marked by *filled circle*. *Inset*: structure of bisulfated CS disaccharide. MS^2 by CID using He as a collision gas. Fragmentation amplitude within 0.40–0.80 V (reproduced from ref. (7) with permission from Springer).

7. For superior stages of fragmentation (see Note 22), keep all the settings and selections made at previous stages (MS including) and proceed further as indicated at steps 3–5.

4. Notes

1. If it is not possible to start the procedure immediately, freeze the brain tissue at −20°C.
2. $NaBH_4$ is toxic. Prepare it carefully under proper ventilation.
3. Carbazole solution is stable for 12 weeks at 4°C in the dark.
4. NaN_3 is highly toxic. Prepare it carefully under proper ventilation.
5. Perform this step under ventilation because of the $NaBH_4$ toxicity.
6. Perform a hole in the lid of the Falcon tube. This will prevent accumulation of gases, which may damage the tube during the process.

7. Perform this step under proper ventilation because of the high toxicity of HNO_2.
8. Keep always the reagents and samples on ice during manipulation, prior to the start of enzymatic digestion.
9. Enzyme/buffer solutions should be stored at −20°C, even though their usage 1 month after enzyme reconstruction is to be avoided.
10. If not used immediately, the digestion product should be stored overnight at 4°C.
11. At this stage, centrifugation is necessary for eliminating possible particles, which may cause clogging of the chip nozzle.
12. With these settings and MeOH as a solvent, 5 μL aspirated volume will last for about 50 min of spraying; higher values of nitrogen back pressure and/or ESI voltage result in increasing of ESI flow rate and speed of sample consumption.
13. Aspiration of air after sample is strongly recommended as it prevents sample dripping from the pipette tip during the pipette course from the well to the chip nozzle.
14. Always load in the well a higher sample volume than that to be aspirated.
15. If no ESI signal is detected, proceed with the following steps, one step at the time: (1) modify first the chip position with respect to the MS entrance orifice by fine adjustment "forward/back," "right/left" in the ChipSoft; (2) increase slightly the nitrogen back pressure; (3) alter slightly the ESI voltage.
16. If the ESI signal is not constant or is lost after only a few minutes of spray, this is an indication that the chip nozzle was clogged. In this case, proceed with the following steps, one step at the time, each operation to be applied if the previous one did not retrieve the signal: (1) increase slightly the nitrogen back pressure; (2) select the option "more sample" in the ChipSoft; (3) select the option "next nozzle" in the ChipSoft.
17. If the ESI signal is lost after a long period of constant signal, this is an indication that the sample in the pipette tip was entirely spent. In this case, for measurement continuation, select the option "more sample" without selecting "stop delivery."
18. By choosing "stop delivery," the tip will be discarded and a fresh tip and nozzle will be used when "deliver sample" will be further selected. By choosing "more sample," the same pipette tip will be conducted to the well for aspirating more sample and afterward engaged for spraying through the same nozzle. This way, the unnecessary change of the nozzle and tip for the same sample may be avoided.
19. A broader isolation window (above 2u) has as a direct consequence the elevation of the number of ions preserved in the trap for

sequencing. Therefore, broad windows appear initially to be beneficial to the total ion current. However, this diminished selectivity may result in a simultaneous isolation and subsequent fragmentation of other ions (i.e., ions exhibiting *m/z* values close to that of the chosen precursor).

20. Avoid decreasing the RF amplitude below 0.30–0.40 V. Below these values the excitation energy will not be enough for the precursor ion to undergo fragmentation.
21. Avoid increasing the RF amplitude above 0.80 V. Superior values may result in the loss of sulfate groups and exclusive generation of desulfated fragment ions.
22. In CID MS^n, the intensity of the signal corresponding to the fragment ions diminishes approximately 10 times after each sequencing event. No further fragmentation should be attempted after the intensity dropped below 100 counts/s.

Acknowledgments

The authors are grateful to the Romanian National Authority for Scientific Research (Grants No. PN-II-RU-TE-2011-2-0008 and PN- II-ID-PCE-2011-3-0047) and German Society for Research (DFG SE1431/1-1 and IRTG1549/1 Molecular and Cellular Glycosciences) for financial support.

References

1. Bielik, A. M. and Zaia J. (2010) Extraction of Chondroitin/Dermatan Sulfate Glycosaminoglycans from Connective Tissue for Mass Spectrometric Analysis. *Methods Mol. Biol.* **600**, 215–225.
2. Sisu, E., Flangea, C., Serb, A., and Zamfir, A. D. (2011) Modern developments in mass spectrometry of chondroitin and dermatan sulfate glycosaminoglycans. *Amino Acids.* **41**, 235–256.
3. Nielsen, T.C., Meikle, P.J., Hopwood, J.J., and Fuller, M. (2008) Minimum substrate requirements of endoglycosidase activities toward dermatan sulfate by electrospray ionization-tandem mass spectrometry. *Glycobiology.* **18**, 1119–1128.
4. Zamfir, A. D., Flangea, C., Sisu, E., Serb, A. F., Dinca, N., Bruckner, P., and Seidler, D. G. (2009) Analysis of novel over- and under-sulfated glycosaminoglycan sequences by enzyme cleavage and multiple stage MS. *Proteomics.* **9**, 3435–3444.
5. Zamfir, A. D., Seidler, D., Kresse, H., and Peter-Katalinic, J. (2003) Structural investigation of chondroitin/dermatan sulfate oligosaccharides from human skin fibroblast decorin. *Glycobiology.* **11**, 733–742.
6. Zamfir, A. D., Seidler, D., Schonherr, E., Kresse, H., and Peter-Katalinić, J. (2004) On-line sheathless capillary electrophoresis/nanoelectrospray ionization-tandem mass spectrometry for the analysis of glycosaminoglycan oligosaccharides. *Electrophoresis.* **25**, 2010–2016.
7. Flangea, C., Schiopu, C., Sisu, E., Serb, A., Przybylski, M., Seidler, D. G., and Zamfir, A. D. (2009) Determination of sulfation pattern in brain glycosaminoglycans by chip-based electrospray ionization ion trap mass spectrometry. *Anal. Bioanal. Chem.* **395**, 2489–2498.
8. Seidler, D. G., Peter-Katalinić, J., and Zamfir, A. D. (2007) Galactosaminoglycan function and oligosaccharide structure determination *Scient. World J.* **19**, 233–241.
9. Bao, X., Muramatsu, T., and Sugahara, K. (2005) Demonstration of the Pleiotrophin-binding oligosaccharide sequences isolated

from chondroitin sulfate/dermatan sulfate hybrid chains of embryonic pig brains. *J. Biol. Chem.* **280**, 35318–35328.

10. Purushothaman, A., Fukuda, J., Mizumoto, S., ten Dam, G. B., van Kuppevelt, T. H., Kitagawa, et al. (2007) Functions of chondroitin sulfate/dermatan sulfate chains in brain development. Critical roles of E and iE disaccharide units recognized by a single chain antibody GD3G7. *J. Biol. Chem.* **282**, 19442–19452.
11. Amon, S., Zamfir, A. D., and Rizzi, A. (2008) Glycosylation analysis of glycoproteins and proteoglycans using capillary electrophoresis-mass spectrometry strategies. *Electrophoresis.* **29**, 2485–2507.
12. Volpi, N., Maccari, F., and Linhardt, R. J. (2008) Capillary electrophoresis of complex natural polysaccharides. *Electrophoresis.* **29**, 3095–3106.
13. Zaia, J. (2009) On-line separations combined with MS for analysis of glycosaminoglycans. *Mass Spectrom. Rev.* **28**, 254–272.
14. Zamfir, A. D., Flangea, C., Altmann, F., and Rizzi, A. M. (2010) Glycosylation analysis of proteins, proteoglycans and glycolipids by CE-MS. *Adv. Chromatogr.* **49**, 135–194.
15. Zamfir, A. D., Seidler, D., Kresse, H., and Peter-Katalinic, J. (2002) Structural characterization of chondroitin/dermatan sulfate oligosaccharides from bovine aorta by capillary electrophoresis and electrospray ionization quadrupole time-of-flight tandem mass spectrometry. *Rapid Commun. Mass Spectrom.* **16**, 2015–2024.
16. Flangea, C., Serb, A. F., Schiopu, C., Tudor, S., Sisu, E., Seidler, D. G., and Zamfir, A. D. (2009) Discrimination of GalNAc (4S/6S) sulfation sites in chondroitin sulfate disaccharides by chip-based nanoelectrospray multistage mass spectrometry. *Cent. Eur. J. Chem.* 7, 752–759.
17. Miller, M. J. C., Costello, C. E., Malmström, A., and Zaia, J. (2006) A tandem mass spectrometric approach to determination of chondroitin/dermatan sulfate oligosaccharide glycoforms. *Glycobiology.* **16**, 502–513.
18. Wolff, J. J., Laremore, T. N., Busch, A. M., Linhardt, R. J., and Amster, I. J. (2008) Electron detachment dissociation of dermatan sulfate oligosaccharides. *J. Am. Soc. Mass Spectrom.* **19**, 294–304.
19. Hitchcock, A. M., Yates, K. E., Costello, C. E., and Zaia, J. (2008) Comparative glycomics of connective tissue glycosaminoglycans. *Proteomics.* **8**, 1384–1397.
20. McClellan, J. E., Costello, C. E., O'Connor, P. B., and Zaia, J. (2002) Influence of charge state on product ion mass spectra and the determination of 4S/6S sulfation sequence of chondroitin sulfate oligosaccharides. *Anal. Chem.* **74**, 3760–3771.
21. Zaia, J., Miller, M. J. C., Seymour, J. L., and Costello, C. E. (2007) The role of mobile protons in negative ion CID of oligosaccharides. *J Am Soc Mass Spectrom.* **18**, 952–960.
22. Zaia, J. and Costello, C. E. (2001) Compositional analysis of glycosaminoglycans by ESI-MS. *Anal. Chem.* **73**, 233–239.
23. Seidler, D. G., Breuer, E., Grande-Allen, K. J., Hascall, V. C., and Kresse, H. (2002) Core protein dependence of epimerization of glucuronosyl residues in galactosaminoglycans. *J.Biol.Chem.* **277**, 42409–42416.
24. Gu, K., Linhardt, R. J., Laliberte, M., Gu, K., and Zimmermann, J. (1995) Purification, characterization and specificity of chondroitin lyases and glucuronidase from *Flavobacterium heparinum. Biochem. J.* **312**, 569–577.
25. Zaia, J. (2004) Mass spectrometry of oligosaccharides. *Mass Spectrom. Rev.* **23**, 161–227.
26. Almeida, R., Mosoarca, C., Chirita, M., Udrescu, V., Dinca, N., Vukelic, Z., et al. (2008) Coupling of fully automated chip-based electrospray ionization to high-capacity ion trap mass spectrometer for ganglioside analysis. *Anal. Biochem.* **378**, 43–52.

Chapter 11

Use of Neutrons Reveals the Dynamics of Cell Surface Glycosaminoglycans

Marion Jasnin

Abstract

Cell surface glycosaminoglycans (GAG), such as heparan sulfate (HS) and heparin, are key multifunctional cell regulators, which are involved in numerous molecular events associated with tumor growth, metastasis, pathogen attachment, and immune response. GAG dynamically bind and regulate the activities of many signaling proteins such as growth factors, chemokines, and cytokines. GAG-binding interactions with proteins rely on the coupling between the geometry, flexibility, and rigidity of the polysaccharide chain. Understanding GAG dynamics at the molecular level can therefore provide fundamental insights into GAG function in a cellular context.

Elastic incoherent neutron scattering is a powerful tool for the exploration of fast molecular motions in biological macromolecules. Recently, the technique was used to evaluate HS flexibility and rigidity on different timescales between the picosecond (ps) and the nanosecond (ns). Here, neutron spectroscopy experimental procedures are presented, with emphasis on the practical details necessary to prepare samples, run neutron scattering experiments, and extract the dynamics parameters from the data.

Key words: Glycosaminoglycan, Heparan sulfate, Elastic incoherent neutron scattering, Picosecond–nanosecond (ps–ns) timescale, Atomic mean square displacement, Flexibility, Resilience

1. Introduction

1.1. Elastic Incoherent Neutron Scattering

The interaction between the neutrons and the atoms in a sample leads to coherent and incoherent scattering. In this chapter on spectroscopy studies, we consider only the incoherent component: a single nucleus is observed with no interference between its scattered wave and the wave scattered by other nuclei in the sample. For this reason, we shall refer to the process as single-particle neutron spectroscopy.

The method presented here, called elastic incoherent neutron scattering (EINS), consists of observing the change in direction of

Françoise Rédini (ed.), *Proteoglycans: Methods and Protocols*, Methods in Molecular Biology, vol. 836,
DOI 10.1007/978-1-61779-498-8_11,

the elastically scattered neutrons (scattered without energy change). The wavelength range of thermal neutrons (1–10 Å) used for the experiments matches that of atomic and molecular distances, while the corresponding energy range (~meV, with associated timescale: ~ps–ns) matches that of thermal atomic fluctuations.

Because the incoherent scattering of hydrogen (^{1}H) is more than one order of magnitude higher than the scattering power of any other atom, including deuterium (D = ^{2}H), EINS reports mainly the motions of hydrogens. On the ps–ns timescale, hydrogens reflect the motions of the groups to which they are bound. By measuring natural abundance GAG in D_2O, GAG motions can be explored free from the water contribution. The dynamics of a specific component of a complex—for example, of a protein in complex with a GAG—can also be enhanced by taking advantage of selective isotopic hydrogen–deuterium (H–D) labeling.

1.2. Elastic Temperature Scan

A neutron spectrometer performs within a given momentum transfer, Q, range and with a given energy resolution. Q defines the length-scale of the motions, whereas the energy resolution provides the timescale (see Note 1). Most of the EINS experiments consist of recording an elastic temperature scan: the elastic intensity, I_{el}, is measured as a function of Q in a wide range of temperatures, including below the freezing point of water if no bulk water is present in the sample.

1.3. Gaussian Approximation

Several analytical models are used for the analysis of the experimental data from protein samples and the extraction of the atomic fluctuations as a function of temperature. The description of the different models is beyond the scope of this chapter but has been well reviewed by Gabel et al. (2).

In the case of GAG samples, we chose the model-free, Gaussian approximation approach (see Note 2), which is an analogy of the Guinier analysis used in small angle scattering (3) (see Note 3), in order to extract the atomic mean square displacement (MSD) of the motions. Following the Gaussian approximation, I_{el} can be written at small Q values as follows:

$$I_{el}(Q) = I_{el}(0)\exp(-<u^2>Q^2/6) \tag{1}$$

$<u^2>$ is the MSD of the motions; the approximation is valid for $\sqrt{(<u^2>Q^2)} < \sqrt{2}$.

The elastic scan can give rise to an analysis in terms of force constant: the inverse slope of $<u^2>$ vs. T, $(d<u^2>/dT)^{-1}$, has the dimension of an elastic force constant, N/m. This parameter was therefore named effective force constant, or resilience, $<k'>$ (5) (see Note 4).

1.4. Analysis of GAG Dynamics

Neutron spectroscopy can provide a full set of experimental average MSD in different timescales from the ps to the ns, as recently presented in our work on HS (6). The MSD values reveal the average flexibility of the polysaccharide chain at a specific timescale. The amplitudes are on absolute scale and can be compared directly to the results of molecular dynamics (MD) simulations to obtain local dynamics information. For example, it would permit to identify the contribution of ring C–H group dynamics, side group motions with respect to the attached ring, intramolecular hydrogen bond dynamics, and fast interconversion between iduronate ring conformations (7, 8).

The other functionally relevant parameter extracted from neutron scattering data is the resilience of the structure (5). In our work on HS (6), we have shown that HS polysaccharides correspond to rather rigid chains with respect to "soft" protein structures. Higher HS resilience may result from the presence of short side groups compared to long, flexible amino acid side chains, as well as from rotational restrictions around the glycosidic linkage, which would participate in rigidifying the overall structure of the chain. The differences in dynamics may differ for different proteins and GAG, and further neutron scattering work on GAG and their respective binding proteins is essential to provide a better understanding of GAG dynamics in a cellular context.

In the following paragraphs, we describe the experimental procedures to prepare GAG samples for neutron spectroscopy experiments, to run an EINS experiment with the spectrometers accessible in large facility centers, and extract dynamics information from the data.

2. Materials

2.1. Neutron Beamline and Spectrometers

Because the neutron fluxes generated in research centers and the measured intensities are low, the experiments require instruments with beam sizes of a few cm^2, in order to illuminate the sample optimally (typically 10^6 neutrons/cm^2/s for a good spectrometer). The data must be collected during a long time, going from a couple of hours up to a few days.

Neutron spectrometers are available via beam time proposal submission in research centers such as the Institut Laue Langevin (ILL) (http://www.ill.eu/) in Grenoble, France; the Spallation Neutron Source (SNS) (http://neutrons.ornl.gov/) in Oak Ridge, United States; and the Australian Nuclear Science and Technology Organisation (ANSTO) (http://www.ansto.gov.au/) in Lucas Heights, Australia. More information on the neutron sources available around the world can be found in Note 5.

Table 1
Characteristics of spectrometers from the ILL facility

Spectrometer	Wavelength (Å)	Measured *Q*-range ($Å^{-1}$)	Associated length-scale 1/*Q* (Å)	Energy resolution (μeV)	Associated timescale up to
IN6	5.12	0.4–2.1	0.5–2.5	90	10 ps
IN13	2.23	0.5–2.3	0.4–2	8	100 ps
IN16	6.27	0.5–1.9	0.5–2	0.9	1 ns

Two types of spectrometers are suitable for the application of single-particle neutron spectroscopy to biological samples: time-of-flight (TOF) spectrometers and backscattering spectrometers. TOF spectrometers have energy resolutions in the order of 10–100 μeV, corresponding to timescales of about 100–10 ps. Backscattering spectrometers have energy resolutions of the order of 1–10 μeV, with associated timescales of 1 ns to 100 ps. The momentum transfer, *Q*, range has also to be taken into account when selecting the spectrometer, as it provides information on the length-scale of the motions. Table 1 provides the characteristics of a few spectrometers accessible at the ILL.

2.2. Sample Holder

A typical sample holder is made up of a flat aluminum base containing a $4 \times 3 \times 0.03$ cm^3 hole, which is sealed with indium wire (diameter of 1 mm) to an aluminum lid of selectable thickness. The thickness of the lid is chosen to get a sample transmission of about 90%; it corresponds to a chamber thickness between 0.2 and 0.5 mm depending on the chemical composition of the sample. The sample holders can be provided by the neutron facilities.

2.3. Sample Amount

The amount of sample must be relatively large: a few hundreds of milligrams are required. For TOF and backscattering experiments, the sample need not be crystalline or even monodisperse; hydrated powders or solutions can be used.

Natural abundance GAG such as HS or heparin can be purchased directly as a lyophilized powder, for example from Celsus or Sigma Laboratories.

In the case of selectively labeled samples, the deuteration laboratory (DLAB) at the ILL in Grenoble can help the user with the sample preparation (see http://www.ill.eu/sites/deuteration/index.htm for further information).

2.4. Sample Hydration

In order to reduce strongly the scattering contribution arising from the solvent, the sample needs to be hydrated by D_2O. It is recommended to work with hydrated powders rather than solution samples (see Note 6). To perform experiments on a wide range of temperature including below the freezing point of water, powders should not be hydrated more than one hydration layer.

2.5. Complementary Samples

Accurate information can be extracted from data on absolute scale. This can be achieved by measuring additionally the following samples:

- An empty cell container (see Subheading 2.2) in the case of hydrated powders.
- A buffer sample, an empty cell container, and a vanadium sample in the case of solutions. The buffer consists in pure D_2O for GAG samples.

3. Methods

3.1. Applying for Beam Time

The first step to perform a neutron scattering experiment is to apply for beam time during the few yearly rounds of the neutron facilities (see Note 7). The spectrometer(s) must be carefully selected in respect to its (their) associated time- and length-scale windows.

See Note 8 in the case of selectively labeled samples.

3.2. Sample Preparation

This step determines the success of the EINS experiment. The protocol described below for HS samples has been used over the last decades for the preparation of hydrated protein powders and can be applied to other types of GAG or GAG in complex with proteins.

HS hydrated powders can be prepared as follows:

1. Take up HS powder in D_2O and lyophilize it. Repeat this step three times.
2. Weigh each piece of the flat aluminum sample holder (lid, base, and screws) as well as the indium wire. Put the indium wire on the associated part of the base and weigh the combined part. Place the HS powder on the 4×3 cm^2 compartment, weigh, and dry over silica gel for one to several days up to reach a constant weight. The procedure removes the remaining water molecules and defines the weight of a dehydrated sample with 0 g of D_2O/g of HS. The optimal dry weight for HS samples is about 150–200 mg.
3. Hydrate the dry sample over vapor pressure of D_2O by using a desiccator, until the weight reaches the amount of water corresponding to the first hydration layer. For HS, we estimate that it corresponds to about 0.43 g D_2O/g of HS.
4. Seal the sample with the appropriate lid (in this case 2.6 mm thickness), using associated aluminum screws, which gives a sample chamber of 0.4 mm thickness.

5. Measure the weight of the sealed sample and keep it at 8°C until the experiment (see Note 9).

3.3. Neutron Scattering Experiments

The neutron measurements are performed with the support of a "local contact" from the neutron facility, who is responsible for the instrument. The local contact provides assistance during the experiment and can also be contacted prior to the proposal submission deadline, in order to discuss the experiment and its feasibility as well as the accurate estimation of the requested beamtime.

3.3.1. Instrument Settings

The instrument is configured by the local contact, who selects the requested incident wavelength and energy resolution of the spectrometer (if several).

3.3.2. Sample Mounting and Data Collection

For an EINS experiment on an hydrated powder performed on a wide range of temperature:

1. Mount the sample on the instrument stick and place it in the associated cryostat at room temperature at an angle of 135° with respect to the incident neutron beam.
2. Cool down to 20 K (−253°C). Note that the cooling can take a couple of hours.
3. Perform the EINS measurements at given temperatures from 20 K (−253°C) to 310 K (−37°C) (see Note 10). On backscattering spectrometers such as IN13 and IN16 at the ILL, elastically scattered neutrons are recorded directly; on TOF instruments such as IN6 at the ILL, the full energy-range (elastically and inelastically scattered neutrons) is recorded, and the elastic intensity is extracted by integrating over a few channels in a time window around the elastic peak of the measured intensity. The evaluation of the counting time depends on the incident neutron flux, the amount of sample and its scattering power, and the measured temperature.

3.4. Data Reduction

The extraction of data on absolute scale is achieved by correcting for:

1. The variation of the incident beam flux: the data are normalized to the monitor that counts the number of incident neutrons.
2. The scattering signal arising from the empty cell container (in the case of hydrated powders, and for the vanadium sample) or from the buffer sample (in the case of solution samples).
3. Detector efficiency. The intensity of the sample has to be normalized by the intensity of the sample measured at 20 K (in the case of hydrated powders) or of a vanadium sample measured at room temperature (in the case of solutions).
4. The multiple scattering effects in the sample. As a first approach, the multiple scattering effects can be neglected by using a sample transmission above 90%.

Different softwares for data reduction are available, depending on the neutron center and the instrument. The local contact can help the user with the handling of the programs and the analysis. At the ILL, data can be analyzed, for example, using the LAMP program and implemented routines (10). Examples of scripts used for data analysis can be found in the supplementary material of Wood et al. (11).

3.5. Data Analysis

1. The MSD, $\langle u^2 \rangle$, of motions that are localized within the time and length window of the spectrometer can be obtained from the Q-dependence of the elastic intensity, I_{el}, according to the Gaussian approximation, valid for $\sqrt{(\langle u^2 \rangle Q^2)} < \sqrt{2}$ (see Subheading 1). $\langle u^2 \rangle$ is extracted at different temperatures, from a linear fit to the plot of $\ln(I_{el}(Q))$ vs. Q^2.
2. The resilience, $\langle k' \rangle$, is extracted from the inverse slope of the $\langle u^2 \rangle$ vs. T plot.

4. Notes

1. A neutron spectrometer offers a window in space and time, given by the momentum transfer range (inversely proportional to the space range) and the energy resolution and range (inversely proportional to the time range). The energy resolution of the spectrometer is the smallest experimental energy change that can be measured and corresponds to the longest time-lapse, over which the experiment is sensitive. For further reading, see Serdyuk et al. (1).
2. Further information on the Gaussian approximation can be found in the "Neutron Spectroscopy" section of Serdyuk et al. (1).
3. The Guinier approximation has been well reviewed in a previous issue of *Methods in Molecular Biology*, which describes the small angle neutron scattering technique for molecular biology (4).
4. $\langle k' \rangle$ is a purely experimental parameter with the dimensions of a force constant. It is very useful in comparative studies of different systems, but it is a measure of the mean force constant only in the case of atoms in a harmonic potential.
5. Neutron sources available around the world have been listed on the NIST Center for Neutron Research website, http://www.ncnr.nist.gov/nsources.html. Among them are the ISIS-Rutherford-Appleton Laboratories (http://www.isis.stfc.ac.uk/index.html) in Chilton, United Kingdom; the FRM-II (http://www.frm2.tum.de/en/index.html) in Garching, Germany; and the ISSP Neutron Scattering Laboratory (http://www.issp.u-tokyo.ac.jp/labs/neutron/index-e.html) in Tokai, Japan.

6. Care must be exercised with solution samples, for which the solvent, even D_2O, and macromolecular self-diffusion can contribute to the scattering signal; the time and space window of the self-diffusion contribution depends on the size of the macromolecule and sample concentration.
7. The proposals are evaluated by a scientific review committee; they include the description of the proposed work, of the samples, and an estimation of the time required to perform the measurements with respect to the instrument characteristics. The time between the submission deadline and the experiment takes about 6 months.
8. In the case of selectively labeled samples measured at the ILL in Grenoble, the user can submit a proposal to the DLAB facility (http://www.ill.eu/sites/deuteration/index.htm) at any time of the year, which will be evaluated by a committee. If the proposal is accepted, the DLAB will provide assistance with the sample preparation. Further information on deuterium labeling is provided in the previous issue of *Methods in Molecular Biology* (9).
9. Because the neutron experiments are performed under vacuum, it is important to control the sample weight before and after each experiment, to verify that the sample did not lose any water during the experiment and therefore conserved its hydration level.
10. For the empty cell and the vanadium samples, the measurements need not be performed on the whole temperature range but only at room temperature.

Acknowledgments

The author thanks Giuseppe Zaccai, Moeava Tehei, and Shoh Asano for helping in the critical reading of the manuscript.

References

1. Serdyuk, I. N., Zaccai, N. S., and Zaccai, G. (2007) Methods in Molecular Biophysics: Structure, Dynamics, Function. Cambridge University Press, Cambridge.
2. Gabel, F., Bicout, D., Lehnert, U., Tehei, M., Weik, M., and Zaccai G. (2002) Protein dynamics studied by neutron scattering. *Q Rev Biophys* **35**, 327–367.
3. Guinier, A., and Fournet, G. (1955) Small Angle Scattering of X-rays. John Wiley, New York.
4. Whitten, A. E., and Trewhella, J. (2009) Small-angle scattering and neutron contrast variation for studying bio-molecular complexes. *Methods Mol Biol* **544**, 307–323.
5. Zaccai, G. (2000) How soft is a protein? A protein dynamics force constant measured by neutron scattering. *Science* **288**,1604–1607.
6. Jasnin, M., van Eijck, L., Koza, M. M., Peters, J., Laguri, C., Lortat-Jacob, H., and Zaccai, G. (2010) Dynamics of heparan sulfate explored by neutron scattering. *Phys Chem Chem Phys.* **12**, 3360–3362.
7. Mulloy, B. and Forster, M.J. (2000) Conformation and dynamics of heparin and heparan sulfate. *Glycobiology* **10**, 1147–1156.

8. Almond, A., DeAngelis, P. L., and Blundell, C.D. (2005) Dynamics of hyaluronan oligosaccharides revealed by ^{15}N relaxation. *J Am Chem Soc.* **127**, 1086–1087.
9. Meilleur, F., Weiss, K. L., and Myles, D. A. (2009) Deuterium labeling for neutron structure-function-dynamics analysis. *Methods Mol Biol* **544**, 281–292.
10. Richard, D., Ferrand, M., and Kearley, G. J. (1996) LAMP, the Large Array Manipulation Program. http://www.ill.fr/data_treat/lamp/front.html.
11. Wood, K., Caronna, C., Fouquet, P. et al (2007) A benchmark for protein dynamics: Ribonuclease A measured by neutron scattering in a large wavevector-energy transfer range. *Chem Phys* **345**, 305–314.

Chapter 12

Following Protein–Glycosaminoglycan Polysaccharide Interactions with Differential Scanning Fluorimetry

Katarzyna A. Uniewicz, Alessandro Ori, Timothy R. Rudd, Marco Guerrini, Mark C. Wilkinson, David G. Fernig, and Edwin A. Yates

Abstract

Studies of the structural changes invoked in proteins by the binding of the glycosaminoglycan (GAG) polysaccharide portion of proteoglycans are of increasing importance to research in a wide range of fields, from biochemistry and molecular biology to biotechnology and medicine. One important aspect is the degree of stabilisation or destabilisation induced in a protein by the binding of these anionic materials, and this can affect enzyme activity, the stability of complexes, folding and the formation of aggregates, including those in neurodegenerative processes. A simple method, able to determine the effect of interactions with GAG polysaccharides on protein stability is described, based on the propensity of a fluorescent dye—Sypro™ Orange—to present differentiable fluorescence emission spectra following contact with exposed core amino acid residues. The method requires only commonly available and inexpensive equipment and is suitable for a multi-well format, allowing multiple readings to be made simultaneously.

Key words: Glycosaminoglycans, Interactions, Proteins, Stabilisation and conformation

1. Introduction

Glycosaminoglycans (GAGs) are an important class of linear, anionic polysaccharides, which are found throughout the animal kingdom and, in evolutionary terms, are thought to have arisen at the same time as multicellular life. GAGs are involved in the major processes of cell growth, division and replication and have become the focus of much research interest because of their involvement in developmental processes and in several disease-related processes of major importance, including ageing, cancer and neurodegenerative disorders. The structure of GAGs varies between family members, but all comprise a disaccharide repeating structure containing an amino

Françoise Rédini (ed.), *Proteoglycans: Methods and Protocols*, Methods in Molecular Biology, vol. 836,
DOI 10.1007/978-1-61779-498-8_12,

sugar. Furthermore, hyaluronate (HA) is the only non-sulphated form, comprising a uronate residue (D-GlcA) linked to an *N*-acetylated amino sugar (D-GlcNAc). The other GAG members either contain *O*-sulphates (chondroitin sulphates, dermatan sulphate and keratan sulphate) or both *O*- and *N*-sulphation (heparan sulphate and heparin). Consequently, all GAG family members are anionic. The GAGs are also of major importance currently in medicine because heparin is widely used as an anticoagulant, but many other potential medical roles for GAGs are under scrutiny (1).

The GAGs exert their biological activities through interactions with a wide variety of proteins, ranging from serine proteases, such as antithrombin (AT) (2) and the Alzheimer's disease beta secretase (BACE-1) (3), to multi-protein assemblies such as the fibroblast growth factor-receptor signalling complexes (FGF:FGFRs) (4–6). In all of these cases, understanding the relationship between the structure of the polysaccharides, both in terms of primary sequence (encompassing substitution patterns) and conformation, is sought after, but highly challenging. The results of many investigations support the view that the substitution pattern of the polysaccharide does not correlate well with activity, so the search for consensus sequences, akin to those proposed for proteins, has not been successful. However, the activity of GAGs does not correspond well with simple charge density either. It now appears that a more subtle principle underlies the structure–activity relationship between GAGs and proteins, one that relies on the complex relationships between substitution patterns, conformation and dynamics. These relationships are beginning to be unravelled in the case of the structurally most complex GAG members, heparan sulphate and heparin (7–9).

The study of GAG–protein interactions is also of interest for the fundamental understanding of the systems and is implicated in changes in enzyme activity as well as the formation of aggregates, including those associated with neurodegenerative disorders. It is also of huge potential importance for the design and optimisation of pharmaceutical agents. However, the study of protein–GAG polysaccharide interactions is challenging for several reasons. First, with the exception of HA, which is homogeneous, the GAGs have both heterogeneous sequences and are polydisperse. Second, the large physical size of the GAGs and their associated low mobility and anisotropy, as well as the potential for multiple binding sites, introduces additional experimental limitations and challenges.

GAG–protein interactions have been studied with a range of techniques, from surface-based approaches (e.g. surface plasmon resonance and ELISA techniques) (10), spectroscopic approaches (e.g. FTIR, NMR, CD and VCD) (11–14) to calorimetric methods (ITC and DSC) (12). Most of these, however, are limited by the need for specialised equipment and expertise and are unsuitable for multiple readings, so do not lend themselves to high-throughput applications or screening.

The interaction of GAG polysaccharides with proteins may alter the activity of the target protein directly (e.g. antithrombin (15)) or of complexes formed from several proteins (e.g. FGF signalling complexes (16–20)). The method described here is a development of a fluorescent shift assay described by Niesen et al. (21), in which a fluorescent dye serves as an indicator of the degree of protein folding/unfolding, differential scanning fluorimetry (DSF). DSF relies on a change in fluorescence of a sensitive fluorescent dye, Sypro™ Orange, which exhibits altered emission properties when bound to exposed core amino acid residues (21,22). DSF provides melting temperature information, as well as the shape of the melting curves reflecting both the enthalpic and entropic characteristics of the interaction process.

2. Materials

2.1. Protein and Polysaccharide Preparation

Protein and polysaccharide samples should be purified before use to the highest available standard and made up in appropriate buffers to convenient concentrations. It is recommended, especially for protein samples, that optimal storage conditions are defined and strictly followed to avoid variations in protein behaviour. Particular care should be paid to the possibility of protein aggregation, which is evident in the form of the melting curves (e.g. see Fig. 1).

2.2. Instrumentation

The experiments are performed on a 7500 Fast real Time PCR system (Applied Biosystems, Warrington, Cheshire, UK, software version 1.4.0) and the samples are subjected to the heating cycle as described (22). The assay uses the default setting for detection of NED/TAMRA/Cy3, which is assigned to filter C. This avoids the need for calibration, which ideally would use λ_{ex} 492 nm/λ_{em} 610 nm, and operates using its approximation; λ_{ex} 560 nm/λ_{em} 582 nm.

2.3. Plate Preparation

Prepare the protein–polysaccharide mixtures in standard polypropylene vials and subsequently add to the wells of a Fast Optical 96-Well Reaction Plate and seal with Optical Adhesive Film (Applied Biosystems).

2.4. General Considerations

1. The suitability of the assay and possibility of its application require preliminary exclusion of any non-specific interactions between the dye and the polysaccharide ligands. Additionally, the use of buffer components with fluorescent characteristics or likely to interact with the dye should be avoided (see Note 1).
2. The final concentration of protein in the assay (usually in the range 5–10 μM) should be established in the preliminary experiments. For experiments in which a range of GAG

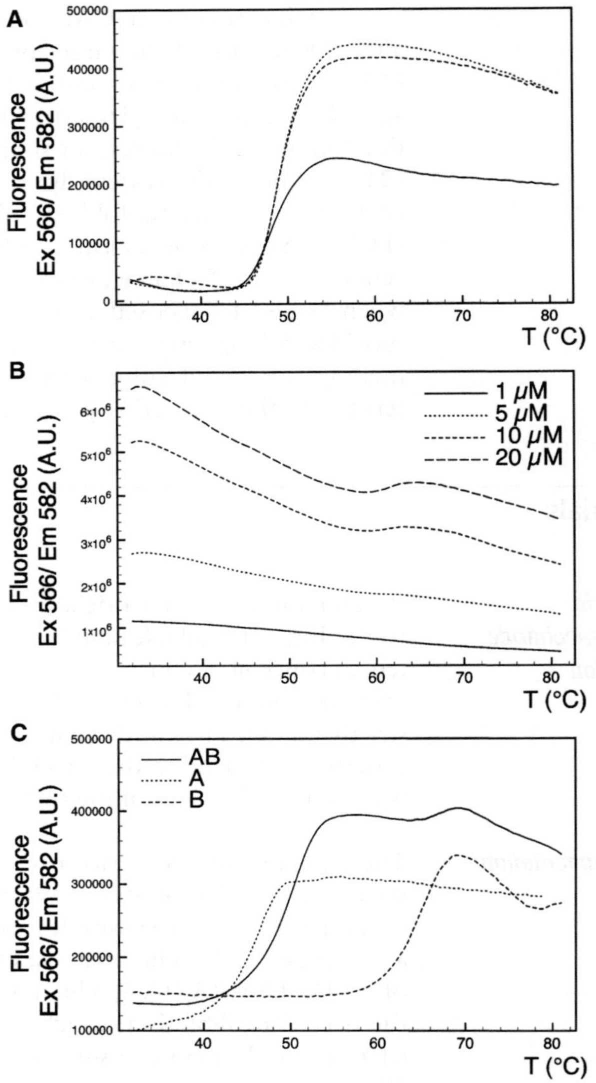

Fig. 1. The shapes of melting curves can be modified by a number of considerations or problems. (**a**) Three repeats, one of which (*solid line*) is distinct to the other two. This may have arisen from uneven mixing of the Sypro™ Orange dye resulting in lower amounts of the dye in one out of three samples within the same repeat. (**b**) The shape of the melting curves reflects protein aggregation following heating during the thermal cycle. (**c**) The melting curves of two proteins (**a** *dotted line*; **b** *dashed line*) and the melting curve of the fusion protein composed of (**a**) and (**b**) (**a**, **b** *solid line*) exhibiting an influence on its melting curve from both of its constituents.

polysaccharides are required, the nominal molecular weight of the polysaccharide can be used, but care should be exercised in cases of high polydispersity, especially when later interpreting the results (see Note 2).

3. Various molar/weight ratios of the polysaccharides vs. the protein might be required to enable detection of differences in ther-

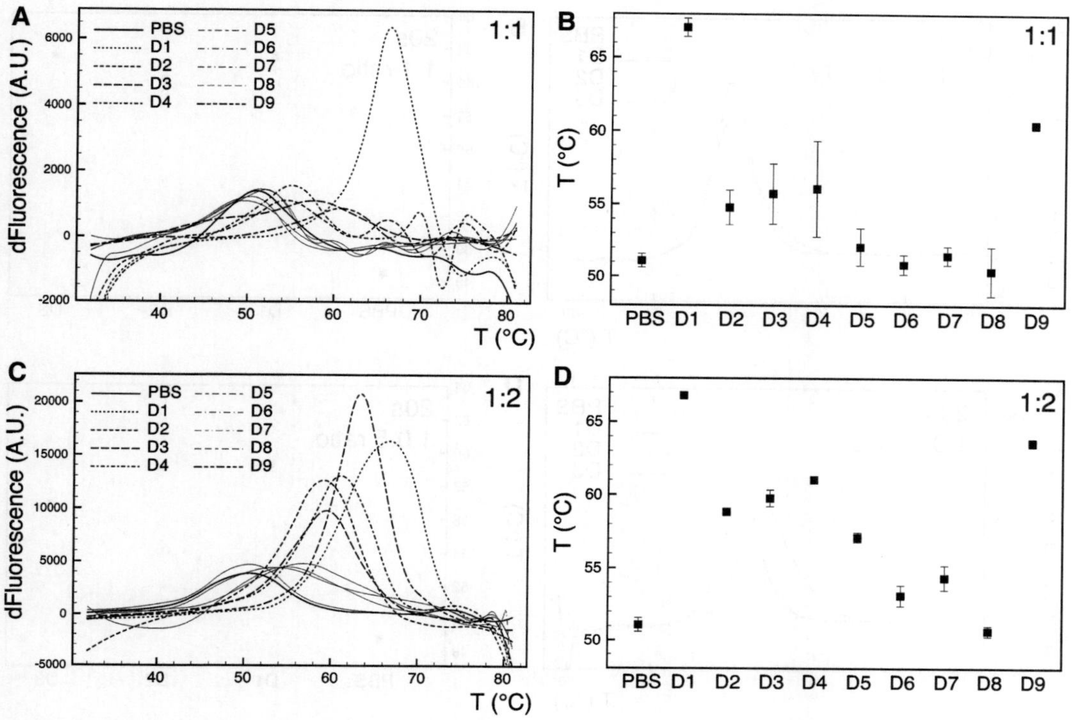

Fig. 2. Appearance of melting curves (**a**, **c**: with a range of polysaccharides) when a protein (FGF) alone has no distinct melting curve, but interaction with some polysaccharides (e.g. D1 in **a**) induces one. The ratio of protein: polysaccharide in (**a**) is 1:1 and in (**b**) is 1:2. The melting temperatures are plotted in (**b**, **d**), with *error bars* derived from several repeats, illustrating both well-defined values of Tm (e.g. D1 in **b**) and poorly defined values (e.g. D4 in **b**).

mostabilising effect (e.g. see Fig. 2). These should be defined in a set of preliminary experiments, each of them including the internal control including the internal positive (protein and the reference polysaccharide) and negative (protein alone) control.

4. The thermal settings of the RT-PCR cycle can be adjusted to obtain maximum information on the thermostabilising effects of the tested polysaccharides (see Note 3 and Fig. 3).

3. Methods

3.1. Differential Scanning Fluorimetry Experimental Measurements

1. Mix the GAG polysaccharide samples (10% v/v), protein solution (10% v/v) and Dulbecco's phosphate-buffered saline (Gibco-Europe, Paisley, UK) (70% v/v) in a vial kept on ice. The final volume per well is 10 μL; however, it is recommended to prepare a mixed solution of all compounds at the final volume of 35 μL, sufficient to prepare three aliquots of 10 μL per well (see Note 4).

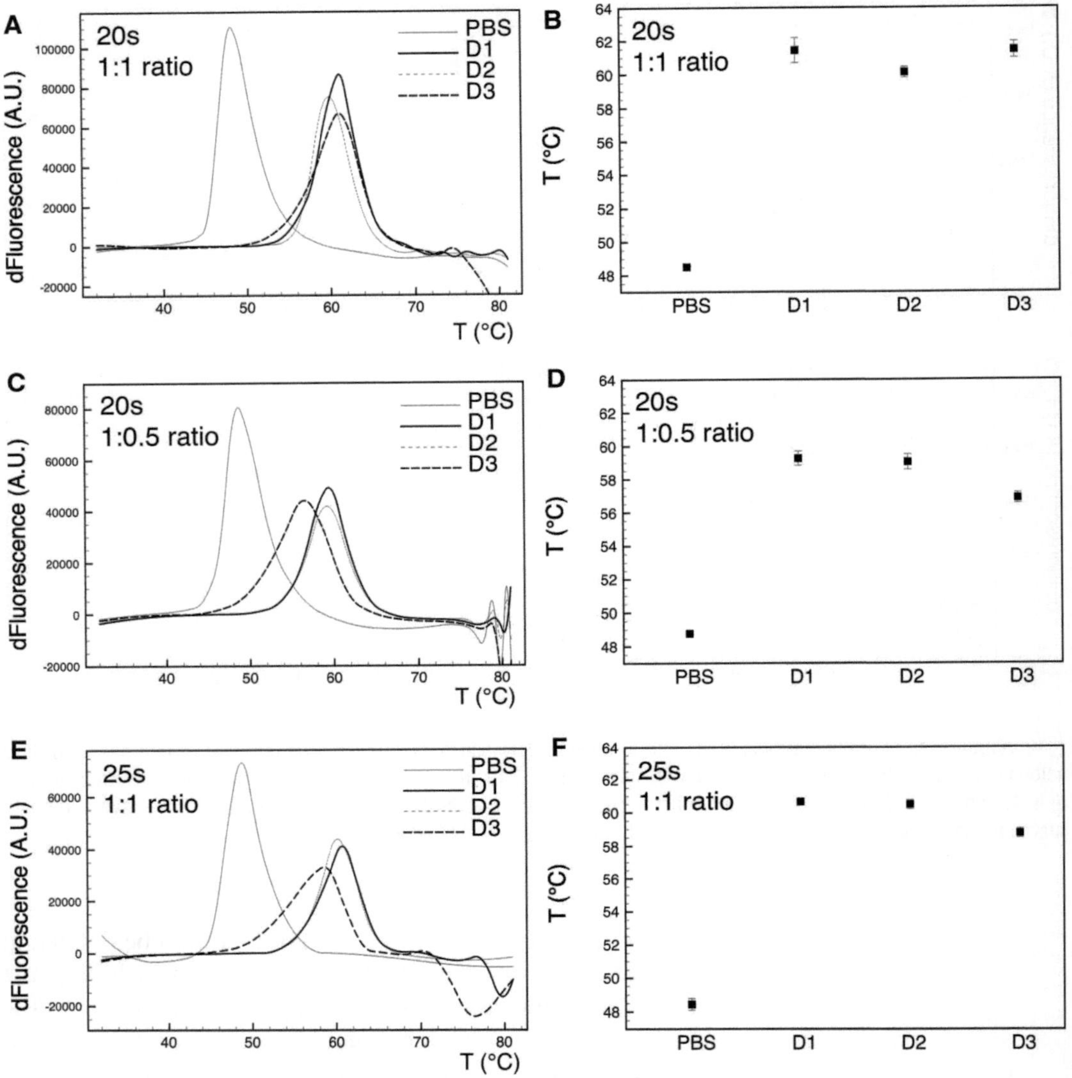

Fig. 3. Optimisation of the Sypro™ Orange fluorescent shift assay involving altered temperature gradient step of 20 s (**a–d**) and 25 s (**e**, **f**) and altering the ratios of protein: polysaccharide (10 μM of each in 1:1 ratio in (**a**, **b**, **e**, **f**) or 1:0.5 ratio in (**c**, **d**)). Note that following optimisation the melting curves (**e**) give rise to well-defined and reproducible values of Tm (**f**).

2. Dilute the fluorescent dye Sypro™ Orange 5,000× (Invitrogen, Paisley, UK; 100× in HPLC grade water) and add to the protein–polysaccharide mixture at 10% v/v. Carefully mix the solution to avoid introducing any air bubbles using one of the two mixing protocols (see Note 5) throughout.
3. Then, carefully seal the plate with Optical Adhesive Film (Applied Biosystems), paying attention not to leave any marks on the film that could affect the reading and immediately analyse in the RT-PCR instrument (see Notes 6 and 7).

4. The heating cycle is a 120-s pre-warming step at 31°C and subsequent gradient between 32 and 81°C in 99 steps of 0.5°C, each of 20 s duration. Data are collected using the existing TAMRA dye setting (λ_{ex} 560 nm and λ_{em} 582 nm).

3.2. Data Analysis

1. Data are analysed by applying an exponential correlation function approximation of a first derivative for each melting curve employing Plot v.0.997 software for Mac OS X (plot.micw.eu) (Origin for Windows). The maxima are used to determine the mean melting temperature (Tm) and the standard deviation (SD) of each Tm, which should be based on at least three repeats. For comparison of the stabilisation effects between polysaccharides, all data should be normalised to a standard polysaccharide (e.g. heparin). For normalisation, each compound can be characterised by the difference between the Tm value of the protein in PBS (TmPBS) and in the presence of the polysaccharide derivative (TmX) according to Eq. 1

$$\text{TmX - TmPBS} \tag{1}$$

2. Obtain a measure of the stabilisation compared to heparin from Eq. 2, where TmHep refers to the mean Tm of the protein in the presence of heparin

$$\text{TmX - TmPBS/TmHep - TmPBS} \tag{2}$$

Set TmPBS to a nominal value of 0 and TmHep to 1, to give a normalised scale with which to compare stabilisation potency.

4. Results

A selection of typical denaturation (melting) curves is shown in Fig. 4, together with their first derivatives. The first derivatives are approximated by an exponential correlation function. The melting temperature (Tm) is identified by the peak position of the first derivative on the temperature scale. Note how the value of Tm can change with conditions, notably in Fig. 4a, b for IL-6, here varying as a function of the amount of protein present. One important type of result is shown in Fig. 2. Here, proteins which alone do not give clear melting curves (see Fig. 2a, c) can be induced to do so following interaction with some polysaccharides (their Tm values for several repeats are plotted in Fig. 2b, d). Note, however, that many polysaccharides do not have this ability (see Fig. 2a) and that, in some cases (especially D4 in Fig. 2b), precise values for Tm may be difficult to obtain. A sharp melting curve, with precise Tm values, indicates that the two-state model is a good approximation.

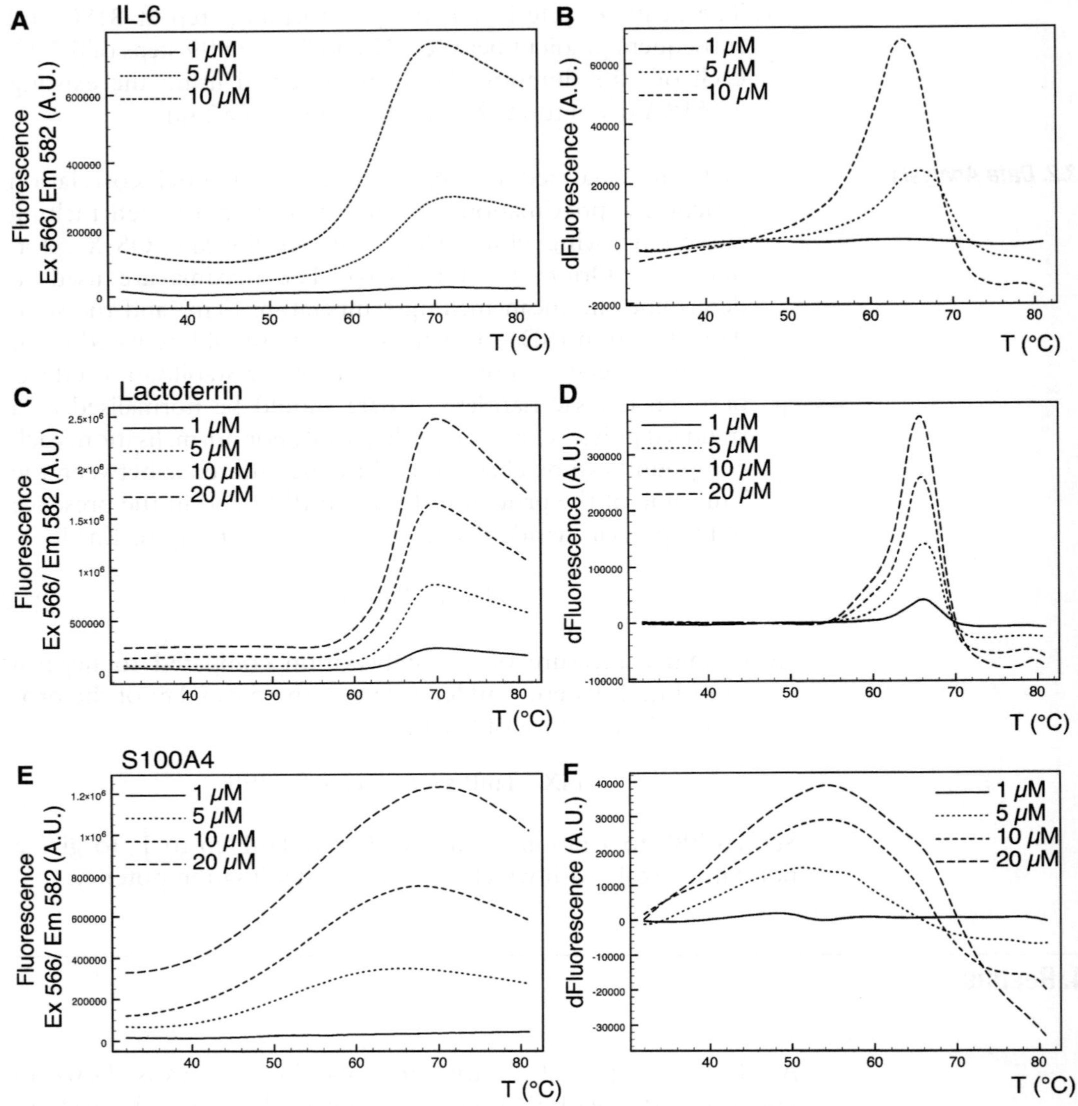

Fig. 4. Denaturation (melting) curves for three proteins (**a**, **c** and **e,** respectively, IL-6, lactoferrin and S100A4) and their respective first derivative plots (**b**, **d**, **f**). The melting curves of each protein were recorded with varying concentrations of varying concentrations of protein.

The assay can be optimised by modifying the molar ratios of the protein and polysaccharide as well as the characteristics of the thermal cycle. A range of experimental values are shown in Fig. 3, for a FGF protein and three polysaccharides with a standard thermal cycle involving 20 s ramp time (melting curves, Fig. 3a, c; Tm values, Fig. 3b, d) and at 1:1 ratio of protein:polysaccharide (see Fig. 3a, b) or 1:0.5 ratio (see Fig. 3c, d). Figure 3e, f show the melting curve and plot of Tm values for a thermal cycle involving a ramp temperature gradient of 25 s at a ratio of 1:1. There are a number of situations in which recording the melting temperature of a protein polysaccharide complex may be problematic. As mentioned above, if the two-state model represents a good approximation, the melting curve should be sharp, leading to a well-defined and precise Tm value. Figure 1a shows three repeats in which one of the repeats (solid line) has a much lower signal. This is probably caused by poor mixing of the dye. If the protein begins to form aggregates as the temperature is raised, irregular melting curves may be observed, as in Fig. 1b, in the presence of different levels of polysaccharide. In Fig. 1c, the case of a fusion protein consisting of two dissimilar components is given. Note that, in this case, the melting curve of the fusion protein combines characteristics of both of its constituent proteins.

5. Conclusions

The method allows the rapid determination of the protein stabilisation characteristics of GAG polysaccharides, or other polyanions, to be determined and analysed, in terms of melting temperatures. The approach requires only widely available equipment and is also suitable for screening applications since it is compatible with multiwell formats, automation and requires relatively modest sample quantities. Furthermore, the process requires only 30 min for completion. Interpretation of the melting curves requires some caution however, because melting curves reflect both enthalpic and entropic contributions and the melting temperature is only one relevant parameter. The form of the melting curve should also be noted. When the two-state model is a good approximation, the melting curve is well-defined, giving rise to a sharp and reproducible value of Tm in the first derivative plot.

6. Notes

1. It has been confirmed (23) that Sypro™ Orange dye does not interact directly with heparin or its de-sulphated derivatives. This means that the thermal cycle of the polysaccharide in the presence of the Sypro™ dye produced a flat line of insignificant fluorescence in comparison with the signal derived from the sample containing protein. Likewise, it would be advantageous to check that the particular polysaccharides under examination do not interact with the dye.

2. Initial tests to determine optimal concentration of the protein to obtain clear denaturation curves are recommended. Comparative tests of 2, 5, 10 and 25 μM samples should be made. Only afterwards, the optimal lowest concentration that provides clear denaturation curves can be selected for the further experiments. Importantly, some proteins might have a propensity to produce clear denaturation curves only after addition of a stabilising polysaccharide (e.g. see Fig. 2) and this possibility should be investigated whenever no clear denaturation curve is recognised. When dealing with GAG polysaccharides, it is often difficult to determine the molarity of a solution precisely because of the inherent sequence diversity and polydispersity of the molecules within the sample. It may be more profitable to compare results in terms of concentrations (μg/mL), but disparities in the molecular weight of the constituent repeating units should also be borne in mind. If the assay is to be conducted with defined oligosaccharide samples, then molarity should be used.
3. The assay was designed to contain an internal control and allow multiple ligand comparison on the basis of relative stability-inducing property of each ligand compared to the denoted control. Such a design allows modification of the thermal cycle, provided the internal control is included in each experiment. Hence, it is possible to change the window of the gradient in order to produce a better-fitted cycle to the properties of the protein. Modifications of ramp length (e.g. 30 s instead of 20 s) or temperature gradient (e.g. 1°C instead of 0.5°C) are possible to improve detection.
4. Variants of the assay are possible. For example, studying the effect of cations on a given polysaccharide–protein interaction. In this case, the buffer should be checked for relevant cations. Divalent cation-free buffer can also be obtained. At the stage when protein and polysaccharide are mixed together (see Subheading 3.1, step 1), it is possible to introduce an additional incubation step. The temperature of this step is optional. Whether the samples are to be kept on ice or on the bench at room temperature needs to be decided and followed throughout the assay.
5. Following dilution in HPLC grade water, the dye should be used within 1–2 h, because it soon starts to precipitate. Although addition of an aged dye solution might still allow detection of denaturation curves, it is recommended to maintain optimal conditions by using fresh Sypro™ Orange solution in the wells of a Fast Optical 96-Well Reaction Plate (Applied Biosystems). The 96-well plates can be spun to eliminate bubbles. There are two methods of adding Sypro™ Orange: (1) Mix protein, polysaccharide, buffer and add 9 μL of the mixture into each well,

then add 1 μL diluted dye into each well. The dye is not stable over long periods and should be added last and variation in the three readings can be observed if the order of addition is changed. (2) Mix all four solutions together, including the dye and add 10 μL to each well. If this is done too slowly, however, the signal may weaken.

6. Before the final addition of the test solutions into the wells of the 96-well plate, it is important to mix them efficiently, as judged by the uniform orange colour of the liquid; however, it should be done gently enough to avoid air bubble formation. If some air bubbles appear, it is possible to remove them using a syringe needle. Otherwise, a gentle vortex of the entire plate can be applied; however, this requires careful support of the whole plate and observation of the menisci of the samples. Be aware that the surrounding temperature might also affect the viscosity of the samples, as well as promoting bubble dispersal. Vortexing may also cause some undesired splashing.
7. It is important to minimise the time between the addition of the dye to the sample and the start of the thermal cycle. Ideally, experiments should be designed to prevent extended incubation of the samples with the dye and allow immediate analysis following plate preparation.

Acknowledgements

The European Commission (Marie Curie Early Stage Training Fellowships) is thanked for funding (KA and AO) and North West Cancer Research Fund for support (DGF).

References

1. Esko, J. D. and Selleck, S. B. (2002) Order out of chaos: Assembly of ligand binding sites in heparan sulfate. *Annu Rev Biochem.* **71**, 435–471.
2. Hovingh, P., Piepkorn, M., and Linker, A. (1986) Biological implications of the structural, antithrombin affinity and anticoagulant activity relationships among vertebrate heparins and heparan sulphates. *Biochem J.* **237**, 573–581.
3. Patey, S. J., Edwards, E. A., Yates, E. A., and Turnbull, J. E. (2006) Heparin derivatives as inhibitors of BACE-1, the Alzheimer's beta-secretase, with reduced activity against factor Xa and other proteases. *J Med Chem.* **49**, 6129–6132.
4. Guimond, S., Maccarana, M., Olwin, B. B., Lindahl, U., and Rapraeger, A. C. (1993) Activating and inhibitory heparin sequences for FGF-2 (basic FGF). Distinct requirements for FGF-1, FGF-2, and FGF-4. *J Biol Chem.* **268**, 23906–23914.
5. Pye, D. A., Vives, R. R., Turnbull, J. E., Hyde, P., and Gallagher, J. T. (1998) Heparan sulfate oligosaccharides require 6-O-sulfation for promotion of basic fibroblast growth factor mitogenic activity. *J Biol Chem.* **273**, 22936–22942.
6. Sugaya, N., Habuchi, H., Nagai, N., Ashikari-Hada, S., and Kimata, K. (2008) 6-O-sulfation of heparan sulfate differentially regulates various fibroblast growth factor-dependent signalings in culture. *J Biol Chem.* **283**, 10366–10376.
7. Seyrek, E., Dubin, P. L., and Henriksen, J. (2007) Nonspecific electrostatic binding

characteristics of the heparin-antithrombin interaction. *Biopolymers* **86**, 249–259.

8. Rudd, T. R. and Yates, E. A. (2010) Conformational degeneracy restricts the effective information content of heparan sulfate. *Mol Biosyst.* **6**, 902–908.
9. Rudd, T. R., Uniewicz, K. A., Ori, A., Giumond, S. E., Skidmore, M. A., Gaudesi, D., et al. (2010) Comparable stabilisation, structural changes and activites can be induced in FGF by a variety of HS and non-GAG analogues: implications for sequence-activity relationships. *Org Biomol Chem.* DOI: 10.1039/c1030ob00246a.
10. Delehedde, M., Lyon, M., Gallagher, J. T., Rudland, P. S., and Fernig, D. G. (2002) Fibroblast growth factor-2 binds to small heparin-derived oligosaccharides and stimulates a sustained phosphorylation of p42/44 mitogen-activated protein kinase and proliferation of rat mammary fibroblasts. *Biochem J.* **366**, 235–244.
11. Gettins, P. (1987) On the domain structure of antithrombin III. Tentative localization of the heparin binding region using ^{1}H NMR spectroscopy. *Biochemistry* **26**, 4403–4408.
12. Guzman-Casado, M., Cardenete, A., Gimanez-Gallego, G., and Parody-Morreale, A. (2001) Myo-inositol hexasulphate and low molecular weight heparin binding to human acidic fibroblast growth factor: A calorimetric and FTIR study. *Int J Biol Macromol.* **28**, 305–313.
13. Svensson, G., Linse, S., and Mani, K. (2009) Chemical and thermal unfolding of glypican-1: Protective effect of heparan sulfate against heat-induced irreversible aggregation. *Biochemistry* **48**, 9994–10004.
14. Zakrzewska, M., Wiedlocha, A., Szlachcic, A., Krowarsch, D., Otlewski, J., and Olsnes, S. (2009) Increased protein stability of FGF1 can compensate for its reduced affinity for heparin. *J Biol Chem.* **284**, 25388–25403.
15. Busby, T. F., Atha, D. H., and Ingham, K. C. (1981) Thermal denaturation of antithrombin III. Stabilization by heparin and lyotropic anions. *J Biol Chem.* **256**, 12140–12147.
16. Copeland, R. A., Ji, H., Halfpenny, A. J., Williams, R. W., Thompson, K.C., Herber, W. K., et al. (1991) The structure of human acidic fibroblast growth factor and its interaction with heparin. *Arch Biochem Biophys.* **289**, 53–61.
17. Culajay, J. F., Blaber, S. I., Khurana, A., and Blaber, M. (2000) Thermodynamic characterization of mutants of human fibroblast growth factor 1 with an increased physiological half-life. *Biochemistry* **39**, 7153–7158.
18. Fan, H., Li, H., Zhang, M., and Russell Middaugh, C. (2007) Effects of solutes on empirical phase diagrams of human fibroblast growth factor 1. *J Pharm Sci.* **96**, 1490–1503.
19. Prestrelski, S. J., Fox, G. M., and Arakawa, T. (1992) Binding of heparin to basic fibroblast growth factor induces a conformational change. *Arch Biochem Biophys.* **293**, 314–319.
20. Vemuri, S., Beylin, I., Sluzky, V., Stratton, P., Eberlein, G., and Wang, Y. J. (1994) The stability of bFGF against thermal denaturation. *J Pharm Pharmacol.* **46**, 481–486.
21. Niesen, F. H., Berglund, H., and Vedadi, M. (2007) The use of differential scanning fluorimetry to detect ligand interactions that promote protein stability. *Nat Protoc.* **2**, 2212–2221.
22. Uniewicz, K. A. and Fernig, D. G. (2008) Neuropilins: A versatile partner of extracellular molecules that regulate development and disease. *Frontier Biosci.* **13**, 4339–4360.
23. Uniewicz et al. (2010) Comparable stabilisation, structural changes and activities can be induced in FGF by a variety of HS and non-GAG analogues: implications for sequence-activity relationships. *Org Biomol Chem.* **8**, 5390–5397.

Chapter 13

In Vivo Scintigraphic Imaging of Proteoglycans

Elisabeth Miot-Noirault, Aurélien Vidal, Philippe Auzeloux, Caroline Peyrode, Jean-Claude Madelmont, and Jean-Michel Chezal

Abstract

In this chapter, we present the methods developed in our lab for the scintigraphic imaging and direct quantitative evaluation of proteoglycan (PG) distribution in vivo. These methods relate to (1) the synthesis and radiolabeling of the NTP 15-5 with ^{99m}Tc, (2) preclinical scintigraphic imaging using laboratory animals, and (3) quantitative analysis of scintigraphic images.

Key words: Proteoglycans, Scintigraphy, ^{99m}Tc-NTP 15-5 radiotracer, Quantitative imaging

1. Introduction

The recent application of scintigraphic imaging, both in planar and tomographic configurations (named as single photon emission computed tomography or SPECT), to small animal produces functional images at the millimeter scale and opens wide in vivo access to many living animal models of human biology and disease (1–4). The basis of scintigraphy is the use of radiotracers as bifunctional agents containing a vector (which has an affinity for a target in vivo) and a radioisotope that confers external detection. Scintigraphic techniques are able to detect very low concentrations of radiotracers in vivo and offer noninvasive and functional in vivo access to specific targets, with a high sensitivity for deep tissue imaging and signal tissue quantitation. In such context, targeting proteoglycans (PG) for scintigraphic imaging of cartilage is the strategy developed by and used in our group for many years (5). This strategy consists of using the quaternary ammonium function (that exhibits a high affinity for PG) as a selective carrier of polyazamacrocycle structure able to chelate ^{99m}Tc (5, 6). Due

Françoise Rédini (ed.), *Proteoglycans: Methods and Protocols*, Methods in Molecular Biology, vol. 836,
DOI 10.1007/978-1-61779-498-8_13,

to its short period (6 h), its 140 keV emission γ rays optimal for external detection with gamma cameras, low cost, and availability, ^{99m}Tc is the most commonly used isotope in nuclear medicine, accounting for about 90% of all diagnostic nuclear medicine scans worldwide (7).

To that end, the radiotracer ^{99m}Tc-*N*-[(3-triethylammonio) propyl]-1,4,7,10,13-pentaazacyclopentadecane (^{99m}Tc-NTP 15-5)was selected on the basis of (a) a high affinity for PG in vitro and in vivo and (b) a high stability of the ^{99m}Tc-complex (5, 6, 8). When intravenously administered to healthy animals (and in many animal species such as mouse, rat, guinea pig, and rabbit), ^{99m}Tc-NTP 15-5 was observed to accumulate within articular cartilage (about 2.18 ± 0.41% and 5.5 ± 1.7% of injected dose/g of cartilage in rat and guinea pig, respectively) and evidenced a high cartilage/surrounding tissues ratio with a high in vivo stability (8–10). From many experimental results in animal models of osteoarthritis and PG-expressing tumors, ^{99m}Tc-NTP 15-5 imaging was strongly related to in vivo PG distribution and is therefore expected to be a powerful SPECT radiotracer for the molecular imaging of the degenerative and tumoral pathologies of cartilage in nuclear medicine (8–13).

For the evaluation of cartilage PG, other biophysical methods such as high-frequency ultrasound and Gd-DTPA-enhanced magnetic resonance imaging have been assessed for their relevance and sensitivity (14, 15). High-frequency ultrasound was shown to provide information about PG content in cartilage specimens in vitro but seems difficult to transpose in vivo. In Gd-DTPA-enhanced MRI, T1 mapping was demonstrated to provide an estimate of the local concentration of the contrast agent, which accumulates in cartilage inversely to negative charges of glycosaminoglycans (GAGs) (15). Nevertheless, this method provides an indirect measurement of PG content, requires critical factors such as the dose of contrast agent (and nonnegligible-associated toxicity), and is known to be several orders of magnitude less sensitive as compared with nuclear imaging techniques (2, 15–17).

With the recent advances in the extracellular matrix biology understanding, PG appear as major and key partners for many organ integrity and function and might represent interesting targets for the assessment (as well as treatment) of degenerative and malignant pathological processes (18–21). In such context, scintigraphic imaging with the radiolabeled-specific ligand ^{99m}Tc-NTP 15-5 would be a powerful tool for a direct in vivo quantitative evaluation of PG in preclinical research, allowing in the same animals PG changes in relation to pathological processes and/or in response to therapies.

2. Materials

2.1. Chemicals

1,4,7,10,13-Pentaazacyclopentadecane (15-5) can be purchased from 3B Scientific Corporation (USA) or Amfinecom Incorporation (USA). Additional chemicals are purchased from standard commercial sources (e.g., Aldrich-, Fluka-, or Sigma (France)) and are used without further purification.

2.2. Radiochemical Labeling of NTP 15-5 with Technetium-99m

1. Synthetic scheme of *N*-[(3-triethylammonio)propyl]-1,4,7,10,13-pentaazacyclopentadecane (NTP 15-5) is presented in Fig. 1.
2. 3-(Bromopropyl)triethylammonium bromide is synthesized from dibromopropane according to reported procedures (22, 23). Briefly, freshly distilled triethylamine (7.00 mL, 50.0 mmol) in 20 mL of anhydrous acetone is added dropwise to a solution of dibromopropane (20.19 g, 100.0 mmol) in anhydrous acetone (35 mL). The mixture is stirred 22 h at reflux. After cooling to room temperature, the resulting precipitate is filtered, washed three times with acetone (20 mL), and dried under vacuum to give 10.21 g of a white solid. The crude 3-(bromopropyl)triethylammonium bromide is crystallized in an absolute ethanol/diethylether mixture and finally isolated after filtration and drying under reduced pressure as a white solid (9.39 g, 30.0 mmol, 62%).

   mp = 153–154°C

   ^{1}H NMR (200 MHz, DMSO-d_6) δ: 1.17 (t, J = 7.0 Hz, 9H, $N^+(CH_2$–$C\underline{H}_3)_3$); 2.15 (m, 2H, Br–CH_2–$C\underline{H}_2$–CH_2–N); 3.23 (m, 2H, Br–CH_2–CH_2–$C\underline{H}_2$–N); 3.26 (q, J = 7.0 Hz, 6H, $N^+(C\underline{H}_2$–$CH_3)_3$); 3.61 (t, J = 6.2 Hz, 2H, CH_2Br).

   ^{13}C NMR (50 MHz, DMSO-d_6) δ: 7.1 (3 × CH_2–$\underline{C}H_3$); 24.4 (Br–$\underline{C}H_2$–CH_2–CH_2–N); 30.7 (CH_2Br); 52.2 (3 × $\underline{C}H_2$–CH_3); 54.6 (Br–CH_2–CH_2–$\underline{C}H_2$–N^+).

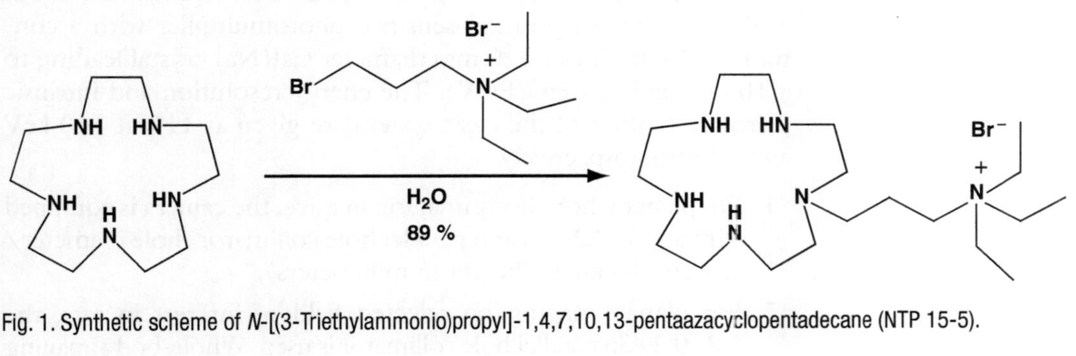

Fig. 1. Synthetic scheme of *N*-[(3-Triethylammonio)propyl]-1,4,7,10,13-pentaazacyclopentadecane (NTP 15-5).

3. *N*-[(3-Triethylammonio)propyl]-1,4,7,10,13-pentaazacyclopentadecane (NTP 15-5) is synthesized from 1,4,7,10,13-Pentaazacyclopentadecane (5) (500 mg, 2.32 mmol) in solution in distilled water (10 mL). 3-(Bromopropyl)triethylammonium bromide (704 mg, 2.32 mmol) is added and the mixture was heated overnight at 90°C under argon, and then evaporated under reduced pressure. This crude product was purified by reverse phase column chromatography eluted with water and obtained after addition of 10 mL of 6 N HCl and evaporation, as a pale yellow solid (1.10 g, 89%).

   ^{1}H NMR (200 MHz, D_2O) δ: 1.25 (t, *J*=7.0 Hz, 9H, $N^+(CH_2C\underline{H}_3)_3$); 2.00 (m, 2H, $N\text{–}CH_2\text{–}C\underline{H}_2\text{–}CH_2\text{–}N^+$); 2.88 (m, 2H, $N\text{–}C\underline{H}_2\text{–}CH_2\text{–}CH_2\text{–}N^+$); 3.19 (m, 2H, $N\text{–}CH_2\text{–}CH_2\text{–}C\underline{H}_2\text{–}N^+$); 3.22 (m, 4H, CH_2 macrocycle); 3.29 (q, *J*=7.0 Hz, 6H, $(C\underline{H}_2\text{–}CH_3)_3$); 3.51 (m, 4H, CH_2 macrocycle); 3.75 (m, 12H, 6×CH_2 macrocycle).

   ^{13}C NMR (50 MHz, D_2O) δ: 10.2 (3×$CH_2\text{–}\underline{C}H_3$); 19.5 ($N\text{–}CH_2\text{–}\underline{C}H_2\text{–}CH_2\text{–}N^+$); 46.2 (2×$CH_2$ macrocycle); 46.6 (4×CH_2 macrocycle); 48.5 (2×CH_2 macrocycle); 52.0 ($N\text{–}\underline{C}H_2\text{–}CH_2\text{–}CH_2\text{–}N^+$); 52.5 (2×$CH_2$ macrocycle); 56.9 (3×$\underline{C}H_2\text{–}CH_3$); 57.5 ($N\text{–}CH_2\text{–}CH_2\text{–}\underline{C}H_2\text{–}N^+$).

   ESI-MS *m/z*: 357 $(M)^+$.

4. Technetium-99m generator.

   Technetium-99m (^{99m}Tc) is produced through the decay of ^{99}Mo beta emitter which has a half life of 66 h, allowing it to be transported over long distances. This radionuclide system made ^{99m}Tc available for clinical application. Radiochemistry units receive ^{99}Mo/^{99m}Tc generators and use the ^{99m}Tc elution (as sodium pertechnetate solution, $Na^{99m}TcO_4$) to label a variety of chemical species (kit formulation) for the majority of the studies they perform.

2.3. Gamma Camera for Scintigraphic Imaging

The gamma camera used in our lab is dedicated to small-animal imaging (γIMAGER, Biospace, France) and allows planar and dual-head SPECT acquisitions. The gamma camera consists of a R 3292 Hamamatsu position-sensitive photomultiplier with a continuous 4 mm thick × 120 mm diameter CsI(Na) crystal leading to a 10 cm field of view (FOV). The energy resolution and intrinsic planar resolution of the basic system are given as 11% at 140 keV and <2 mm, respectively.

1. For planar whole-body imaging in mice, the camera is equipped with a 1.3/0.2/20 and parallel hole collimator (hole diameter/septum thickness/height in millimeters).
2. For planar imaging on selected field of interest in rats, the 1.3/0.2/35 parallel hole collimator is used. Whole-body imaging of a rat is not possible in only one acquisition, but could be

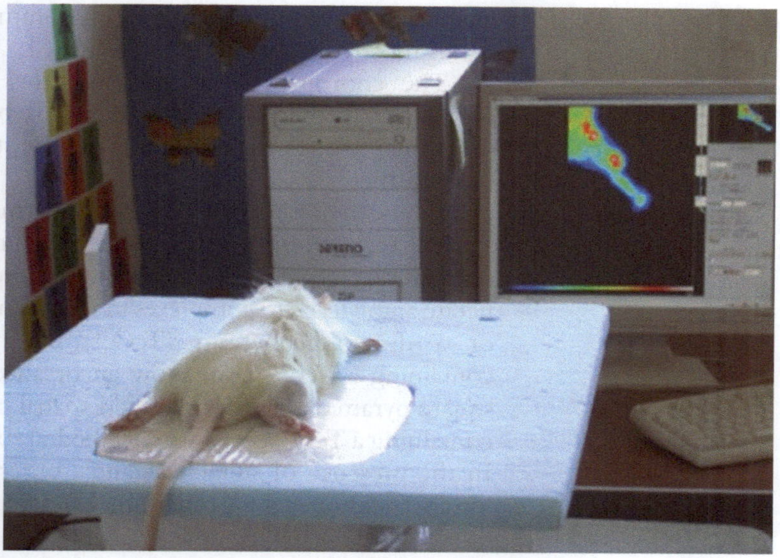

Fig. 2. Support dedicated to planar acquisition in rats. This support allows the positioning of the field of interest over the collimator of the gamma camera.

obtained by multiple sequential acquisitions on successive field of interest. For rats, a support could be used to allow a positioning of the field of interest over the collimator of the camera (see Fig. 2).

3. For dual-head SPECT acquisition in mice and rats, the 1.8/0.2/20 parallel hole collimator is used. In this configuration, detectors are fixed and the animal rotates, with the rotation (1 turn/min) synchronized to data acquisition.
4. For a higher resolution imaging in mice and rats, in both planar and dual-head SPECT acquisition, pinhole collimators with 1-mm size aperture are used.
5. For rabbits or larger animals, a clinical research gamma camera with a larger FOV should be used. It should nevertheless be mentioned that the intrinsic resolution of the clinical gamma camera (working with CsI(Na) crystal) is usually higher than 2 mm.

3. Methods

3.1. Technetium-99m Labeling of NTP 15-5

Technetium can exist in nine oxidation states varying from (+VII) to (−I). The highest oxidation state (+VII) is occupied by a pertechnetate anion (TcO_4^-) which is eluted from the generator. The chemical reactivity of the pertechnetate anion is negligible; it does not bind directly to any ligand. Thus, for radiolabeling of molecules, reduction to lower oxidation states in the presence of a suitable ligand is a prerequisite. Tin (+II) is commonly used as a reductant and is mostly used as tin chloride ($SnCl_2$).

The coordination metal complexes of technetium (central metal) are formed by means of bonds between technetium and functional groups such as amine, amide, thiol, phosphine, oxime, or isonitrile.

In the majority of ^{99m}Tc radiopharmaceuticals, the oxidation state of technetium is (+V). Indeed, when strong reducing agents (such as tin chloride) are used in the presence of ligands, the pertechnetate anion leads to complexes in which an oxo-technetium (TcO^{3+}) or a di-oxo-technetium (TcO_2^+) core is formed. Complexes containing a TcO^{3+} core show an octahedral six-coordinated or a square pyramidal five-coordinated spatial configuration. Complexes containing a TcO_2^+ core form octahedral six-coordinated complexes. In the presence of suitable ligands, other cores and complexes of lower oxidation states (+IV, +III, +I) may be achieved.

3.1.1. Complexation of ^{99m}Tc with NTP 15-5 (see Fig. 3)

1. In a glass vial, dissolve 1.0 mg of NTP 15-5 in an aqueous solution of NaOH 2 N (100 μL) (Solution A). It should be noted that the NTP 15-5 radiolabeling must take place in strong alkaline media to avoid protonation of macrocyclic amines.
2. Dissolve 2.0 mg of $SnCl_2$ in 100 μL of water (100 μL, WFI quality) (Solution B).
3. Add the *Solution B* to the *Solution A* and shake vigorously.
4. To the resulting reaction mixture, add the solution of pertechnetate (TcO_4^-) (185–3,700 MBq in 0.1–5.0 mL); (for radioprotection, see Note 1).
5. Seal the glass vial and heat the reaction mixture at 70°C for 15 min.

3.1.2. Neutralization

1. Dissolve 35 mg of NaH_2PO_4, $2H_2O$ in water (200 μL, WFI quality) (Solution C).
2. Add the *Solution C* to the reaction mixture at room temperature.
3. Control the pH of the resulting solution. It should be between 7.0 and 8.0.

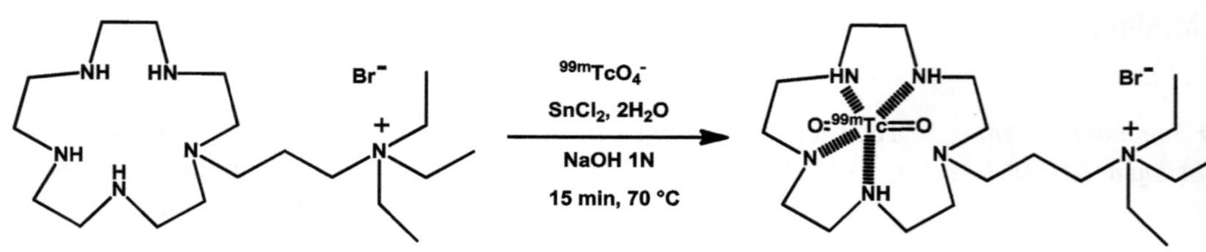

Fig. 3. Radiolabeling of NTP 15-5 with ^{99m}Tc.

3.1.3. Examination of the Yield of Complexation of ^{99m}Tc with NTP 15-5 Technique: Thin Layer Chromatography

The main impurities in ^{99m}Tc pharmaceutical preparations are free pertechnetate ($^{99m}TcO_4^-$) and reduced, hydrolyzed technetium (colloidal ^{99m}Tc). These two ^{99m}Tc species may be separated from ^{99m}Tc pharmaceuticals by simple thin layer chromatography (TLC) procedures.

The migration properties of free pertechnetate are influenced by the choice of different mobile and stationary phase. Colloids do not migrate in most TLC systems.

1. Apply an aliquot (2–10 μL) of the crude reaction mixture to a Whatman KC18 reversed phase TLC plate (5 × 20 cm).
2. Place the plate in an appropriate chromatographic tank containing a MeOH/CH_3CN/THF/$AcONH_4$ (1 N) (3/3/2/2, v/v/v/v) mixture.
3. Elute the plate over a path of 18 cm.
4. Determine the distribution of radioactivity using a suitable radiodetector (e.g., AMBIS 101, Scanalytics, USA). There should be three radioactive spots on the TLC: The first one at Rf = 0.0 corresponds to colloids, the second one at the solvent front (Rf = 1.0) corresponds to unreduced technetium (TcO_4^-). The addition of these two spots should represent less than 5% of the total radioactivity. The third spot at Rf = 0.4 corresponds to ^{99m}Tc-NTP 15-5 and should represent more than 95% of the total radioactivity.

3.1.4. Formulation of ^{99m}Tc-NTP 15-5 for Intravenous Injection

Dilute with saline (0.9% aq. NaCl) the solution containing ^{99m}Tc-NTP 15-5 to the desired concentration for intravenous injection.

3.2. In Vivo Scintigraphic Acquisitions

Static and dynamic acquisitions are performed to obtain, respectively, the biodistribution and pharmacokinetic data of the radiotracer ^{99m}Tc-NTP 15-5 in both nontarget and target organs, as a reflection of in vivo PG distribution.

3.2.1. Animal Handling and Ethics

All the animals should be handled and cared in accordance with the European Directive 86/809/EEC. Protocols should be performed under the authorization of the Veterinary Services and/or animal ethic committee. Animal experimentation should only be conducted by trained personnel under the authority of a researcher with registered certification for animal laboratory science. The number of animals is defined by the user; to our opinion, a minimal of four animals per each acquisition is a prerequisite.

3.2.2. Dynamic Planar Imaging for In Vivo Pharmacokinetic Studies in Rodents

This protocol is used to provide distribution and pharmacokinetic data of the ^{99m}Tc-NTP 15-5 radiotracer in target and nontarget organs, as well as in whole body in mice.

This protocol should be used when a new study (e.g., a study using a new animal model) is performed: this would help to define the best parameters for acquisition (such as delay between radiotracer administration and acquisition), in order to obtain the highest target to surrounding tissue signal.

1. Start anesthesia of the animals by placing them in a chamber filled with 3% isofluorane in a flow of medical-grade oxygen. Then maintain animals under 1.5–3% isofluorane inhalation during all the duration of experimentation by introducing the nose of the animals into coaxial breathing nozzles; the anesthetic mixture expired by the animals should be reaspirated in the nozzle and carried back to carbon cartridge (Minerve, France).
2. During the whole experimentation, control the respiration rate and temperature of the animals at regulatory intervals.
3. Place the animals with the field of interest over the collimator of the gamma camera. Animals are placed in decubitus ventral or dorsal, depending on the field of interest to image. Use, if necessary, a dedicated support for large rodents as described in Fig. 2.
4. After cleaning the tail of the animals with medical grade 70% alcohol, introduce a sterile Teflon catheter equipped with a sharp edge needle (24G for rat and 29G for mice) and filled with sterile physiological serum with 0.01% heparin into the tail vein.
5. Fix the catheter to the tail of the animals with adhesive tape.
6. Prepare the file for acquisition containing all the parameters for a LIST mode acquisition with a 15% window set at 140 keV, so that you will only have to press the starting button simultaneously with the starting of injection. The duration of LIST mode acquisition is related to the aim of the study: from our experience, for animals anesthetized with isofluorane inhalation, a 3-h duration acquisition could be performed. If your acquisition system does not work in LIST mode, enter the parameters chosen for a routine dynamic acquisition (number of time group, number of frames per group, and duration of frames for each group).
7. Calibrate a syringe containing the ^{99m}Tc-NTP 15-5 radiotracer using a dose calibrator (Capintec, USA). We usually inject 25 and 30 MBq/animal in mice and rats, respectively, in a volume lower than 250 μL for a 25 g mouse and 500 μL for a 300–350 g rat. For each animal, note precisely the radioactivity $A_{(Full)}$ and volume of each "full syringe" as well as the time of measurement.
8. Introduce the syringe into a tungsten-shielded protector system.

9. Connect the tungsten-shielded 1-mL calibrated syringe containing ^{99m}Tc-NTP 15-5 radiotracer to the catheter.
10. Inject the radiotracer and press simultaneously the acquisition starting button.
11. Rinse immediately the catheter with 200 μL of 0.9% sterile physiological serum.
12. Remove carefully the catheter from the vein.
13. For each animal, measure the radioactivity of each empty syringe $A_{(\text{Empty})}$ and each catheter $A_{(\text{catheter})}$, using the same dose calibrator and note precisely the time of counting.
14. After radioactive decay correction (see Note 2), calculate the injected dose (ID) to each animal by using the formula:

$$\text{ID (MBq)} = A_{(\text{Full})} - A_{(\text{Empty})} - A_{(\text{catheter})}$$

15. Make sure that animals do not move during the whole duration of acquisition.
16. When the acquisition in LIST mode format is ended, save the file and postprocess it by using the Gammavision+ software (Biospace, France). The sample time for the sequence of time frames is defined by the user. Choice of sample time is based on the kinetics and clearance of radiotracer in both target and nontarget organs. From a "radioactive counting statistics" point of view, sample time should be at least 10 s to achieve good signal-to-noise ratio.

3.2.3. Static Planar Acquisition for In Vivo Targeting Studies in Rodents

This protocol is used to provide biodistribution data of the ^{99m}Tc-NTP 15-5 radiotracer in target and nontarget organs, at selected time points.

1. Use a contention box dedicated to intravenous injection in vigil mice or rats.
2. Carefully clean the tail of the rodents with medical disinfectant such as grade 70% alcohol.
3. Calibrate a syringe containing the ^{99m}Tc-NTP 15-5 radiotracer using a dose calibrator (Capintec, USA), in order to inject 25 and 30 MBq/animal in mice and rats, respectively. For each animal, note precisely the radioactivity $A_{(\text{Full})}$ and volume of each "full syringe," as well as the time of measurement.
4. Introduce the syringe into a tungsten-shielded protector system.
5. Inject in about 30 s, ^{99m}Tc-NTP 15-5 radiotracer directly into the tail vein by using an insulin-type syringe equipped with a 24 or 29G needle for mice and rats, respectively. For each animal, note precisely the time of injection. Acquisition of

anesthetized animals should be started after a minimal delay of 15 min after IV administration of tracer.

6. After injection, replace the vigil animals in their cage until acquisition.
7. During this delay, measure the radioactivity of each empty syringe $A_{(\text{Empty})}$, using the same dose calibrator and note precisely the time of counting.
8. After radioactive decay correction, calculate the injected dose (ID) to each animal by using the formula:

$$\text{ID (MBq)} = A_{(\text{Full})} - A_{(\text{Empty})}$$

9. Anesthetize the animals by placing them in a chamber filled with 3% isofluorane in a flow of medical-grade oxygen and maintain animals under 1.5–3% isofluorane inhalation during all the duration of acquisition as described in Subheading 3.2.2, step 1. Alternatively, for short duration acquisition, one may use a mixture of ketamine and xylazine (4:1 ratio) delivered by intraperitoneal route.
10. During anesthesia, observe the physiological state of the animals, in terms of regulation of respiration rate and temperature.
11. For static acquisition, place the field of interest (or whole body for mice) of anesthetized animals over the collimator of the gamma camera and use, if necessary, a dedicated support for large rodents as described in materials. Animals are placed in decubitus ventral or dorsal, depending on the field of interest to image. Keep in mind that the field of interest you want to image should be as close as possible to the detector.
12. Start the acquisition with a 15% window centered on the 140-keV photopeak of ^{99m}Tc. The choice of the duration of acquisition should allow a high target to nontarget ratio. From our experience, this is generally achieved for the range of injected activity by using 5–10 min duration acquisition for each FOV.
13. Such acquisition could be repeated in the same animals, at different selective time points over a 24-h period after radiotracer injection. This could be a powerful way to obtain in vivo biodistribution parameters in target and nontarget organs, as well as elimination parameters. Nevertheless, keep in mind that the signal in the images decreases in an exponential way in relation to the physical period of the radionuclide (i.e., 6 h for ^{99m}Tc) and that duration of acquisition should be adapted to counting statistics.

3.2.4. SPECT Acquisition in Rodents

The protocol for SPECT acquisition is the same as for static acquisition, except for the configuration of the gamma camera.

1. In our lab, we use the dual-head detection system of the γIMAGER (Biospace, France) in which the animals are

positioned on the vertical rotating platform, with acquisition being synchronized to rotation. Alternatively, other SPECT imaging devices can be used, in which detectors are around the animal.

2. Repeat steps 1–8 in Subheading 3.2.3.
3. For anesthesia, prefer ketamine/xylazine (4:1 ratio) by intraperitoneal route.
4. Place and fix the animal on the rotating platform.
5. Start acquisition, with a 15% window centered on the 140 keV for at least a 20-min duration (for the range of injected activity).
6. At the end of acquisition, perform image reconstruction using the dedicated software interface. Tomographic reconstruction is performed using a validated 3D ordered-subset expectation maximization (OSEM) algorithm with at least three iterations. The distance between axis of rotation and collimator should also be considered. It is possible to apply filters, with filter order defined by the user.
7. SPECT acquisition could also be reconstructed for dynamic analyses. The sequence of the time frames is defined by the user, but should be at least 1 min.
8. In our lab, we use AMIRA software (Biospace, France) to read and analyze the 3D SPECT volume. Alternatively, similar analysis could be obtained using other free software after DICOM (Digital Imaging and Communication in Medicine) export.

3.3. Image Analysis for In Vivo Quantitative Assessment of PG Distribution

3.3.1. Analysis of Static Planar Images for Expressing Tracer Uptake as Percentage of Injected Dose or Percentage of Whole-Body Activity

1. In our lab, image analysis is performed using the Gammavision+ software (Biospace, France). Alternatively, similar analysis could be obtained using other free software after DICOM export, such as Amide (24).
2. For analysis of static planar images, regions of interest (ROI) are drawn in two dimensions to delineate uptake areas in order to determine radioactivity accumulation. The use of activity profile for ROI placement is recommended to assure an easy and reproducible positioning of the ROI for serial images in the same animal over time and among all the animals imaged.
3. For each ROI, size in pixel and in mm^2, total activity in counts per minute (cpm), average count in cpm per pixel, and average count in cpm/mm^2 are obtained. These results are displayed as tables.
4. Export the table as a spreadsheet, and use excel functions for calculations.
5. Normalize all activity values to the dose injected to each animal.

6. Apply correction for radioactive decay, by taking into account the delay between injection and acquisition (see Note 2).
7. For expressing tracer uptake in selected organs, several methodologies could be used:
   - Expressing as a percentage of injected dose per organ (%ID/organ).
   - Expressing as percentage of whole-body activity (% WBA). This is only for mice, when whole-body imaging is obtained in one acquisition.

3.3.2. Analysis of Static Planar Images for Expressing Target to Background Ratio

The target to background ratio (TBR) is a semiquantitative parameter which is commonly used in imaging, to provide information about the specificity of tracer uptake in targets, as well as comparison between animals and scans.

TBR parameter is of particular interest for the longitudinal monitoring of the same animals over time, as well as for evaluating response to therapeutic intervention.

1. Delineate fixed-sized ROIs over target and background (BKG) patterns; (usually muscle is chosen as background) (an example is shown in Fig. 4). Use activity profile to ensure a reproducible positioning of the ROI for serial images in the same animal over time and among all the animals imaged. For each ROI, take into account surfacic activity, i.e., average cpm per pixel.
2. Export quantitation table as a spreadsheet and use functions for calculations.

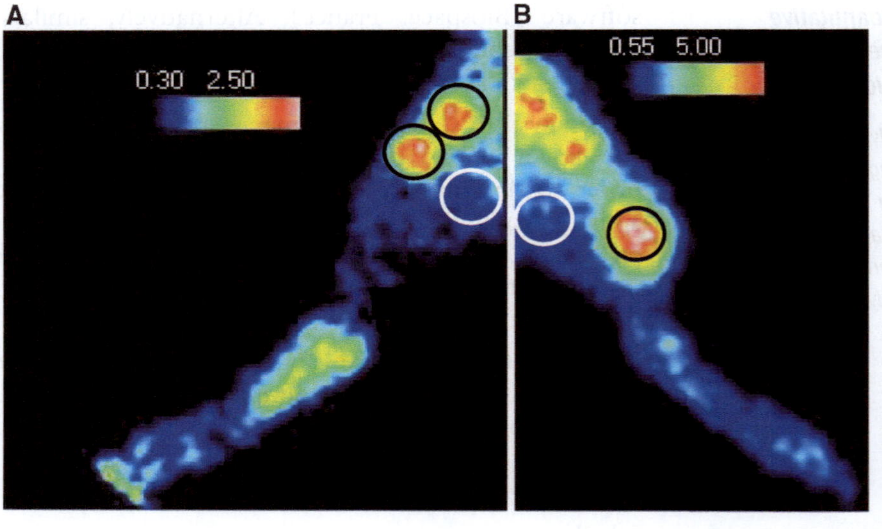

Fig. 4. Quantitative analysis for TBR determination of ^{99m}Tc-NTP 15-5 scintigraphic images in a chondrosarcoma-bearing rat. Fixed-sized ROIs were delineated as illustrated (**a**) over cartilage and muscle uptake for the contralateral paw and (**b**) over tumoral and muscle uptake for the primary tumor.

3. Normalize all values of surfacic activity to the dose injected to each animal.
4. Apply correction for radioactive decay as explained in Note 2.
5. Express TBR as:

$$\text{TBR} = \text{average cpm per pixel within target} \\ / \text{average cpm per pixel within BKG.}$$

6. Save systematically the ROI for each animal.
7. For serial examination of the same animals over time, reuse the same ROIs.
8. Use TBR value for evaluating quantitatively PG distribution in vivo in targets, as well as PG changes in targets in relation to pathology evolution and/or response to therapy.

3.3.3. Analysis of Dynamic Planar Images for Expressing Pharmacokinetic Parameters as Time Activity Curves

1. Delineate fixed-sized ROIs in two dimensions using Gamma Vision+ software (Biospace, France) in order to delineate structures which evidenced tracer uptake and also over BKG.
2. Use the dedicated software interface to obtain for each ROI time–activity curves.
3. Calculate pharmacokinetic parameters from the time–activity curves using classical pharmacokinetic softwares or home-made spreadsheets.

3.3.4. Analysis of SPECT Images

1. In our lab, we use AMIRA software, but similar analysis could be obtained using other free software such as Amide, after DICOM export.
2. Draw ROIs in three dimensions (also called VOI for volume of interest). For such segmentation method, several approaches could be chosen by the user: manual or semiautomatic segmentation.
3. The total activity in the VOI is then given.
4. Normalize activity value to the injected dose, apply correction for radioactive decay, see Note 2.
5. If pharmacokinetic analysis is needed, use the dedicated software interface to obtain for each VOI time–activity curves.
6. For TBR analysis, it is preferable to delineate ROIs in two dimensions on successive selected sections reconstructed from the 3D volume and in the three axes of the animals. Proceed as described in Subheading 3.3.2, steps 1–8.

4. Notes

1. Radiosyntheses using ^{99m}Tc are performed in a 2.5-cm-lead shielded cell (Medisystem, France). ^{99m}Tc is manipulated using tungsten syringe shield (5-mm thick tungsten body).
2. Radioactivity decays in time according to the equation:

$$A_t = A_0 e^{-\lambda t}$$

with A_t = radioactivity at time of measurement and A_0 = radioactivity at the reference time (for such protocols of in vivo imaging, reference time is the time of radiotracer administration).

$$\lambda = \text{radioactive constant: } \lambda = \ln 2 / T$$

with T = physical period of radioisotope and equals 6 h for ^{99m}Tc.

So, for radioactive decay correction, calculate A_0 by using the equation with T = 6 h and t = delay (in hours) between injection and measurement.

Acknowledgments

In our group, studies related to the in vivo scintigraphy of proteoglycans for the imaging of cartilage and its degenerative and tumoral pathologies are supported by the Regional Fund for Innovation (FRI 2) OSEO, the Contrat de Projet Etat Region (CPER), the Fondation pour la Recherche Médicale (FRM), the Ligue Contre Le Cancer, and the Institut National du Cancer.

Studies aiming at validating ^{99m}Tc-NTP 15-5 radiotracer as a candidate for clinical imaging of cartilage pathologies in nuclear medicine are conducted in close collaboration with Cyclopharma Laboratoires.

The authors would like to thank Prof. D. Heymann, Prof. F. Gouin, and Dr. F. Rédini from UMR S957 INSERM from Nantes University for their collaboration and expertise on chondrosarcoma.

References

1. Biswal, S., Resnick, D. L., Hoffman, J. M., and Gambhir, S. S. (2007) Molecular imaging: integration of molecular imaging into the musculoskeletal imaging practice. *Radiology* **244**, 651–671.
2. Vanderheyden, J. L. (2009) The use of imaging in preclinical drug development. *Q J Nucl Med Mol Imaging* **53**, 374–381.
3. Weissleder, R. (2006) Molecular imaging in cancer. *Science* **312,** 1168–1171.
4. Meikle, S. R., Kench, P., Kassiou, M., and Banati, R. B. (2005) Small animal SPECT and its place in the matrix of molecular imaging technologies. *Phys Med Biol.* **50**, R45–R61.
5. Madelmont, J. C., Giraud, I., Nicolas, C., Maurizis, J. C., Rapp, M., Ollier, M., et al. (2001) Novel quaternary ammonium derivatives, method for preparing same and pharmaceutical use. *Patent WO/2001/000621 PCT/FR2000/001731.*
6. Maurizis, J. C., Rapp, M., Nicolas, C., Ollier, M., Verny, M., and Madelmont, J. C. (2000) Disposition in rats of N-pyridinium-propyl-cyclam, N-triethy-lammonium-propyl-cyclam, and N-[triethylammonium]-3-propyl-[15]ane-N5, potential cartilage imaging agents. *Drug Metab Dispos.* **28**, 418–422.
7. Banerjee, S. R., Maresca, K. P., Francesconi, L., Valliant, J., Babich, J. W., and Zubieta, J. (2005) New directions in the coordination chemistry of ^{99m}Tc: a reflection on technetium core structures and a strategy for new chelate design. *Nucl Med Biol.* **32**, 1–20.
8. Ollier, M., Maurizis, J. C., Nicolas, C., Bonafous, J., De Latour, M., Veyre, A., and Madelmont, J. C. (2001) Joint scintigraphy in rabbits with ^{99m}Tc-N-[3-(triethylammonio) propyl]-15ane-N5,a new radiodiagnostic agent for articular cartilage imaging. *J Nucl Med.* **42**, 141–145.
9. Miot-Noirault, E., Vidal, A., Pastoureau, P., Bonafous, J., Chomel, A., Sarry, L., et al. (2007) Early detection and monitoring of cartilage alteration in the experimental meniscectomized guinea pig model of osteoarthritis by ^{99m}Tc-NTP 15–5 scintigraphy. *Eur J Nucl Med Mol Imaging* **34**, 1280–1290.
10. Miot-Noirault, E., Vidal, A., Auzeloux, P., Madelmont, J. C., Maublant, J., and Moins, N. (2008) First In Vivo SPECT Imaging of Mouse Femorotibial Cartilage Using ^{99m}Tc-NTP 15-5. *Mol Imaging* 7, 263–271.
11. Miot-Noirault, E., Gouin, F., Vidal, A., Rapp, M., Maublant, J., Askienazy, S., et al. (2009) First preclinical imaging of primary cartilage neoplasm and its local recurrence using ^{99m}Tc-NTP 15–5 radiotracer. *J Nucl Med.* **50**, 1541–1547.
12. Lamoureux, F., Picarda, G., Garrigue-Antar, L., Baud'huin, M., Trichet, V., Vidal, A., et al. (2009) Glycosaminoglycans as potential regulators of osteoprotegerin therapeutic activity in osteosarcoma. *Cancer Res.* **69**, 526–36.
13. Sarda-Mantel, L. and Le Guludec, D. (2009) Molecular imaging of cartilage; invited perspectives. *J Nucl Med.* **50**, 1391–1393.
14. Wang, Q., Zheng, Y. P., Qin, L., Huang, Q. H., Lam, W. L., Leung, G., Guo, X., and Lu, H. B. (2008) Real-time ultrasonic assessment of progressive proteoglycan depletion in articular cartilage. *Ultrasound Med Biol.* **34**, 1085–1092.
15. Nieminen, M. T., Rieppo, J., Silvennoinen, J., Töyräs, J., Hakumäki, J. M., Hyttinen, M. M., et al. (2002) Spatial assessment of articular proteoglycans with Gd-DTPA-Enhanced T1 imaging. *Magn Reson Med.* **48,** 640–648.
16. Laurent, D., Wasvary, J., Rudin, M., O'Byrne, E., and Pellas, T. (2003) *In vivo* assessment of macromolecular content in articular cartilage of the goat knee. *Magn Reson Med.* **49**, 1037–1046.
17. Eckelman, W. C., Reba, R. C., and Kelloff, G. J. (2008) Targeted imaging: an important biomarker for understanding disease progression in the era of personalized medicine. *Drug Discov Today* **13**, 748–759.
18. Hardingham, T. E. and Fosang, A. J. (1992). Proteoglycans : many forms and many functions. *FASEB J* **6**, 861–870.
19. Lamoureux, F., Baud'huin, M., Duplomb, L., Heymann, D., and Rédini, F. (2007) Proteoglycans : key partners in bone cell biology. *Bioessays* **29**, 758–771.

20. Schaefer, L., and Schaefer, R. M. (2010) Proteoglycans: from structural compounds to signaling molecules. *Cell Tissue Res.* 339, 237–246.
21. Borges, L. F., Touat, Z., Leclercq, A., Zen, A. A., Jondeau, G., Franc, B., et al. (2009) Tissue diffusion and retention of metalloproteinases in ascending aortic aneurysms and dissections. *Hum Pathol.* **40**, 306–313.
22. Jeon, W. S., Kim, E., Ko, Y. H., Hwang, H., Lee, J. W., Kim, S.-Y., et al. (2004) Molecular Loop Lock: A Redox-Driven Molecular Machine Based on a Host-Stabilized Charge-Transfer Complex. *Angew Chem. Int. Edit.* **44**, 87–91.
23. Botero Cid, M. H., Holzgrabe, U., Kostenis, E., Mohr, K., and Traenkle, C. (1994) Search for the Pharmacophore of Bispyridinium-Type Allosteric Modulators of Muscarinic Receptors. *J Med Chem.* **37**, 1439–1445.
24. Loening, A. M. and Gambhir, S. S. (2003) AMIDE: a free software tool for multimodality medical image analysis. *Mol Imaging* **2**, 131–137.

Part III

PGs: Multifunctional Cell Regulators

Chapter 14

Serglycin: The Master of the Mast Cell

Elin Rönnberg and Gunnar Pejler

Abstract

Serglycin is a proteoglycan composed of a relatively small (~17 kDa) core protein to which sulfated glycosaminoglycans of either heparin, heparan sulfate or chondroitin sulfate types are attached. Serglycin is expressed in many cell types, including in particular cells of hematopoietic origin. To study the function of serglycin, we have used a serglycin knockout mouse strain. A striking finding was that the mast cell population was severely affected by the absence of serglycin, as evidenced by distorted granule morphology and defective staining with cationic dyes. Moreover, the absence of serglycin caused a dramatic effect on the ability of mast cells to store a number of granule compounds, including several mast cell-specific proteases as well as biogenic amines. Hence, serglycin has a major function in maintaining mast cell secretory granule homeostasis.

Key words: Serglycin, Proteoglycans, Mast cells, Granules, Secretion

Abbreviations

MC	Mast cell
BMMC	Bone marrow-derived mast cell
mMCP	Mouse mast cell protease
CS	Chondroitin sulfate
PG	Proteoglycan
GAG	Glycosaminoglycan
SCF	Stem cell factor
MC-CPA	Mast cell carboxypeptidase A
HPRT	Hypoxanthine guanine phosphoribosyl transferase
NDST	*N*-deacetylase/*N*-sulfotransferase
GalNAc	*N*-acetyl-galactosamine
C4ST	Chondroitin 4-sulfotransferase
GalNAc4S6ST	GalNAc(4S)-6-O-sulfotransferase
C6ST	Chondroitin 6-sulfotransferase
GlcUA	Glucuronic acid
IdoUA	Iduronic acid
WT	Wild type

Françoise Rédini (ed.), *Proteoglycans: Methods and Protocols*, Methods in Molecular Biology, vol. 836,
DOI 10.1007/978-1-61779-498-8_14, © Springer Science+Business Media, LLC 2012

1. Introduction

Mast cells (MCs) are hematopoietic cells that circulate as progenitors in blood, and develop into mature MCs after entry into tissues and exposure to local growth factors, in particular stem cell factor (SCF) (reviewed in ref. 1,2). MCs have previously been recognized mostly for their well-established role in allergic conditions, including allergic asthma (3). However, more recent findings, obtained through studies utilizing various MC-deficient mice, have implicated MCs in a variety of other disorders such as arthritis (4), multiple sclerosis (5), diabetes (6), abdominal aortic aneurysm formation (7), atherosclerosis (8), and cancer (9). MCs are therefore emerging as major players in multiple disorders. Therapeutic regimens that interfere with MC function may therefore constitute promising, future strategies for treatment of numerous diseases. However, MCs are also known to possess a number of beneficial functions, especially in the context of innate immunity to, e.g., bacteria and parasites (10, 11); reviewed in ref. 12.

A distinguishing feature of MCs is their remarkably high content of electron-dense secretory granules. These contain large amounts of numerous preformed compounds, including biogenic amines (histamine, serotonin), certain preformed cytokines (e.g., TNF), lysosomal compounds such as cysteine cathepsins, serglycin proteoglycan (PG), as well as a number of MC-specific proteases, the latter being of either chymase, tryptase or carboxypeptidase A (MC-CPA) type (2, 13). Serglycin, like all PGs, is built up of a protein "core" to which sulfated and thereby negatively charged glycosaminoglycans (GAGs) are attached via a specific linkage region (14–16). The GAG components of serglycin can either be of heparin/heparan sulfate or chondroitin sulfate (CS) type, with heparin/heparan sulfate built up of repeating $(GlcUA/IdoUA\text{-}GlcNAc)_n$ units, while CS is composed of repeating $(GlcUA\text{-}GalNAc)_n$ units.

Among the various PG species, serglycin was the first one to be cloned. It was originally cloned from rat yolk sac (17, 18), but it was soon after discovered that serglycin is highly expressed in MCs (19–21) (Fig. 1a). MC granules have for a long time been known to contain large amounts of PGs, mainly of heparin and CS type, and it has also been known that the MC granule PGs are of extraordinary high anionic charge density, due to a high level of sulfation of the heparin/CS chains. Therefore, it may be assumed that the highly sulfated PGs present in MC granules are of serglycin type, and this notion was confirmed following successful targeting of the serglycin gene (22). Notably, targeting of the serglycin gene caused an almost complete abrogation of total PG synthesis in MCs, indicating that serglycin is by far the dominating PG species in this cell type. A typical feature of MCs in general is their strong staining with cationic dyes such as Toluidine blue and May-Grünwald–Giemsa (Fig. 2a), presumably explained by strong binding of these

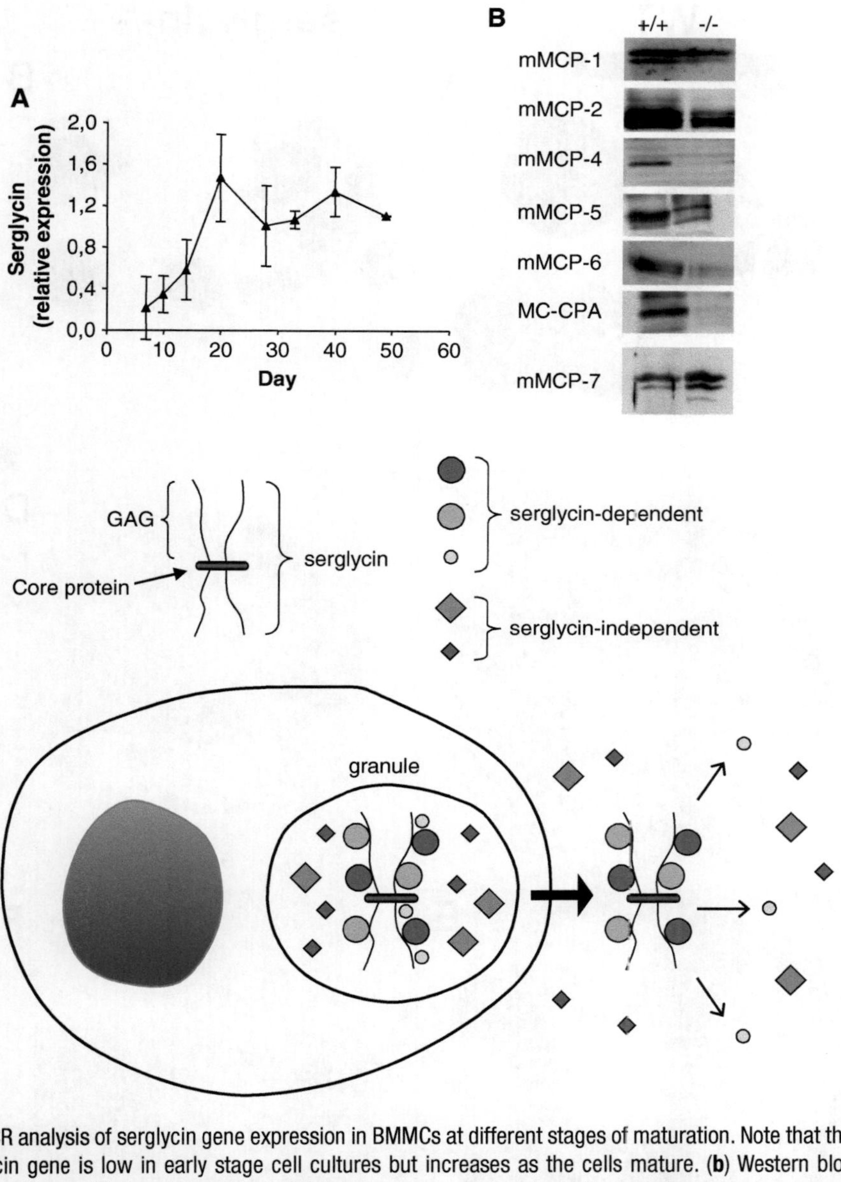

Fig. 1. (**a**) qPCR analysis of serglycin gene expression in BMMCs at different stages of maturation. Note that the expression of the serglycin gene is low in early stage cell cultures but increases as the cells mature. (**b**) Western blot analysis of BMMCs from wild type (WT) and serglycin$^{-/-}$ BMMCs. All antisera were of rabbit origin. For the generation of mouse MC protease (mMCP)-5, mMCP-6 and MC-CPA antisera, full-length recombinant proteins were used as antigens. For antisera toward mMCP-1, mMCP-2, and mMCP-4, the following peptides (designed by Lars Hellman, Uppsala University, Sweden) coupled to Keyhole limpet hemocyanin were used as antigens: SYGDSHGKPP (mMCP-1); HNINKNEPTQQI (mMCP-2); KKAKETPSVNV (mMCP-4). Note that mMCP-4, -5, -6 and MC-CPA storage is critically dependent on serglycin, that mMCP-2 storage is partially serglycin-dependent, and that the storage of mMCP-1 and -7 is unaffected by the absence of serglycin. (**c**) Schematic image showing the presence of serglycin within MC secretory granule. Certain compounds (e.g., mMCP-4, -5, -6, MC-CPA, histamine) are stored in complex with serglycin, whereas the storage of other compounds is serglycin-independent. Upon MC degranulation, serglycin is released. After release, certain compounds remain attached to serglycin (*large circles*; e.g., mMCP-4, MC-CPA), whereas other compounds (*small circles*; e.g., histamine) are detached from serglycin after release and exposure to neutral pH. (**a**)Reproduced from ref. 24 with permission from Elsevier. (**b**) Reproduced from ref. 14 with permission from Portland Press Ltd. (**c**) Reproduced from ref. 23 with permission from Wiley InterScience.

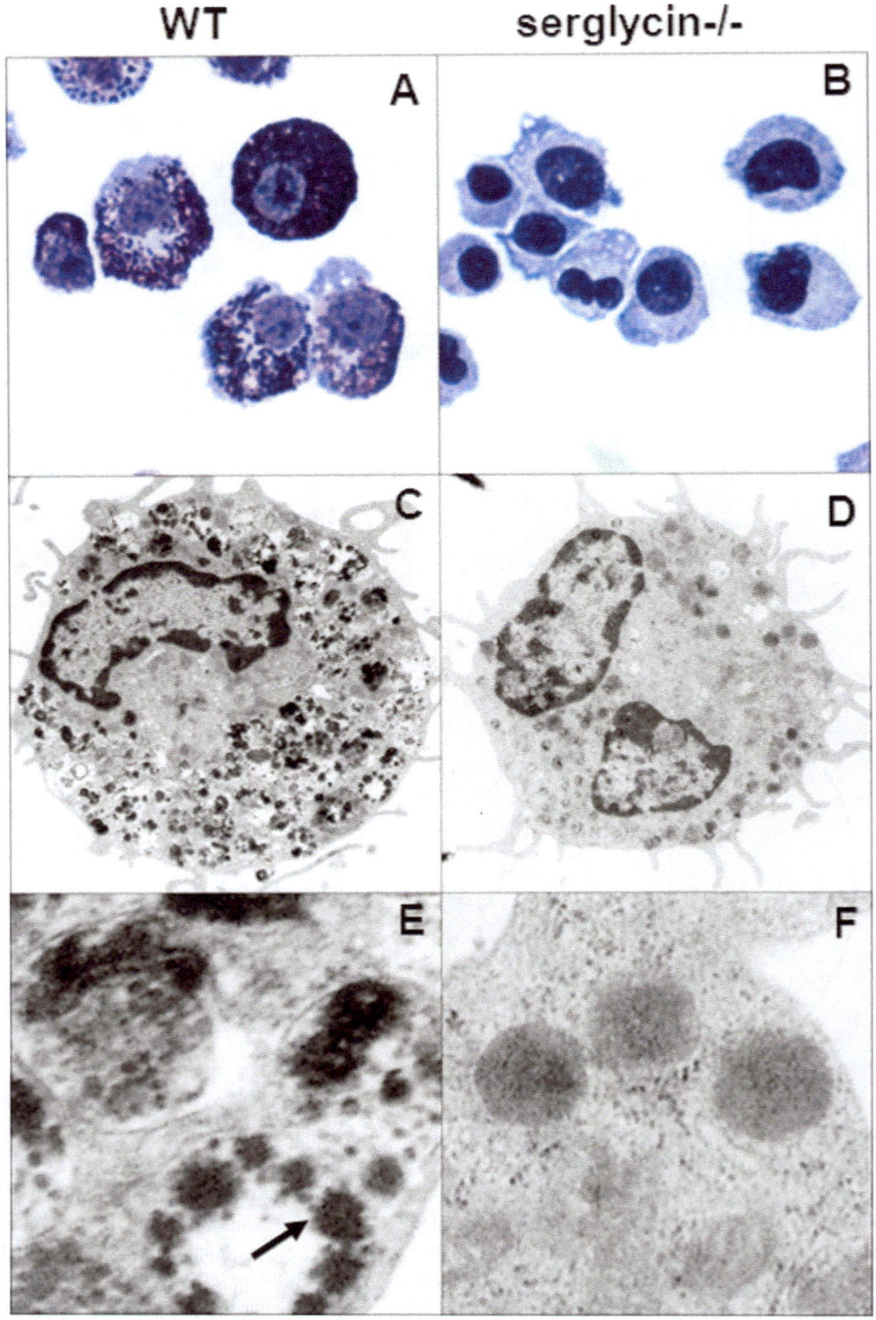

Fig. 2. (**a**, **b**) May-Grünwald–Giemsa-stained cytospin slides showing WT (**a**) and serglycin$^{-/-}$ (**b**) BMMCs. Note the strong granular staining of WT cells (**a**), whereas the serglycin$^{-/-}$ cells show dramatically reduced staining intensity, with no apparent staining of granules (**b**). (**c**, **d**) Transmission electron microscopy analysis of WT (**c**) and serglycin$^{-/-}$ (**d**) BMMCs. Original magnification 6,000×. Granules, in approximately equal numbers, are present in both WT and serglycin$^{-/-}$ cells, but the granule morphology is affected by the absence of serglycin. (**e**, **f**) Granules of WT and serglycin$^{-/-}$ BMMCs in larger magnification, showing the presence of dense core formation in WT granules only (**e** *arrow*), whereas the granules of serglycin$^{-/-}$ cells instead are filled with evenly distributed material of amorphous character (**f**).

to the highly negatively charged serglycin PG. In strong support of this notion, serglycin$^{-/-}$ MCs showed a virtually complete inability to stain with these dyes (22) (Fig. 2b) accompanied by defective dense core formation (23) (Fig. 2c–f), indicating a key role for serglycin in generating the morphological features typical for MCs of all species.

Considering the profound impact of serglycin deficiency on MC granule morphology, we also assessed whether the storage of the various MC granule compounds was affected by the lack of serglycin. Indeed, we have found that the storage of several (but not all) MC-specific proteases is severely affected by the absence of serglycin (22, 23) (Fig. 1b, c), and we have also demonstrated a key role for serglycin in mediating the storage of the biogenic amines, histamine, and serotonin (24). Notably, serglycin-deficiency has also been shown to affect the storage of granule compounds in various types of cells other than MCs (25–27).

In this chapter we describe the methods that we use for studying the role of serglycin in maintaining secretory granule homeostasis in MCs.

2. Materials

2.1. Mast Cell Culture

2.1.1. Reagents and Tools

1. Sterile scissors and tweezers.
2. 70% Ethanol.
3. Phosphate-buffered saline (PBS; SVA, Uppsala, Sweden).
4. 5-mL sterile syringes.
5. 0.4-mm needles.
6. Sterile Petri dishes.
7. Sterile plastic centrifuge tubes (Falcon type).
8. DMEM cell culture medium (SVA).
9. Heat-inactivated fetal bovine serum (FBS; Invitrogen, Carlsbad, CA).
10. Penicillin–Streptomycin sulfate (SVA).
11. L-glutamine (SVA).
12. WEHI-3B cells.
13. SCF, IL-3, IL-9, TGF-β (Peprotech, Rocky Hill, NJ) (aliquoted and stored at –20°C).

2.1.2. Preparation of WEHI-3B-Conditioned Media

1. Grow WEHI-3B cells in DMEM supplemented with 10% heat-inactivated FBS, 60 μg/mL Penicillin, 50 μg/mL Streptomycin sulfate, 2 mM L-glutamine.
2. Change medium every third to fourth day.

3. When the cells have expanded, keep them in the same medium for about 1 week or until the medium turns orange/yellow. Centrifuge the cells at 300 × *g* for 10 min and transfer the medium to a flask or tubes, freeze, and store at −20°C.
4. Resuspend the cells in new medium and culture them until it is time to collect conditioned medium again.
5. Before use, sterile-filter the WEHI-3B-conditioned media.

2.1.3. Preparation of Bone Marrow-Derived MC Culture Media

1. Supplement DMEM with 10% heat-inactivated FBS, 60 μg/mL Penicillin, 50 μg/mL Streptomycin sulfate, 2 mM L-glutamine.
2. For generation of CTMC-like bone marrow-derived MCs (BMMCs), also supplement medium with either 30% WEHI-3B-conditioned media (sterile filtered) or 5 ng/mL IL-3 + 25 ng/mL SCF.
3. For generation of MMC-like BMMCs, supplement instead with 1 ng/mL IL-3 + 25 ng/mL SCF + 5 ng/mL IL-9 + 1 ng/mL TGF-β.

2.2. Proteoglycan Analysis

2.2.1. Sulfate Labeling

1. Tyrode's buffer: 130 mM NaCl, 5 mM KCl, 1.4 mM $CaCl_2$, 1 mM $MgCl_2$, 5.6 mM glucose, 10 mM HEPES, 0.1% BSA (pH 7.4).
2. Medium with reduced sulfate content: DME/F-12 Base Medium (1.3 μg/L sulfate; Sigma) supplemented with 5% FBS, 60 μg/mL Penicillin, 50 μg/mL Streptomycin sulfate.
3. Carrier-free $^{35}SO_4^{2-}$ (PerkinElmer, Boston, USA).

2.2.2. Purification of ^{35}S-Labeled PGs

1. Lysis buffer: PBS, 2 M NaCl, 0.5% Triton X-100 (protease inhibitors may be added; however, since serglycin is protease-resistant this is not essential; see Note 1 under Subheading 3.2.2).
2. Buffer I: 50 mM Tris–HCl (pH 8.0), 0.1 M NaCl, 0.1% Triton X-100.
3. Buffer II: 50 mM NaAc (pH 4.0), 0.15 M NaCl, 0.1% Triton X-100.
4. Elution buffer: PBS/2 M NaCl, 0.5% Triton X-100.
5. Columns (BioRad, 2 mL Poly-Prep columns).
6. Anion exchange resin (DEAE-Sephacel; GE Healthcare, Uppsala, Sweden).

2.2.3. Release of Free GAG Chains (Alkali Treatment)

1. 4 M NaOH.
2. 4 M HCl.
3. pH paper.

2.2.4. Size Determination of ^{35}S-Labeled PGs and Released GAG Chains

1. Running buffer: 50 mM Tris–HCl (pH 7.4), 1 M NaCl, 0.1% Triton X-100.
2. Superose 6 and/or Superose 12 columns (GE Healthcare, Uppsala, Sweden).

2.2.5. Determination of Anionic Charge Density (Anion Exchange Chromatography)

1. 50 mM NaAc (pH 4.0), 0.05 M LiCl.
2. 50 mM NaAc (pH 4.0), 2 M LiCl.
3. Anion exchange column (e.g., HiTrap™ DEAE FF anion exchanger; GE Healthcare).

2.3. Real-Time Quantitative RT-PCR

1. ABI Prism 7900 HT (Applied Biosystems, Carlsbad, CA).
2. iQ™ SYBR® Supermix (BioRad, CA).
3. DNAse/RNAse-free H_2O.
4. 384-Well microtiter plates (Applied Biosystems, Foster City, CA).
5. Primers (see Table 1).

2.4. SDS–PAGE–Western Blot

1. ClearPage™ SDS–PAGE gels (VWR International, West Chester, PA).
2. ClearPage SDS–PAGE running buffer (VWR International).

Table 1

Target	Sequence	Amplicon size (bp)	Efficiency[a] (58°C)
Hprt, *fw* Hprt, *rev*	5′-GAT TAG CGA TGA TGA ACC AGG TTA-3′ 5′-GAC ATC TCG AGC AAG TCT TTC AGT C-3′	133	1.09
Serglycin, *fw* Serglycin, *rev*	5′-GCA AGG TTA TCC TGC TCG GAG-3′ 5′-GGT CAA ACT GTG GTC CCT TCT C-3′	101	0.96
NDST-1, *fw* NDST-1, *rev*	5′-CTG CAC TCC TGG ACC AAC CT-3 5′-ACA GGG ATC CTG CCA AAG C-3′	111	1.17
NDST-2, *fw* NDST-2, *rev*	5′GTG GCT GAT GTT GAG GCT TTG-3′ 5′-ATC CTC CTC TTC TGT CCC GG-3′	120	1.04
C4ST-1, *fw* C4ST-1, *rev*	5′-CCA AAG TAT GTT GCA CCC AGT C-3′ 5′-CTG GTC CCG TCT CAT CTG GT-3′	154	1.15
C4ST-2, *fw* C4ST-2, *rev*	5′-CGC TAG GTC CGT CTC CCA G-3′ 5′-CAG ATA GAA GTG GGC GGT GC-3′	193	1.15
GalNAc4S6ST, *fw* GalNAc4S6ST, rev	5′-GGC TTT TCA GGT CAC CTA GCA-3′ 5′-GAC ATT ATG GGT TCC TCG TTG A-3′	102	0.90
C6ST, *fw* C6ST, *rev*	5′-CAT ATC CAG GGT CTC CGA CAA G-3′ 5′-CAA GAG AGA TGC ATT CTC CGA TAA G-3′	103	1.05
MC-CPA *fw* MC-CPA, *rev*	5′-TGA CAG GGA GAA GGT ATT CCG-3′ 5′-CCA AGG TTG ACT GGA TGG TCT-3′	185	1.02
mMCP-6, *fw* mMCP-6, *rev*	5′-CAT TGA TAA TGA CGA GCC TCT CC-3′ 5′-CAT CTC CCG TGT AGA GGC CAG-3′	116	1.05

All primers were designed for qPCR using the software *Primer Express* Version 1.0

fw forward primer; *rev* reverse primer

[a]Efficiency = $10^{(-1/\text{slope})}-1$

3. PageRuler™ Plus Prestained Protein Ladder (Fermentas International Inc., ON, Canada).
4. 5× SDS–PAGE sample buffer: 250 μL 20% SDS, 600 μL glycerol, 100 μL mercaptoethanol, 50 μL bromphenol blue (BFB).
5. Methanol.
6. Odyssey blocking buffer (Li-cor, Lincoln, NE).
7. Tween-20 (Merck, Darmstadt, Germany).
8. Primary antibodies.
9. Blotting Paper (VWR International).
10. PVDF-FL membranes (Millipore, Bedford, MA).
11. Black Western incubation box (Li-cor).
12. Odyssey Infrared Imager (Li-cor).
13. Secondary antibodies (depending on primary antibody species; all from Li-cor): Li-cor IRDye 800 CW Donkey Anti-Rabbit IgG; Li-cor IRDye 680 CW Donkey Anti-Mouse IgG; Li-cor IRDye 800 CW Donkey Anti-Goat IgG; Li-cor IRDye 800 CW Donkey Anti-Rat IgG.
14. Transfer buffer: 4.84 g Tris, 22.5 g glycine, 400 mL methanol, 1 mL 20% SDS. Dilute to 2 L with H_2O
15. 10× Tris-buffered saline (TBS): Dissolve the following in 800 mL H_2O: 12.1 g Tris, 87 g NaCl. Set pH to 7.4. Dilute to 1 L with H_2O 1 ml 20% SDS.

2.5. Staining Procedures

2.5.1. May-Grünwald–Giemsa staining

1. May-Grünwald solution (Merck).
2. Giemsa solution (Merck).
3. VectaMount™ Permanent mounting medium (Vector Laboratories Inc, Burlingame, CA).

2.5.2. Toluidine Blue Staining

1. Toluidine blue stock: mix 1 g Toluidine blue O (Sigma-Aldrich, Stockholm, Sweden), 100 mL 70% ethanol.
2. 1% NaCl (in H_2O). Make this solution fresh every time. Adjust pH to 2.0–2.5 using glacial acetic acid or HCl.
3. Working solution: 5 mL Toluidine blue stock + 45 mL 1% NaCl (pH 2.0–2.5). Mix well. The pH should be around 2.3 and less than 2.5. Make the solution fresh and discard after use. A pH higher than 2.5 will result in a less distinct staining.
4. VectaMount™ Permanent mounting medium (Vector Laboratories Inc, Burlingame, CA).

2.5.3. Chloroacetate Esterase Assay

1. Pararosanilin solution: mix 0.5 g pararosanilin (Sigma), 20 mL H_2O, and 2.5 mL of concentrated HCl. Warm gently and filter. Refrigerate.

2. Nitrosylated pararosanilin. Mix 0.1 mL of the pararosanilin solution and 0.1 mL of 4% sodium nitrite (fresh). After mixing, let stand for 1 min. Use immediately.
3. Sorenson Stock A: 2.73 g Na_2HPO_4 + 250 mL H_2O.
4. Sorenson Stock B: 2.27 g KH_2PO_4 + 250 mL H_2O.
5. Sorenson working buffer: 41 mL Sorenson Stock A + 9 mL Sorenson Stock B.
6. Incubating medium. Combine in the following order: 10 mg Naphtol-ASD chloroacetate, 1 mL *N,N*-dimethylformamide, 35 mL Sorenson's working buffer, and 0.2 mL nitrosylated pararosanilin. Mix well, filter, and use immediately.

2.6. MC Activation Procedures

2.6.1. IgE-Mediated MC Activation

1. Tyrode's buffer (see Subheading 2.2.1).
2. Mouse anti-TNP IgE (R35-72, 0.5 mg/mL; Pharmingen, San Diego, CA).
3. TNP-OVA (7 mg/mL solution; Biosearch Technologies, Novato, CA).

2.6.2. Calcium Ionophore-Mediated MC Activation

1. Tyrode's buffer (see Subheading 2.2.1).
2. A23187 (2 mM stock solution in DMSO; store at −20°C).

2.7. β-Hexosaminidase Assay

1. Tyrode's buffer (see Subheading 2.2.1).
2. Tyrode's buffer, 1% Triton X-100,
3. p-Nitrophenyl-*N*-acetyl-β-D-glucosaminide (Sigma).
4. 0.05 M citric acid.
5. 0.05 M tri-sodium citrate.
6. 0.05 M Na_2CO_3.
7. 0.05 M $NaHCO_3$.
8. Citrate buffer: Add 49.5 mL of 0.05 M citric acid to 50.5 mL of 0.05 M tri-sodium citrate. Check to see that the pH is 4.5.
9. Na_2CO_3 buffer: Add 60 mL of 0.05 M Na_2CO_3 to 40 mL of 0.05 M $NaHCO_3$. Check to see that the pH is 10.0.

3. Methods

3.1. Bone Marrow-Derived MCs

Different protocols can be used, depending on which type of BMMCs is wanted.

For connective tissue MC (CTMC)-like BMMCs, bone marrow cells can be grown in either WEHI-3B-conditioned media or with a combination of recombinant IL-3 and SCF. If recombinant cytokines are used, cells grow and mature faster but it is more expensive. For generation of mucosal MC (MMC)-like BMMCs, use recombinant IL-3 + SCF + IL-9 + TGF-β (28).

BMMC culture

1. Kill the mice.
2. Wet the body with 70% ethanol.
3. Dissect the hind legs, femur and tibia, and remove the skin.
4. Put the bones in sterile 50-mL plastic tubes (Falcon type) on ice.
5. From now on, work in a sterile hood.
6. Transfer the bones to a Petri dish.
7. Remove all tissue surrounding the bones.
8. Cut the ends of the bones away.
9. Use a syringe filled with PBS, flush out the bone marrow into a sterile plastic tube (Falcon type).
10. Centrifuge, 10 min, 300 × *g*.
11. Resuspend the cells in 10 mL PBS and centrifuge again; 10 min, 300 × *g*.
12. Resuspend cells in culture medium, approximately 500,000 cells/mL (usually about 50 mL/mouse).
13. Transfer the cells to a culture flask.
14. Count the cells every third/fourth day, centrifuge cells for 10 min at 300 × *g*, resuspend to 500,000 cells/mL in fresh medium, and place them in a new flask. Discard the cells that are attached to the bottom of the old flask. In the early phase cultures the number of cells will decrease, but after a while they should start proliferating. After approximately 3 weeks, mature BMMCs are obtained.

3.2. Proteoglycan Analysis

3.2.1. Biosynthetic Labeling of Proteoglycans

1. Culture BMMCs as described under Subheading 3.1. For each labeling experiment, preferably use at least ~20×10^6 cells.
2. Wash the cells 3 × (300 × *g*) in Tyrode's buffer. Resuspend the cells in medium with reduced sulfate content.
3. Add 0.5 mCi of carrier-free $^{35}SO_4^{2-}$ and incubate for desired time (e.g., 1, 4, 24 h).
4. Centrifuge the cells (300 × *g*; 10 min), recover the cell supernatant (for analysis of secreted PGs), and wash the cell pellet (3× with Tyrode's buffer). Cell pellets and cell supernatants can be frozen (−20°C) until use.

3.2.2. Purification of ^{35}S-Labeled PGs

1. Solubilize the labeled cell pellets in 500 μL lysis buffer, 30 min, 4°C.
2. Dilute the lysates with H_2O, 0.5% Triton X-100 to 0.1 M NaCl, e.g., add 9.5 mL H_2O, 0.5% Triton X-100 to 500 μl of sample in lysis buffer (see Note 2).

3. Pack columns with ~300 μL anion exchange resin (4°C).
4. Equilibrate the columns with ~10 volumes of buffer I.
5. Load the column with your sample. Wash with 10 volumes of buffer I and 10 volumes of buffer II.
6. Elute the column with 20 × 300 μL of elution buffer. Collect 300 μL fractions in separate tubes. Analyze 10 μL from each fraction for ^{35}S-radioactivity, using scintillation counting.

3.2.3. Release of Free GAG Chains (Alkali Treatment, see Note 3)

1. Add 1/7 volume of 4 M NaOH to your sample, e.g., 50 μL 4 M NaOH to 350 μL of sample. Make sure that the sample is in a non-buffered solution! The final NaOH concentration is thus 0.5 M.
2. Incubate over night at 4°C.
3. Neutralize with 4 M HCl. Normally, somewhat less volume of 4 M HCl than 4 M NaOH is required for neutralization. Check that the pH is neutral using pH paper.

3.2.4. Size Determination of ^{35}S-Labeled PGs and Released GAG Chains

1. Equilibrate a Superose 6 (for intact PGs) or Superose 12 (for free GAG chains) column with Running buffer.
2. Load the sample (~5,000 cpm or more) in minimal volume (preferably <500 μL; depending on loop size).
3. Run the column at a flow rate of 0.5 mL/min.
4. Collect fractions of 250–500 μL. Analyze each fraction (total volume) for ^{35}S-radioactivity using scintillation counting.

3.2.5. Determination of Anionic Charge Density (Anion Exchange Chromatography)

1. Equilibrate an anion exchange column with equilibrating buffer.
2. Load the sample (≥5,000 cpm) and wash with ~10 column volumes of equilibrating buffer. Elute the column with a LiCl gradient increasing from 0.05 to 2 M, in 50 mM NaAc (pH 4.0) at a flow rate of 0.5 mL/min.
3. Collect fractions of 1 mL and analyze for ^{35}S-radioactivity by scintillation counting.

3.3. Primer Efficiency and qPCR

3.3.1. Primer Efficiency

1. Dilute cDNA: 1:1, 1:10, 1:100.
2. Make a master mix containing, for an individual well: 5.0 μL SYBR® Supermix, 0.2 μL forward primer (10 μM; 200 nM final concentration), 0.2 μL reverse primer (10 μM; 200 nM final concentration), 3.6 μL H_2O. Add 9 μL of your master mix and 1 μL cDNA in triplicates to a 384-well microtiter plate.
3. Centrifuge the plate (300 × *g*, 2 min).
4. Run the real-time quantitative RT-PCR using the following cycling conditions. Step 1: 95°C (10 min); step 2: 95°C (30 s); step 3: 58°C (30 s); step 4: 72°C (20 s). Repeat steps 2–4, 40× + dissociation stage.

5. Plot the C_T values against the log of the concentration of the samples (0, −1, −2 in this example). The slope of the curve should be from −3.1 to −3.6. The efficiency is calculated by: efficiency = $10^{(-1/\text{slope})}-1$. If the correct efficiency is not achieved, try another temperature, or another primer concentration.
6. Check that there is only one peak in the dissociation curve. If more than one peak is seen, this indicates a contamination, primer dimers, or that the primers are nonspecific.

3.3.2. Real-Time Quantitative RT-PCR

1. When a satisfactory primer efficiency and dissociation curve has been obtained, run the qPCR in triplicates using the same conditions.
2. The results are calculated according to User Bulletin #2: ABI PRISM 770 Sequence Detection System (P7N 4303859). In short, first the Ct value of your reference gene (in our case HPRT) is subtracted from the Ct value of the target gene, ΔCt = (tar−ref). Then subtract the ΔCt of the control sample (control or any other chosen sample that you want to relate all your other samples to) from the ΔCt of all other samples $\Delta\Delta Ct = \Delta Ct - \Delta Ct_{control}$. To get the relative amounts expressed, use the formula $2^{-\Delta\Delta Ct}$.

3.4. SDS–PAGE–Western Blot

1. Sample preparation. Use 20 μL of sample and add 5 μL 5× SDS–PAGE sample buffer. Boil 5 min in a heating block (100°C).
2. Set up a ClearPage™ SDS–PAGE gel (percentage depending on the size of the protein to be detected), fill in space between gels and around according to mark with ClearPage SDS–PAGE running buffer. Load the samples and protein ladder.
3. Run the gel. Step 1: 90 V—until the sample has migrated through the stacking gel. Step 2: 110 V, until the blue line is running out of the gel.
4. Wet transfer. Use PVDF-FL membranes. Wet the dry membrane in methanol before transfer. Note, the membranes are sensitive to scratches and bending; marks will be visible in the scanning. Soak the papers and "sponges" in transfer buffer. Place in the following order, starting from *white* side of holder: sponge, 2× paper, membrane, gel, 2× paper, sponge. All of these components should have been previously soaked in transfer buffer. Place the holder with the black side of the holder facing the black side of the transfer box, and place the ice-block inside the transfer box. Cover with transfer buffer to the top of the cassettes (transfer buffer can be reused). *Transfer conditions*: 200 mA (approximately 1 h).
5. Blocking: Dilute the Odyssey blocking buffer 1:1 in TBS or PBS. Do not use Tween-20 in the blocking buffer. Block for approximately 1 h on a rocking table.

6. Incubation of membranes with primary antibody. Dilute primary antibody (1:100–2,000) in blocking buffer (reuse the blocking buffer from the blocking step); 5 mL for each membrane, place membranes in plastic bags. Incubate on a rocking table, over night at 4°C (or 2 h, room temperature).
7. Washing. 3 × 10 min with TBS, 0.1% Tween-20, followed by 1 × 5 min with TBS.
8. Secondary antibody. Near-infrared (NIR) fluorescent labeled anti-rabbit antibody (or anti-rat or anti-mouse, depending on the primary antibody) diluted 1:1,000 in Odyssey blocking buffer (see Note 4). Use approximately 5–10 mL per membrane. Keep membranes in dark at all times until reading IR.
9. Incubate on a rocking table for approximately 1–2 h.
10. Washing. Wash 2 × 10 min in TBS, 0.1% Tween-20, followed by 1 × 10 min in TBS. Use the "black boxes."
11. Scan membranes using an Odyssey Infrared Imager.

3.5. Staining Procedures

3.5.1. May-Grünwald–Giemsa Staining

1. Prepare cytospin slides using approximately 20–50 × 10^3 cells/slide in culture medium.
2. Stain slides with May-Grünwald (concentrated), 5 min. Solutions can be reused.
3. Stain with May-Grünwald diluted 1:2 in H_2O, 1 min.
4. Stain slides with 2.5% Giemsa (in H_2O), 15 min.
5. Wash slides with H_2O until the water does not turn blue anymore.
6. Let the slides air-dry.
7. Mounting. Add 1 drop (using a Pasteur pipette) VectaMount Permanent mounting medium, and place a coverslip on top.

3.5.2. Toluidine Blue Staining

1. Prepare cytospin slides or deparaffinized (hydrated) tissue sections.
2. Stain cytospin slides or sections in Toluidine blue working solution for 2–5 min.
3. Wash in H_2O, five changes.
4. Dehydrate quickly through 95% and two changes of 100% ethanol.
5. Mounting: Add 1 drop (using a Pasteur pipette) VectaMount Permanent mounting medium, and place a cover slip on top.

3.5.3. Chloroacetate Esterase Assay

1. Use cytospin slides or tissue sections from paraffin-embedded tissues. After deparaffinization of tissue sections, bring them to H_2O.
2. Place samples in incubating medium (prepared as described under Subheading 2.5.3) for 40 min.

3. Wash in running tap water for 5 min.
4. Counterstain.
5. Mounting, as described under Toluidine staining (see Subheading 3.5.2).
6. Inspect by light microscopy.

3.6. MC Activation Procedures

3.6.1. IgE-Mediated MC Activation

1. Centrifuge (300 × *g*; 10 min) 5 × 10^6^ MCs (e.g., BMMCs) and resuspend in 1 mL culture medium.
2. Add 2 μL of anti-TNP IgE (from 0.5 mg/mL stock solution).
3. Incubate for 5 h.
4. Wash the cells 2× with Tyrode's buffer (centrifugation at 300 × *g*; 10 min) and resuspend cells in 1 mL of Tyrode's buffer.
5. Add 400 μL of the cell suspension to two individual wells of a 24-well cell culture plate.
6. Add 2.4 μL TNP-OVA (from a 1:100 dilution of a 7 mg/mL stock).
7. Incubate at 37°C for various times (e.g., 30 min, 1, 2 h).
8. Using a pipette, take up all of the contents from the well (including cells; flush pipette up and down to make sure that all cells are transferred).
9. Centrifuge the cells (300 × *g*; 10 min). Transfer the supernatant (medium fraction) to a separate tube. Store cell pellets and medium fractions at −20°C until use.

3.6.2. Calcium Ionophore-Mediated MC Activation

1. Wash 5 × 10^6^ MCs (e.g., BMMCs) 2× with Tyrode's buffer. Resuspend in 1 mL Tyrode's buffer.
2. Add 400 μL of the cell suspension to two individual wells of a 24-well cell culture plate.
3. Add 4 μL of 10× diluted A23187 (diluted from a 2 mM stock solution; final concentration in the wells: 2 μM) to one of the wells. Add 4 μL of buffer only to the other well.
4. Incubate at 37°C for various time periods (e.g., 30 min, 1, 2 h).
5. Using a pipette, take up all of the contents from the well (including cells; flush pipette up and down to make sure that all cells are transferred).
6. Centrifuge the cells (300 × *g*; 10 min). Transfer the supernatant (medium fraction) to a separate tube. Store cell pellets and medium fractions at −20°C until use.

3.7. β-Hexosaminidase Assay

1. Wash mature MCs (e.g., BMMCs; ~2 × 10^6^ cells) in Tyrode's buffer. After activation of the cells (see Subheadings 3.6.1 and 3.6.2), collect conditioned media (centrifuge at 300 × *g* for

10 min to remove cells; media can be frozen at −20°C). As a control for total β-hexosaminidase content, prepare cell lysates by solubilization in Tyrode's buffer, 1% Triton X-100 (200 μL/10^6 cells).

2. Using 96-well microtiter plates, incubate conditioned media and cell lysates, 20 μL of sample with 80 μL 1 mM p-nitrophenyl-*N*-acetyl-β-D-glucosaminide in 0.05 M citrate buffer (pH 4.5) at 37°C for 1 h.

3. Quench the reactions by addition of 200 μL 0.05 M Na_2CO_3 buffer (pH 10.0).

4. Read the OD at 405 nm.

5. Calculate the extent of degranulation by taking the OD value of the media divided by the total OD of the sample (media + cell fraction).

4. Notes

1. During extraction of serglycin from MCs, high NaCl concentrations are needed to break the strong electrostatic interactions between serglycin and positively charged MC proteases.

2. The procedure for extraction of serglycin will lead to recovery of intact PGs, i.e., GAG chains attached to core protein. If free GAG chains are desired, use the alkali treatment procedure (see Subheading 3.2.3).

3. A distinguishing feature of serglycin, as opposed to other PG species, is its resistance to proteolytic degradation. This property is explained by the dense substitution of the serglycin core protein with GAG chains. Hence, proteolytic digestion will not lead to release of free GAG chains, whereas proteolytic digestion (e.g., with papain or trypsin) is a common procedure to generate free GAG chains from PGs of other species.

4. Do not keep the fluorescently labeled antibodies in light! Use black Western incubation boxes for incubations.

Acknowledgments

The authors of this chapter receive support from The Swedish Research Council, Formas, King Gustaf V's 80-year Anniversary Fund, Torsten and Ragnar Söderberg Foundation, and The Swedish Cancer Foundation.

References

1. Metcalfe, D. D., Baram, D., and Mekori, Y. A. (1997) Mast cells. *Physiol Rev* **77**, 1033–1079.
2. Galli, S. J., Nakae, S., and Tsai, M. (2005) Mast cells in the development of adaptive immune responses. *Nat Immunol.* **6**, 135–142.
3. Yu, M., Tsai, M., Tam, S. Y., Jones, C., Zehnder, J., and Galli, S. J. (2006) Mast cells can promote the development of multiple features of chronic asthma in mice. *J Clin Invest.* **116**, 1633–1641.
4. Lee, D. M., Friend, D. S., Gurish, M. F., Benoist, C., Mathis, D., and Brenner, M. B. (2002) Mast cells: a cellular link between autoantibodies and inflammatory arthritis. *Science* **297**, 1689–1692.
5. Secor, V. H., Secor, W. E., Gutekunst, C. A., and Brown, M. A. (2000) Mast cells are essential for early onset and severe disease in a murine model of multiple sclerosis. *J Exp Med.* **191**, 813–822.
6. Liu, J., Divoux, A., Sun, J., Zhang, J., Clement, K., Glickman, J. N., et al. (2009) Genetic deficiency and pharmacological stabilization of mast cells reduce diet-induced obesity and diabetes in mice. *Nat Med.* **15**, 940–945.
7. Sun, J., Sukhova, G. K., Yang, M., Wolters, P. J., MacFarlane, L. A., Libby, P., et al. (2007) Mast cells modulate the pathogenesis of elastase-induced abdominal aortic aneurysms in mice. *J Clin Invest.* **117**, 3359–3368.
8. Sun, J., Sukhova, G. K., Wolters, P. J., Yang, M., Kitamoto, S., Libby, P., et al. (2007) Mast cells promote atherosclerosis by releasing proinflammatory cytokines. *Nat Med.* **13**, 719–724.
9. Soucek, L., Lawlor, E. R., Soto, D., Shchors, K., Swigart, L. B., and Evan, G. I. (2007) Mast cells are required for angiogenesis and macroscopic expansion of Myc-induced pancreatic islet tumors. *Nat Med.* **13**, 1211–1218.
10. Malaviya, R., Ross, E. A., MacGregor, J. I., Ikeda, T., Little, J. R., Jakschik, B. A., and Abraham, S. N. (1994) Mast cell phagocytosis of FimH-expressing enterobacteria. *J Immunol.* **152**, 1907–1914.
11. Echtenacher, B., Mannel, D. N., and Hultner, L. (1996) Critical protective role of mast cells in a model of acute septic peritonitis. *Nature* **381**, 75–77.
12. Marshall, J. S. (2004) Mast-cell responses to pathogens. *Nature reviews* **4**, 787–799.
13. Pejler, G., Åbrink, M., Ringvall, M., and Wernersson, S. (2007) Mast cell proteases. *Adv Immunol.* **95**, 167–255.
14. Pejler, G., Abrink, M., and Wernersson, S. (2009) Serglycin proteoglycan: regulating the storage and activities of hematopoietic proteases. *Biofactors* **35**, 61–68.
15. Kolset, S. O. and Tveit, H. (2008) Serglycin - Structure and biology. *Cell Mol Life Sci.* **65**, 1073–1085.
16. Bishop, J. R., Schuksz, M., and Esko, J. D. (2007) Heparan sulphate proteoglycans fine-tune mammalian physiology. *Nature* **446**, 1030–1037.
17. Oldberg, A., Hayman, E. G., and Ruoslahti, E. (1981) Isolation of a chondroitin sulfate proteoglycan from a rat yolk sac tumor and immunochemical demonstration of its cell surface localization. *J Biol Chem.* **256**, 10847–10852.
18. Bourdon, M. A., Oldberg, A., Pierschbacher, M., and Ruoslahti, E. (1985) Molecular cloning and sequence analysis of a chondroitin sulfate proteoglycan cDNA. *Proc Natl Acad Sci USA* **82**, 1321–1325.
19. Kjellén, L., Pettersson, I., Lillhager, P., Steen, M. L., Pettersson, U., Lehtonen, P., et al. (1989) Primary structure of a mouse mastocytoma proteoglycan core protein. *Biochem J.* **263**, 105–113.
20. Tantravahi, R. V., Stevens, R. L., Austen, K. F., and Weis, J. H. (1986) A single gene in mast cells encodes the core peptides of heparin and chondroitin sulfate proteoglycans. *Proc Natl Acad Sci USA* **83**, 9207–9210.
21. Stevens, R. L., Fox, C. C., Lichtenstein, L. M., and Austen, K. F. (1988) Identification of chondroitin sulfate E proteoglycans and heparin proteoglycans in the secretory granules of human lung mast cells. *Proc Natl Acad Sci USA* **85**, 2284–2287.
22. Åbrink, M., Grujic, M., and Pejler, G. (2004) Serglycin is essential for maturation of mast cell secretory granule. *J Biol Chem.* **279**, 40897–40905.
23. Braga, T., Grujic, M., Lukinius, A., Hellman, L., Abrink, M., and Pejler, G. (2007) Serglycin proteoglycan is required for secretory granule integrity in mucosal mast cells. *Biochem J.* **403**, 49–57.
24. Ringvall, M., Rönnberg, E., Wernersson, S., Duelli, A., Henningsson, F., Åbrink, M., et al. (2008) Serotonin and histamine storage in mast cell secretory granules is dependent on

serglycin proteoglycan. *J Allergy Clin Immunol.* **121**, 1020–1026.

25. Woulfe, D. S., Lilliendahl, J. K., August, S., Rauova, L., Kowalska, M. A., Abrink, M., et al. (2008) Serglycin proteoglycan deletion induces defects in platelet aggregation and thrombus formation in mice. *Blood* **111**, 3458–3467.

26. Niemann, C. U., Abrink, M., Pejler, G., Fischer, R. L., Christensen, E. I., Knight, S. D., and Borregaard, N. (2007) Neutrophil elastase depends on serglycin proteoglycan for localization in granules. *Blood* **109**, 4478–4486.

27. Grujic, M., Braga, T., Lukinius, A., Eloranta, M. L., Knight, S. D., Pejler, G., and Abrink, M. (2005) Serglycin-deficient cytotoxic T lymphocytes display defective secretory granule maturation and granzyme B storage. *J Biol Chem.* **280**, 33411–33418.

28. Miller, H. R., Wright, S. H., Knight, P. A., and Thornton, E. M. (1999) A novel function for transforming growth factor-beta1: upregulation of the expression and the IgE-independent extracellular release of a mucosal mast cell granule-specific beta-chymase, mouse mast cell protease-1. *Blood* **93**, 3473–3486.

Chapter 15

Analysis of Aggrecan Catabolism by Immunoblotting and Immunohistochemistry

Peter J. Roughley and John S. Mort

Abstract

Aggrecan is essential for the normal function of articular cartilage and intervertebral disc, where it provides the ability for the tissues to withstand compressive loading. This property depends on both the high charge density endowed by its numerous chondroitin sulfate and keratan sulfate chains and its ability to form large molecular aggregates via interaction with hyaluronan. Degradation of aggrecan via the action of proteases takes place throughout life and the degradation products accumulate in the tissue and impair its function. Such degradation is exacerbated in degenerative or inflammatory joint disorders. The use of antibodies recognizing the various regions of aggrecan and the neoepitopes generated upon proteolytic cleavage has shown that matrix metalloproteinases and aggrecanases, members of the ADAMTS family, are responsible for aggrecan degradation, both throughout life and in disease. By using immunoblotting techniques, it is possible to determine the extent of aggrecan degradation and to identify the degradation products that have accumulated in the tissue, and immunohistochemistry allows the location of the aggrecan degradation to be established.

Key words: Aggrecan, Cartilage, Intervertebral disc, Degradation, Matrix metalloproteinase, Aggrecanase, Neoepitope, Immunoblotting, Immunohistochemistry

1. Introduction

1.1. Function of Aggrecan

Aggrecan is a large proteoglycan, possessing a core protein with over 2,000 amino acids to which over 100 sulfated glycosaminoglycan (GAG) chains are attached. It belongs to the hyalectan family of proteoglycans (see Note 1), which are characterized by their ability to interact with hyaluronan (HA) to form proteoglycan aggregates (1), with each aggregate possessing up to 100 aggrecans per HA molecule. The interaction between aggrecan and HA in the aggregates is stabilized by a link protein (see Note 2).

Aggrecan is present in high abundance in all hyaline cartilages and the intervertebral discs, where it provides the osmotic properties

Françoise Rédini (ed.), *Proteoglycans: Methods and Protocols*, Methods in Molecular Biology, vol. 836, DOI 10.1007/978-1-61779-498-8_15, © Springer Science+Business Media, LLC 2012

needed for tissue swelling and the ability to resist compression (2). It is also present at lower abundance in the menisci and in tendons where they pass over bone and fulfills a similar compression resisting role (see Note 3). The osmotic properties of aggrecan depend on the high anionic charge density conferred by its sulfated GAG chains, which may be chondroitin sulfate CS or keratan sulfate (KS). Resistance to compression, however, requires more than a high charge density, as the aggrecan must also be confined within the tissue and not able to freely diffuse from the site under compression. Such localization is provided by the large size of the proteoglycan aggregates. In addition to its role in resisting compression, aggrecan can also inhibit endothelial cell migration (3) and so contribute to the avascular nature of cartilages and their resistance to tumor invasion.

1.2. Structure of Aggrecan

The aggrecan core protein consists of multiple functional domains (4) (see Fig. 1). The amino terminal region consists of two globular domains, termed G1 and G2, which are separated by a short interglobular domain (IGD). The G1 region is responsible for the interaction with HA, and its structure is similar to that of link protein (5), with both possessing HA-binding and protein–protein interaction domains. The G2 region is similar in structure to the HA-binding domains of the G1 region, but is unable to interact with HA (6). At present, its function is unclear. A third globular domain, termed G3, is present at the carboxy terminus of the core protein, where it directs trafficking of the aggrecan through the various organelles of the cell during synthesis and allows interaction with other components of the extracellular matrix following secretion (7). The KS and CS chains reside in a long extended region between the G2 and G3 regions. KS is predominant in the KS-attachment domain adjacent to the G2 region, whereas CS resides closer to the G3 region and occupies two distinct domains termed CS1 and CS2 (see Note 4). At present, it is unclear whether CS and KS chains are solely involved in aggrecan hydration or whether they may also serve additional functional roles.

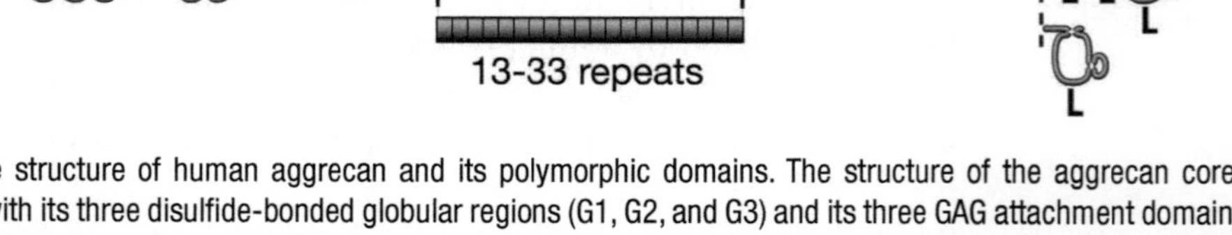

Fig. 1. The structure of human aggrecan and its polymorphic domains. The structure of the aggrecan core protein is depicted with its three disulfide-bonded globular regions (G1, G2, and G3) and its three GAG attachment domains (KS, CS1, and CS2). Aggrecan polymorphism due to a variable number of repeats within the CS1 domain and alternative splicing within the G3 region is indicated. The G3 region is shown in its longest form possessing two EGF-like domains (E), a lectin domain (L) and a Sushi domain (S), and its shortest form possessing only the lectin domain.

In the human, the aggrecan core protein exhibits polymorphic variation in its G3 region and CS1 domain (see Fig. 1) (see Note 5). The G3 region arises from exons in the aggrecan gene that encode for two epidermal growth factor-like (EGF) repeats, one lectin domain and one Sushi domain. Either or both of the EGF domains or the Sushi domain may be absent due to alternative splicing of the gene transcript, and the resulting G3 regions show differences in their functional properties (8). The CS1 domain is derived from a single exon of the aggrecan gene that possesses 19 residue tandem repeats of CS attachment sequences, and each aggrecan allele can exhibit a variable number of the tandem repeats (9). At present, it is unclear whether the number of tandem repeats is of functional consequence, but it has been suggested that aggrecan possessing a low number of repeats is functionally inferior and predisposes individuals to premature articular cartilage or intervertebral disc degeneration (10).

In addition to the polymorphic changes described above, aggrecan also undergoes a variety of structural changes throughout life (11). These changes include variations in CS and KS chain length and sulfation pattern due to altered synthesis and variations in core protein length due to degradation. The extent of the changes can vary with anatomical site, but occur mostly during juvenile growth and to a lesser extent during adult life, though this can vary with disease. Indeed, degenerative and inflammatory disorders of articular cartilage and intervertebral disc are associated with increased degradation of aggrecan in the adult. The changes in aggrecan structure can be caused by biochemical factors such as variations in cytokine or growth factor exposure, or biomechanical factors such as immobilization or overloading. While the functional consequences of the changes in aggrecan glycosylation are unclear, increased degradation of aggrecan is always considered as a detrimental event.

1.3. Degradation of Aggrecan

Degradation of the aggrecan core protein is due to the action of proteases and is most commonly associated with members of the matrix metalloproteinase (MMP) (12) and a disintregrin and metalloproteinase with thrombospondin motifs (ADAMTS) (13) families (see Note 6). ADAMTS4 and ADAMTS5 are the most active members of the ADAMTS family and are commonly called aggrecanases because of their selectivity for aggrecan as a substrate. Proteolytic cleavage of aggrecan can take place throughout the core protein, but is most commonly associated with cleavage in the IGD and CS2 domain. While most proteases are able to cleave within the IGD in vitro (14), cleavage in vivo appears to be a consequence of MMP or aggrecanase action (Fig. 2). The aggrecanases are also able to cleave at specific sites within the CS2 domain (15), whereas the cleavage sites of the MMPs outside of the IGD are less well characterized.

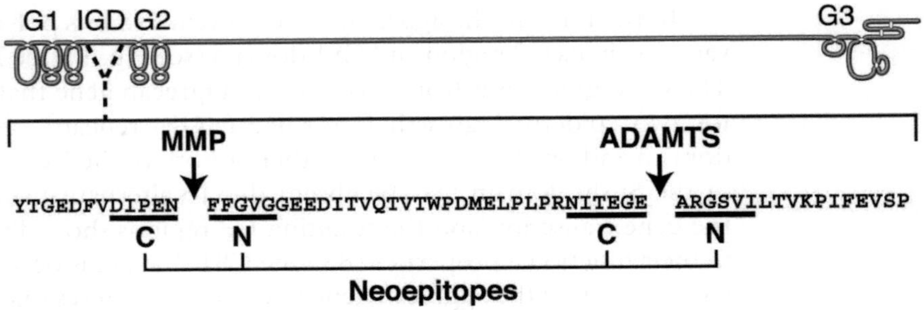

Fig. 2. Proteolytic cleavage sites within the aggrecan IGD. The structure of the aggrecan core protein is depicted with its three disulfide-bonded globular regions (G1, G2, and G3) and its interglobular domain (IGD). The amino acid sequence of the IGD is shown, together with the cleavage sites for matrix metalloproteinases (MMP) and aggrecanases (ADAMTS), and the C and N-terminal neoepitopes generated by such cleavage.

Proteolytic cleavage of aggrecan at a single site results in the generation of two fragments; one aggregating fragment possessing the G1 region and one nonaggregating fragment possessing the G3 region. In the absence of other catabolic events, such as degradation of collagen or HA (see Note 7), the aggregating fragments can be retained in the tissue for decades (16). In contrast, the nonaggregating fragments are free to diffuse within the tissue, and in articular cartilage, they are rapidly lost into the synovial fluid. Such rapid loss does not occur in the intervertebral discs, due to entrapment by the outer fibrous layers of the annulus fibrosus and the adjacent vertebrae. With time, proteolysis results in the aggregating fragments being reduced to the size of free G1 regions which are relatively resistant to proteolysis when bound to HA, and in the discs in reduction in size of the nonaggregating fragments, which eventually will facilitate their diffusion and loss. In normal articular cartilage and intervertebral disc, intact aggrecan molecules have half lives of a few years before they undergo extensive proteolysis, whereas free G1 regions have average retention times of about 20 years. In the disc, the nonaggregating fragments also have average retention times of about 20 years.

1.4. Analysis of Aggrecan Degradation

The ability to specifically detect aggrecan and its degradation products in cartilage and intervertebral disc has been instrumental in identifying the proteolytic mechanisms that operate within the tissues. This specificity has been achieved by the use of immunologic reagents, either for immunoblotting or immunohistochemical analysis. Immunoblotting provides a means of studying the extent of aggrecan degradation following extraction from the tissue and fractionation by gel electrophoresis, whereas immunohistochemistry allows the location of aggrecan degradation within the tissue to be determined. The antibodies used for both techniques can be divided into two groups: conventional antibodies that detect a specific region of aggrecan, and anti-neoepitope antibodies that detect degradation products generated by a specific protease. Anti-neoepitope antibodies specifically recognize the new terminal amino acid sequences generated by proteolytic cleavage of a protein, but not the equivalent amino acid sequence occurring in the intact protein (see Note 8).

The immunoblotting and immunohistochemical techniques allow aggrecan degradation to be studied throughout development and aging of the normal individual and also during the tissue destruction associated with joint diseases. The techniques can also be applied to organ culture models of cartilage and intervertebral disc catabolism and permit the proteases responsible for aggrecan degradation associated with specific inflammatory or mechanical insults to be identified.

2. Materials

2.1. Extraction and Deglycosylation of Aggrecan

1. Guanidinium chloride solution: 4 M guanidinium chloride, 0.1 M Tris–HCl, pH 7.5. Make fresh.
2. Complete Protease Inhibitor Cocktail Tablets (Roche). Dissolve in guanidinium chloride solution as needed.
3. Ethanol: 95% ethanol.
4. Keratanase buffer: 20 mM sodium acetate, pH 6.0. Store at room temperature.
5. Keratanase II (Seikagaku) resuspended at 0.2 U/mL in keratanase buffer. Store frozen at –20°C.
6. Chondroitinase buffer (10×): 1 M Tris, 1 M sodium acetate, pH 7.5. Store at room temperature.
7. Chondroitinase ABC (Seikagaku) resuspended at 10 U/mL in water. Store frozen at –20°C.

2.2. Preparation of Antisera

1. Peptides can be obtained commercially (see Notes 9 and 10), and purification by HPLC to at least 95% is recommended. The identity of the purified product is confirmed by mass spectroscopy, which is usually included with the synthesis by the supplier. Store peptides at –20°C. Peptides used for generating antibodies recognizing aggrecan G1, G2, G3, CS1 and neoepitopes are shown in Table 1.
2. Coupling buffer: 0.1 M potassium phosphate, pH 7.5, containing 1 mM EDTA (adjust with 10% potassium hydroxide).
3. Maleimide-activated KLH (keyhole limpet hemocyanin) (Pierce) (see Note 11).
4. Phosphate-buffered saline (PBS): 6 mM Na_2HPO_4, 4 mM KH_2PO_4, 145 mM NaCl, pH 7.2. Store at 4°C.
5. New Zealand White rabbits (female, 2.5–3.0 kg).
6. Freund's complete and incomplete adjuvant (Difco).

Table 1
Peptide sequences used for generation of anti-human aggrecan antisera

Epitope	Peptide sequence[a]
G1 region[b]	HDNSLSVSIPQPS*GGC* RVLLGTSLTIPCYFIDPMHPVTTAPS TEGRVRVNSAYQDK*GGC* SSRYDAICYTG
G2 region[b]	*CGG*YRPGPTRYSLTFEEA IVSPRTPCVGD VFFATRLEQFT*GGC* EALEFCESHNATAT TGQLYAAWSRGLDK*GGC* PIVTPRPACGGDKPGV TGLPDPLSRHHAF*GGC*
G3 region[b]	GWNKYQGHCYRHFPDR TWVDAERRCREQQSHLSS VEHARTGQKKDRY*GGC* CQPSGHWEEPRIT *CGG*TTYKRRLQKRSSRH
CS1[c]	*G*RIEWPSTPTVGEL*GGC*
MMP-C	*CGG*FVDIPEN
MMP-N[d]	FFGVG*AKKGC*
AGG-C	*CGG*NITEGE
AGG-N	ARGSVI*GGC*

[a] C-terminal *GGC* or N-terminal *CGG* sequences (in italics) are added for coupling when no cysteine residue is present in the peptide
[b] Mixtures of peptides used for generating G1, G2, and G3 region antisera
[c] N-terminal *G* residue added to avoid generation of neoepitope antibody
[d] To improve solubility of the peptide, the normal *GGC* coupling sequence was replaced by *AKKGC*

2.3. Reactivity and Specificity of Antisera

1. ELISA plates, Immulon 2, 96-well flat bottomed.
2. Extended and truncated peptides. In addition to the immunizing neoepitope antigen, a peptide possessing one additional residue extending past the cleavage terminus and a peptide lacking the terminal residue should also be prepared.
3. Maleimide-activated ovalbumin (Pierce) (see Note 11).
4. Coating buffer: 0.01 M sodium bicarbonate, pH 9.6. Store at 4°C.
5. PBS-T: PBS containing 0.05% Tween-20. Store at 4°C.
6. Blocking buffer: PBS-T containing 1% bovine serum albumin. Store at 4°C.

7. Alkaline phosphatase-conjugated goat anti-rabbit IgG (Sigma).
8. Alkaline phosphatase substrate: *p*-nitrophenylphosphate (Sigma). Dissolve one tablet in 10 mL diethanolamine buffer (9.6% v/v diethanolamine in 0.26 mM $MgCl_2$, pH 9.8).

2.4. Purification of Antibodies

1. Sulfolink coupling gel (Pierce).
2. Equilibration buffer: 50 mM Tris–HCl, pH 8.5 containing 5 mM EDTA. Can be stored at room temperature.
3. Glycine buffer: 0.1 M glycine, pH 2.8.
4. Tris buffer: 1 M Tris–HCl, pH 9.5.
5. PBS-azide: PBS containing 0.05% sodium azide. Can be stored at room temperature.

2.5. Electrophoresis and Immunoblotting

1. NuPAGE 4–12% gradient Bis-Tris gels (10 well, 1.5 mm thick) (Invitrogen). Other fixed percentage or gradient gel systems could be used.
2. Sample buffer (4×): NuPAGE LDS sample buffer (Invitrogen).
3. Reducing agent (10×): NuPAGE reducing agent (Invitrogen).
4. Electrophoresis buffer (20×): NuPAGE MOPS SDS running buffer (Invitrogen).
5. Nitrocellulose membranes (0.2 μm) (Bio-Rad).
6. Transfer buffer (20×): NuPAGE transfer buffer (Invitrogen).
7. Skim milk: nonfat dry milk powder (Carnation).
8. Tris-buffered saline/Tween (TBST): 20 mM Tris–HCl, 137 mM NaCl, pH 7.6 containing 0.1% Tween 20.
9. Primary antibodies: Polyclonal anti-peptide antibodies to aggrecan globular regions and anti-neoepitope antibodies to IGD cleavage sites (see Subheading 3.2 for method of preparation).
10. Secondary antibody: Biotinylated anti-rabbit IgG (Amersham).
11. Streptavidin-biotin-horseradish peroxidase (HRP) complex (Amersham).
12. ECL western blotting detection reagents (Amersham).
13. X-ray film: Bioflex MSI film (Clonex Corporation).

2.6. Immunohistochemistry

1. Phosphate buffer: 0.1 M NaH_2PO_4, 0.1 M Na_2HPO_4, pH 7.4. Store at 4°C.
2. PLP solution A: 0.2 M lysine in water adjusted to pH 7.4 with 0.1 M Na_2HPO_4. Can store for 10 days at 4°C.
3. PLP solution B: 1 part PLP solution A and 1 part phosphate buffer. Make fresh.
4. 8% Paraformaldehyde: 8 g paraformaldehyde in 100 mL water. Dissolve paraformaldehyde at 60°C by adding 2 M NaOH dropwise until solution clears. Make fresh.

5. EDTA/Tris: 10% EDTA in 0.1 M Tris, pH 7.4. Store at 4°C.
6. OCT: Optimal Cutting Temperature compound (Tissue-Tek).
7. Gelatin: 5 g gelatin, 50 mg chromium potassium phosphate (chrome-alum) in 100 mL water. Dissolve gelatin at 55°C then add chrome-alum. Allow to cool to room temperature then filter. Should be made fresh, but can be stored for several days at 4°C if coating large numbers of slides.
8. Gelatin-coated slides: Dip microscope slides in gelatin, wipe back of slide, and allow to dry vertically at room temperature. Can be stored in slide box at room temperature for several months.
9. Phosphate-buffered saline (PBS): 6 mM Na_2HPO_4, 4 mM KH_2PO_4, 0.145 M NaCl, pH 7.2. Store at 4°C.
10. 4% Paraformaldehyde: 4 g paraformaldehyde in 100 mL PBS. Make fresh.
11. Chondroitinase ABC and chondroitinase buffer as in Subheading 2.1.
12. Primary and secondary antibodies as in Subheading 2.5.
13. Vectastain Elite ABC kit for rabbit IgG (Vector Laboratories).
14. Incubation buffer: 0.8% bovine serum albumin, 0.1% gelatin, 1% normal goat serum in PBS with 0.05% sodium azide.
15. Diaminobenzidine (DAB) substrate: DAB peroxidase substrate kit (Vector).

3. Methods

3.1. Extraction and Deglycosylation of Aggrecan (see Note 12)

1. Cartilage or intervertebral disc is divided into small pieces (1–2 mm in each dimension) using a scalpel. Tissue can be used freshly for extraction or stored frozen at –20°C.
2. Tissue is suspended in guanidinium chloride extraction solution containing proteinase inhibitors (see Note 13). Between 10 and 20 mL of extraction fluid is used per 1 g tissue. Intervertebral disc requires more extraction fluid than cartilage because of its swelling.
3. The tissue is extracted for 48 h at 4°C with continuous mixing. For large volumes (>10 mL), extraction can be carried out in a conical flask using a magnetic stirrer for mixing. For intermediate volumes (0.5–10 mL), extraction can be carried out in a sealed tube using a rocking platform for mixing. For small volumes (<0.5 mL), extraction can be carried out in microfuge tubes using a vibrating shaker.
4. The extract is separated from the tissue residue by filtration. A syringe plugged with glass wool is ideal. Alternatively, the

extracts can be recovered by centrifuged at 10,000 × *g* for 30 min at 4°C. Extracts can be stored at −20°C.

5. Guanidine is removed from the extracts by precipitation with 9 volumes of ethanol. Usually 100 μL of extract is used and precipitation is carried out in a 1.5-mL microfuge tube. After the ethanol has been added to the extracts, the mixture is left overnight at 4°C to facilitate precipitation, and the precipitate is recovered by centrifugation using a microfuge at maximum speed. The precipitate is washed in cold ethanol, recovered by centrifugation, then redissolved in 100 μL keratanase buffer (see Note 14).
6. The sample is digested for 6 h at 37°C with 10 μL (2 mU) keratanase II (see Note 15).
7. Add 11 μL 10× chondroitinase buffer to the digest, then 2 μL chondroitinase ABC (20 mU), and digest overnight at 37°C (see Note 16).
8. Finally, inactivate both enzymes by incubation of the samples at 100°C for 5 min.

3.2. Preparation of Antisera

1. Maleimide-activated KLH is reconstituted in water at 10 mg/mL, and the peptide is dissolved in water at 10 mg/mL (see Note 17). To 0.6 mL of coupling buffer, add 0.2 mL of activated KLH and 0.2 mL of peptide solution and allow to react at room temperature for 2 h.
2. Remove excess peptide by dialysis of the preparation against PBS at 4°C. Peptide-conjugates can be stored at −20°C.
3. Obtain a preimmune bleed from each rabbit for use as a future control (see Note 18).
4. To a 1.5-mL microfuge tube, add 0.4 mL of KLH-peptide conjugate and 0.4 mL of Freund's complete adjuvant. Cap the tube and agitate as rapidly as possible using a vortex mixer to produce a stiff emulsion. Inject 0.25 mL of the emulsion intramuscularly into each flank of the animal.
5. After 2 weeks, the rabbits are boosted with the same mixture, except that Freund's incomplete adjuvant is used.
6. A second boost, as above, is given after a further 2 weeks.
7. After 10 days, the rabbits are bled out by heart puncture, and the blood collected in nonanticoagulant tubes and allowed to clot overnight at 4°C.
8. Following clot contraction, the serum is collected and centrifuged at low speed (500 × *g*) to remove contaminating red blood cells. Serum should be stored in aliquots at −20°C.

3.2.1. Direct ELISA to Evaluate Antisera Reactivity

1. Peptide-ovalbumin (see Note 19) conjugates are prepared as in step 1 in Subheading 3.2 for KLH-conjugates.
2. Coat ELISA plates with 50 ng of peptide-ovalbumin conjugate in 50 μL of coating buffer. Allow solution to stand at 4°C overnight.
3. Remove coating solution and wash wells with PBS-T.
4. Add 100 μL of blocking buffer and allow to stand at room temperature for 1 h.
5. Remove blocking buffer and add serial antisera dilutions (50 μL) from 1 in 10 to 1 in 100,000 in blocking buffer and incubate at 37°C for 1 h.
6. Remove antibody solutions and wash wells with PBS-T.
7. Add 50 μL of alkaline phosphatase-conjugated goat anti-rabbit IgG (1:1,000 dilution) and incubate at 37°C for 1 h.
8. Remove antibody and wash wells with PBS-T.
9. Add 50 μL of alkaline phosphatase substrate solution and allow color to develop for 15 min (or longer if necessary) at 37°C.
10. Measure absorbance at 405 nm using a plate reader.

3.2.2. Competition ELISA to Determine Specificity of Neoepitope Antisera

1. Coat plates with peptide conjugate as in step 2 in Subheading 3.2.1.
2. Determine dilution of antiserum to give 50% of maximum binding (by following steps in Subheading 3.2.1).
3. Make serial dilutions of competing peptide ranging from 25 to 150,000 ng/mL in this antibody dilution. Incubate for 1 h at room temperature.
4. Add antibody/peptide complexes to the wells. Incubate for 1 h at room temperature.
5. Remove complexes and wash plate with PBS-T, then develop with second step antibody and alkaline phosphatase substrate as above (steps 7–9 in Subheading 3.2.1).
6. Read absorbance at 405 nm and calculate IC_{50} values of inhibition (see Note 20). An example of a competitive ELISA is shown in Fig. 3.

3.3. Purification of Antibodies

1. Pack a 10-mL disposable column with 1.5 mL of Sulfolink resin (previously allowed to reach room temperature).
2. Wash the column with 10 mL of affinity column equilibration buffer, allow excess buffer to drain, and close the exit of the column.
3. Add 1.5 mL of a 2 mg/mL solution of ligand peptide in water to the gel. Cap the column and mix the contents by inverting several times over the course of 15 min. Allow to remain at room temperature for a further 30 min.

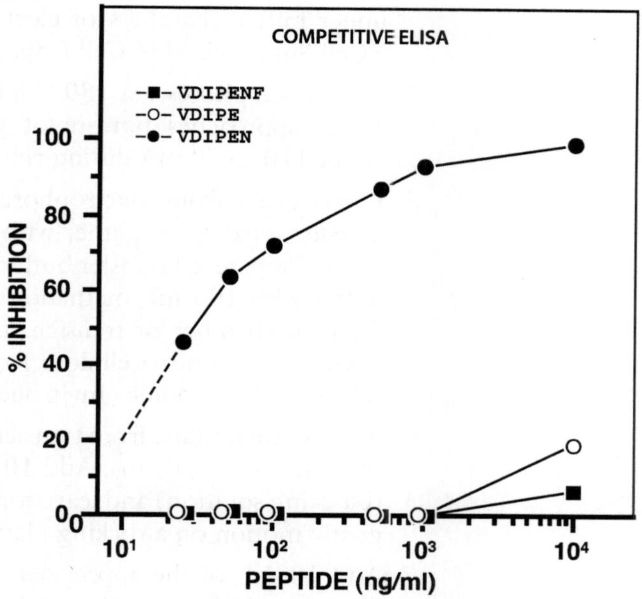

Fig. 3. Analysis of neoepitope specificity. The amino acid sequence specificity of the MMP-C neoepitope antibody was evaluated by a competitive ELISA, using the immunizing peptide (VDIPEN), a peptide with the C-terminal amino acid missing (VDIPE), and a peptide with an additional C-terminal amino acid (VDIPENF). The truncated and extended peptides show a three orders of magnitude lower immunoreactivity.

4. Drain and wash the column with 5 mL of equilibration buffer.
5. Block unreacted sites on the resin by adding 1.5 mL of 50 mM cysteine in equilibration buffer and mix again as above.
6. Drain the column and wash with 10 mL of equilibration buffer followed by 10 mL of PBS-azide by gravity flow.
7. Allow serum to equilibrate to room temperature, then apply 10 mL of serum to the affinity column by gravity flow.
8. Wash the column with 25 mL of PBS-azide.
9. Apply 15 mL of glycine buffer (see Note 21) and collect 1.0 mL fractions into tubes containing 50 μL of Tris buffer.
10. Read absorbance of fractions at 280 nm.
11. Dialyze peak fractions against PBS-azide overnight at 4°C. Store in aliquots at −20°C.

3.4. Electrophoresis and Immunoblotting

1. Mix 10 μL deglycosylated proteoglycan with 7.5 μL 4× sample buffer, 3 μL 10× reducing agent, and 9.5 μL water (final volume 30 μL) in a 1.5-mL microfuge tube. Heat at 70°C for 10 min.
2. Load 10–30 μL of sample into each well of NuPAGE 4–12% gradient gel.
3. Prepare gel running buffer by mixing 50 mL 20× electrophoresis buffer with 950 mL deionized water, and fill lower and

upper buffer chambers of electrophoresis apparatus. We use XCell SureLock Mini Cell from Invitrogen.

4. Run electrophoresis at 180 V for 1–1.5 h (until dye in sample buffer approaches bottom of gel). Current decreases from about 110 to 70 mA during run.
5. Remove gel from electrophoresis apparatus and place in a transfer apparatus together with adjacent nitrocellulose membrane. Prepare gel transfer buffer by mixing 50 mL 20× transfer buffer with 100 mL methanol and 850 mL deionized water, then fill chamber of transfer apparatus. Electroblot proteins from gel onto nitrocellulose for 2 h at 30 V (see Note 22). XCell II Blot Module (Invitrogen) was used.
6. Remove membrane from transfer apparatus and place in square Petri dish (10 × 10 cm). Add 10 mL of 5% skim milk in TBST (blocking solution) and leave for 1 h at room temperature with gentle motion on a rocking platform.
7. Add 10 mL of the appropriate dilution of primary antibody diluted in blocking solution to the membrane and leave overnight at 4°C. Usually, antisera can be used at a 1:1,000 dilution, and purified antibodies at 1 μg/mL. Remove primary antibody, then wash membrane in TBST (4 × 5 min).
8. Add 10 mL of a 1:1,000 dilution of secondary antibody diluted in blocking solution to the membrane and leave for 1 h at room temperature. Remove secondary antibody, then wash membrane in TBST (4 × 5 min).
9. Add 10 mL of 1:750 dilution of streptavidin-biotin-horseradish peroxidase (HRP) complex in TBST and leave for 45 min at room temperature. Remove complex, then wash membrane in TBST (4 × 5 min).
10. Add 2 mL solution 1 plus 2 mL solution 2 of ECL detection reagents to membrane and leave at room temperature for 1 min (see Note 23).
11. Expose membrane to X-ray film in cassette for 30 s to 10 min (depending on antigen abundance) to visualize location of immunoreactive proteins. Film is separated from wet membrane by a clear plastic sheet. Examples of results using anti-G1 and anti-G3 antibodies are shown in Fig. 4 and using anti-neoepitope antibodies are shown in Fig. 5.

3.5. Immunohistochemistry

1. Prepare PLP fixative (100 mL) fresh by mixing PLP solution B (75 mL) with 8% paraformaldehyde (25 mL), then adding 0.214 g sodium periodate (0.01 M) (see Note 24).
2. Fix tissue in PLP fixative for 1 h at room temperature, then place at 4°C. Time of fixation depends on thickness of tissue. Usually, 1–2 days are sufficient.

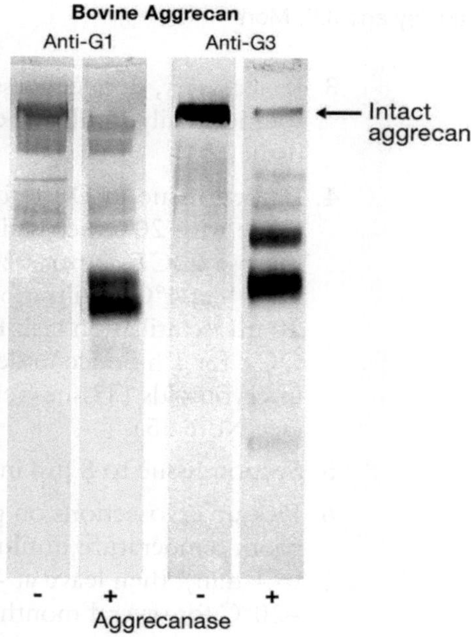

Fig. 4. Immunoblotting of aggrecan using anti-G1 and anti-G3 antibodies. Aggrecan extracted from bovine cartilage was analyzed both before and after aggrecanase (ADAMTS5) digestion. Before digestion, the anti-G1 antibody shows both intact aggrecan and some degradation products, whereas the anti-G3 antibody shows only intact aggrecan. After digestion, degradation products due to cleavage in the IGD are detected by the anti-G1 antibody, and those due to cleavage in the CS2 domain are detected by the anti-G3 antibody.

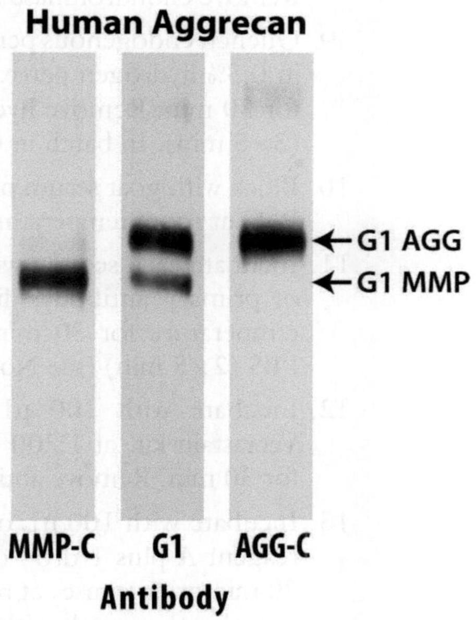

Fig. 5. Immunoblotting of aggrecan using anti-neoepitope antibodies. Aggrecan extracted from adult human cartilage was analyzed using an anti-G1 antibody and antibodies recognizing the C-terminal neoepitopes generated by matrix metalloproteinase (MMP-C) and aggrecanase (AGG-C) cleavage in the IGD. G1 regions of aggrecan compatible with both MMP and aggrecanase cleavage can be identified.

3. If necessary, decalcify tissue in EDTA/Tris at 4°C, changing solution daily until tissue is soft. Check by inserting pin into tissue.
4. Embed tissue in OCT for cryosectioning by infiltrating fixed tissue with 20% sucrose in PBS for 1 h at 4°C, then with 20% sucrose/OCT (2 parts 20% sucrose in PBS plus 1 part OCT) for 1 h at 4°C. Infiltrations can be done on rocker platform in 20 mL scintillation vial. Finally, infiltrate in new 20% sucrose/OCT for 1 h on ice under vacuum in a desiccator then embed in cryomolds (Tissue-Tek) on dry ice. Store blocks at −20°C (see Note 25).
5. Section tissue to 8 μm in cryostat.
6. Pick up cryosections on gelatin-coated slide and allow to sit at room temperature until moisture has evaporated (usually 30 s to 1 min), then leave at −20°C overnight. Or can be stored at −20°C for several months.
7. Postfix sections in 4% paraformaldehyde for 10 min at room temperature, then wash with PBS (2 × 5 min). In batch in Coplin jars.
8. Prepare chondroitinase ABC solution for permeabilization of tissue sections by mixing chondroitinase ABC stock with 9 volumes of chondroitinase buffer (1×). Place 100 μL chondroitinase ABC solution over each section and leave for 1 h at 37°C. Remove chondroitinase then wash with PBS (3 × 5 min).
9. Quench endogenous peroxidase activity by incubating sections in 0.3% hydrogen peroxide in methanol at room temperature for 30 min. Remove hydrogen peroxide, then wash with PBS (3 × 5 min). In batch in Coplin jars.
10. Block with goat serum present in Vectastain kit, diluted 1:66 in PBS, at room temperature for 20 min. Use 100 μL per section.
11. Incubate each section with 100 μL of the appropriate dilution of primary antibody, diluted in incubation buffer, at room temperature for 30 min. Remove antibody, then wash with PBS (2 × 5 min) (see Note 26).
12. Incubate with 100 μL of secondary antibody present in Vectastain kit, at 1:200 dilution in PBS, at room temperature for 30 min. Remove antibody then wash with PBS (2 × 5 min).
13. Incubate with 100 μL of Vectastain ABC complex (1 drop of reagent A plus 1 drop of reagent B in 2.5 mL PBS prepared 30 min prior to use) at room temperature for 30 min. Remove complex then wash with PBS (2 × 5 min).
14. Prepare fresh DAB substrate by mixing component solutions according to Vector instructions. Incubate section with 1–2 drops of DAB substrate (sufficient to cover section) at room temperature for 5 min. Wash sections for 5 min in Coplin jar

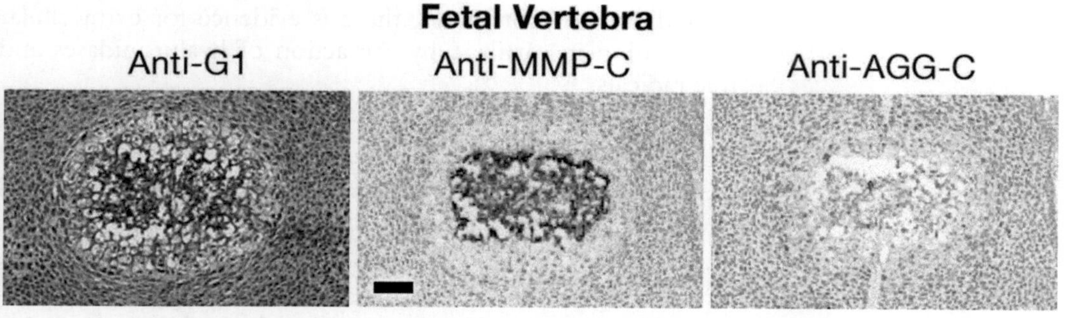

Fig. 6. Immunohistochemical analysis of aggrecan and its degradation products. Fetal mouse vertebrae were analyzed using an anti-G1 antibody and antibodies recognizing the C-terminal neoepitopes generated by matrix metalloproteinase (MMP-C) and aggrecanase (AGG-C) cleavage in the IGD. Aggrecan can be seen throughout the cartilage using the anti-G1 antibody. Cleavage by MMPs is detected only in the center of the vertebra where cartilage is being resorbed during endochondral ossification, but there is no evidence for aggrecanase activity. The bar represents 100 μm.

under running tap water to remove DAB. Air-dry, counterstain if desired, and mount. Examples of results using anti-G1 and anti-neoepitope antibodies are shown in Fig. 6.

4. Notes

1. The hyalectan family includes aggrecan, versican, brevican, and neurocan, all of which possess a structurally similar G1 region (17). They are also known as lecticans or aggregating proteoglycans.
2. Four link protein genes exist and occur paired with the hyalectan genes (18). The cartilage link protein gene occurs paired with the versican gene rather than the aggrecan gene. Versican is also expressed in cartilage, but at a much lower abundance compared to aggrecan.
3. In addition to being present in the cartilaginous tissues, aggrecan is also expressed in brain (19). Its function in brain is unclear, as is its degree of GAG substitution and proteolytic possessing.
4. Aggrecan structure within the GAG attachment regions varies between species, particularly with respect to the number of repeat motifs contained in the KS and CS1 domains (9,20).
5. Polymorphism within the G3 and CS1 regions has only been reported in the human.
6. There is currently no evidence for the involvement of serine proteinases or carboxyl proteinases within the cartilage extracellular matrix in vivo. However, cleavage compatible with the action of calpain, a cysteine proteinase, has been reported.

7. In addition to proteinases, there is evidence for extracellular degradation in cartilage by the action of hyaluronidases and free radicals.
8. Proteolytic cleavage of a protein results in a new N-terminus and C-terminus at each cleavage site, whose amino acid sequences can be used to generate corresponding N-terminal and C-terminal neoepitope antibodies.
9. For preparation of anti-neoepitope antibodies, five or six residues representing the N- or C-terminal of the cleavage site are sufficient. This is extended by the addition of a double glycine linker and a cysteine residue to allow coupling to the carrier protein. Epitopes containing a cysteine residue should be excluded.
10. For continuous epitopes, sequences of up to 25 or more residues can be used. Design will depend on issues such as species specificity and possible cross-reaction with orthologous sequences. Internal naturally occurring cysteine residues can be used for coupling, otherwise an artificial cysteine residue and glycine linker can be appended to either terminus.
11. Carrier proteins can also be activated using the bifunctional coupling reagent N-succinimidyl bromoacetate (21).
12. Analysis of aggrecan by immunoblotting does not require purification from the tissue extracts. Because of the structural heterogeneity of aggrecan, purification will result in the selective loss of some fragments.
13. The extraction can also be carried out in the presence of a mixture of general protease inhibitors: 1 mM phenylmethylsulfonyl fluoride (PMSF), 1 mM iodoacetate (IAA), 1 mM ethylene diamine tetraacetate (EDTA), and 10 μg/mL pepstatin A, which inhibit serine-, cysteine-, metallo-, and aspartic proteases, respectively.
14. For larger samples, dialysis can be used as an alternative to ethanol precipitation for removal of guanidine. In this case, filtered extracts are dialyzed twice at 4°C against 100 volumes of the keratanase buffer.
15. Keratanase II can be replaced by keratanase I or endo-β-galactosidase for digestion of keratan sulfate. The enzymes differ in their ability to cleave disaccharides of keratan sulfate, with keratanase II being able to cleave disulfated disaccharides and endo-β-galactosidase to cleave nonsulfated disaccharides, but all are able to diminish keratan sulfate heterogeneity to permit precise immunoblotting.
16. Chondroitinase ABC can be replaced by chondroitinase AC for digestion of chondroitin sulfate. The enzymes differ in their ability to degrade dermatan sulfate (chondroitin sulfate B), but this is not present on aggrecan.

17. The majority of peptides dissolve in water. We have also had success with dissolving peptides in dimethylformamide or dimethylsulfoxide so that the final concentration of organic solvent in the coupling reaction is no more than 20 or 30%, respectively.
18. Many institutions have standard operating protocols (SOPs) for antibody production or offer a complete service.
19. Unless the antibodies to be analyzed have been affinity-purified, a strong anti-KLH component will be present. This necessitates the use of an unrelated carrier protein for the conjugate used to coat the plates.
20. Such analyses demonstrate that specificity for the terminal residue is not absolute. For example, the bovine aggrecan MMP cleavage neoepitope …VDIPES is still recognized but with lower affinity by an antibody to the human …VDIPEN epitope.
21. In some cases, glycine elution may prove too drastic. Other possibilities are elution with isocyanate, citrate, or proprionic acid. Rapid removal of the eluting agent is necessary in all cases.
22. Nitrocellulose membranes can be replaced by polyvinylidene fluoride (PVDF) membranes for immunoblotting. PVDF membranes are able to bind more protein than nitrocellulose and therefore can give greater detection sensitivity, but nitrocellulose membranes tend to give cleaner background and are preferred if detection is not a problem.
23. ECL detection of immunoreactive proteins can be replaced by color detection using either alkaline phosphatase or horseradish peroxidase-based systems. While color detection is less sensitive, it does have the advantage that one can visualize reaction product directly.
24. PLP is the preferred fixative for immunostaining of aggrecan, as it permits cross-linking of both the protein and carbohydrate components. With a highly glycosylated proteins such as aggrecan, protein fixation alone may not be sufficient to ensure adequate localization within the tissue.
25. Frozen sections can be replaced by paraffin sections if necessary, but frozen sections are preferred for retention of antigenicity. Increased antibody concentrations may be required when using paraffin sections.
26. Antibody reactivity can vary between immunoblotting and immunohistochemistry because of differences in epitope presentation.

Acknowledgments

The authors would like to thank Lisa Lamplugh for technical assistance with the immunoblotting and immunohistochemistry and Guylaine Bédard for the artwork. The work was supported by the Shriners of North America and the Canadian Institutes of Health Research.

References

1. Hascall, V. C. (1988) Proteoglycans: the chondroitin sulfate/keratan sulfate proteoglycan of cartilage. *ISI Atlas Sci. Biochem.* **1**, 189–198.
2. Watanabe, H., Yamada, Y., and Kimata, K. (1998) Roles of aggrecan, a large chondroitin sulfate proteoglycan, in cartilage structure and function. *J. Biochem.* (Tokyo) **124**, 687–693.
3. Johnson, W. E. B., Caterson, B., Eisenstein, S. M., and Roberts, S. (2005) Human intervertebral disc aggrecan inhibits endothelial cell adhesion and cell migration *in vitro*. *Spine* **30**, 1139–1147.
4. Doege, K. J., Sasaki, M., Kimura, T., and Yamada, Y. (1991) Complete coding sequence and deduced primary structure of the human cartilage large aggregating proteoglycan, aggrecan. Human-specific repeats, and additional alternatively spliced forms. *J. Biol. Chem.* **266**, 894–902.
5. Neame, P. J. and Barry, F. P. (1993) The link proteins. *Experientia* **49**, 393–402.
6. Fosang, A. J. and Hardingham, T. E. (1989) Isolation of the N-terminal globular protein domains from cartilage proteoglycans. Identification of G2 domain and its lack of interaction with hyaluronate and link protein. *Biochem. J.* **261**, 801–809.
7. Zheng, J., Luo, W., and Tanzer, M. L. (1998) Aggrecan synthesis and secretion - A paradigm for molecular and cellular coordination of multiglobular protein folding and intracellular trafficking. *J. Biol. Chem.* **273**, 12999–13006.
8. Day, J. M., Olin, A. I., Murdoch, A. D., Canfield, A., Sasaki, T., Timpl, R., Hardingham, T. E., and Aspberg, A. (2004) Alternative splicing in the aggrecan G3 domain influences binding interactions with tenascin-C and other extracellular matrix proteins. *J. Biol. Chem.* **279**, 12511–12518.
9. Doege, K. J., Coulter, S. N., Meek, L. M., Maslen, K., and Wood, J. G. (1997) A human-specific polymorphism in the coding region of the aggrecan gene - Variable number of tandem repeats produce a range of core protein sizes in the general population. *J. Biol. Chem.* **272**, 13974–13979.
10. Roughley, P., Martens, D., Rantakokko, J., Alini, M., Mwale, F., and Antoniou, J. (2006) The involvement of aggrecan polymorphism in degeneration of human intervertebral disc and articular cartilage. *Eur. Cell Mater.* **11**, 1–7.
11. Roughley, P. J. and Mort, J. S. (1986) Ageing and the aggregating proteoglycans of human articular cartilage. *Clin. Sci.* **71**, 337–344.
12. Nagase, H., Visse, R., and Murphy, G. (2006) Structure and function of matrix metalloproteinases and TIMPs. *Cardiovasc. Res.* **69**, 562–573.
13. Jones, G. C. and Riley, G. P. (2005) ADAMTS proteinases: a multi-domain, multi-functional family with roles in extracellular matrix turnover and arthritis. *Arthritis Res. Ther.* **7**, 160–169.
14. Fosang, A. J., Neame, P. J., Last, K., Hardingham, T. E., Murphy, G., and Hamilton, J. A. (1992) The interglobular domain of cartilage aggrecan is cleaved by PUMP, gelatinases, and cathepsin B. *J. Biol. Chem.* **267**, 19470–19474.
15. Tortorella, M. D., Pratta, M., Liu, R. Q., Austin, J., Ross, O. H., Abbaszade, I., et al. (2000) Sites of aggrecan cleavage by recombinant human aggrecanase-1 (ADAMTS-4). *J. Biol. Chem.* **275**, 18566–18573.
16. Sivan, S. S., Tsitron, E., Wachtel, E., Roughley, P. J., Sakkee, N., Van der Ham, F., et al. (2006) Aggrecan turnover in human intervertebral disc as determined by the racemization of aspartic acid. *J. Biol. Chem.* **281**, 13009–13014.
17. Margolis, R. U. and Margolis, R. K. (1994) Aggrecan-versican-neurocan family of proteoglycans. *Methods Enzymol.* **245**, 105–128.

18. Spicer, A. P., Joo, A., and Bowling, R. A., Jr. (2003) A hyaluronan binding link protein gene family whose members are physically linked adjacent to chondroitin sulfate proteoglycan core protein genes - The missing links. *J. Biol. Chem.* **278**, 21083–21091.

19. Schwartz, N. B. and Domowicz, M. (2004) Proteoglycans in brain development. *Glycoconjugate J.* **21**, 329–341.

20. Barry, F. P., Neame, P. J., Sasse, J., and Pearson, D. (1994) Length variation in the keratan sulfate domain of mammalian aggrecan. *Matrix* **14**, 323–328.

21. Mort, J. S. and Roughley, P. J. (2004) Production of antibodies against degradative neoepitopes in aggrecan. *Methods Mol. Med.* **100**, 237–250.

Chapter 16

Heparan Sulfate Proteoglycans as Multifunctional Cell Regulators: Cell Surface Receptors

Jin-ping Li and Dorothe Spillmann

Abstract

Proteoglycans are macromolecules expressed on the cell surfaces and in the extracellular matrix of most animal tissues (Annu Rev Biochem 68:729–777, 1999; Int Rev Cell Mol Biol 276:105–159, 2009). Heparan sulfate proteoglycans (HSPGs) are essential for animal development and homeostasis, and are involved in various pathological processes. The functions of HSPGs are largely exerted through interaction of the heparan sulfate (HS) side chains with different types of ligands, including diverse molecules such as cytokines, enzymes, and pathogens. One of the important roles of cell surface HSPGs is to mediate cytokine-induced cell signaling through interaction with growth factors (GFs) and their cognate receptors. A selective dependence of GFs for different structural features of HS has been demonstrated by applying cell models that are mutated variously in HS structure due to deficiency in enzymes involved in the biosynthesis of HS chains.

Key words: Proteoglycan, Heparan sulfate, Heparin, Cytokine, Cell surface receptor, HS mutant cell lines

1. Introduction

1.1. Role of Heparan Sulfate as a Cellular Co-receptor

Heparan sulfate proteoglycans (HSPGs) are typically composed of a core protein to which a number of heparan sulfate (HS) polysaccharide chains are covalently attached (1, 2). The function of HSPGs is largely through interaction of the HS chains with a variety of extracellular ligands. Latter include components of the extracellular matrix, cell surface receptors and diverse cytokines, enzymes, proteases and their inhibitors, as well as microbial adhesion molecules. Thus, a number of GFs require HS as a co-receptor to interact with the corresponding GF receptors. It has been proposed that the HS chain participates directly in the assembly of a ternary signaling complex of GF, GF receptor, and HS, where HS may even act as a fine-tuner of the signaling process (3). Such a role of

Françoise Rédini (ed.), *Proteoglycans: Methods and Protocols*, Methods in Molecular Biology, vol. 836,
DOI 10.1007/978-1-61779-498-8_16, © Springer Science+Business Media, LLC 2012

HSPGs has been shown prototypically for the fibroblast GF (FGF) family (4). FGF2 (basic FGF) has been the first identified GF to depend on HS for binding to its high affinity receptor (5) followed by the majority of FGFs, most of which have been shown to make use of HS as co-receptors. Since then, an overwhelming number of studies have focused on HS interaction with different combinations of FGFs and FGFRs (6). Resolving the co-crystal structure of ternary complexes of FGF, FGF receptor in complex with heparin fragments has provided insights into the molecular interaction mechanisms (7, 8), yet the structural details that are required for successful interaction are still a matter of debate. A co-receptor role of HS [and occasionally also other glycosaminoglycans (GAGs)] has also been demonstrated for epidermal GF (EGF) (9), glial derived neurotrophic factor (GDNF) (10, 11), hepatocyte GF/ scatter factor (HGF) (12), platelet-derived GF (PDGF) (13), sonic hedgehog (Shh) (14), vascular endothelial GF (VEGF) (15, 16), and for the members of the transforming GF (TGF) superfamily (17), as well as for Wnt (18, 19).

1.2. Enzymes Involved in Biosynthesis of HS

Biosynthesis of HS chains is initiated by formation of a polysaccharide–protein linkage region, with a four-sugar unit (glucuronyl-galactosyl-galactosyl-xylose) attached to a serine residue in the core protein (see Fig. 1) (2). This tetrasaccharide sequence, shared by HS and chondroitin sulfate (CS), is extended by addition of either D-acetylglucosamine (GlcNAc) or D-acetylgalactosamine (GalNAc) and D-glucuronic acid (GlcA) residues, to form HS or CS polysaccharides, respectively. Parallel to the polymerization process, the repeating disaccharide units are modified by a series of reactions catalyzed by different enzymes (see Fig. 1). Gene-targeted elimination of the enzymes involved in biosynthesis of HS results in altered HS structures accompanied with a variety of phenotypes in mice, providing useful models for investigation of HS structure–function relationships, especially in relation to GF-mediated activities (2). Cells expressing no or aberrant HS structures, either derived from the mutant mice or by chemical mutagenesis, have been extensively employed for studying the functions of HSPG on cell surfaces. In the following we give a short background to selected enzymes involved in HS/CS biosynthesis and two enzymes active after completed HS biosynthesis.

UDP-xylose synthase (UXS) catalyzes the formation of UDP-xylose through decarboxylation of UDP-glucuronic acid, providing the essential substrate for initiation of HS and CS chain polymerization.

Xylosyltransferase (XYLT) transfers the UDP-xylose to serine residues in the core protein. Two isoforms of this enzyme have been identified in mouse and human, however, many cells, including wild-type Chinese hamster ovary (CHO) cells, seem to predominantly express XylT2 (20). It should be pointed out that in PGs found

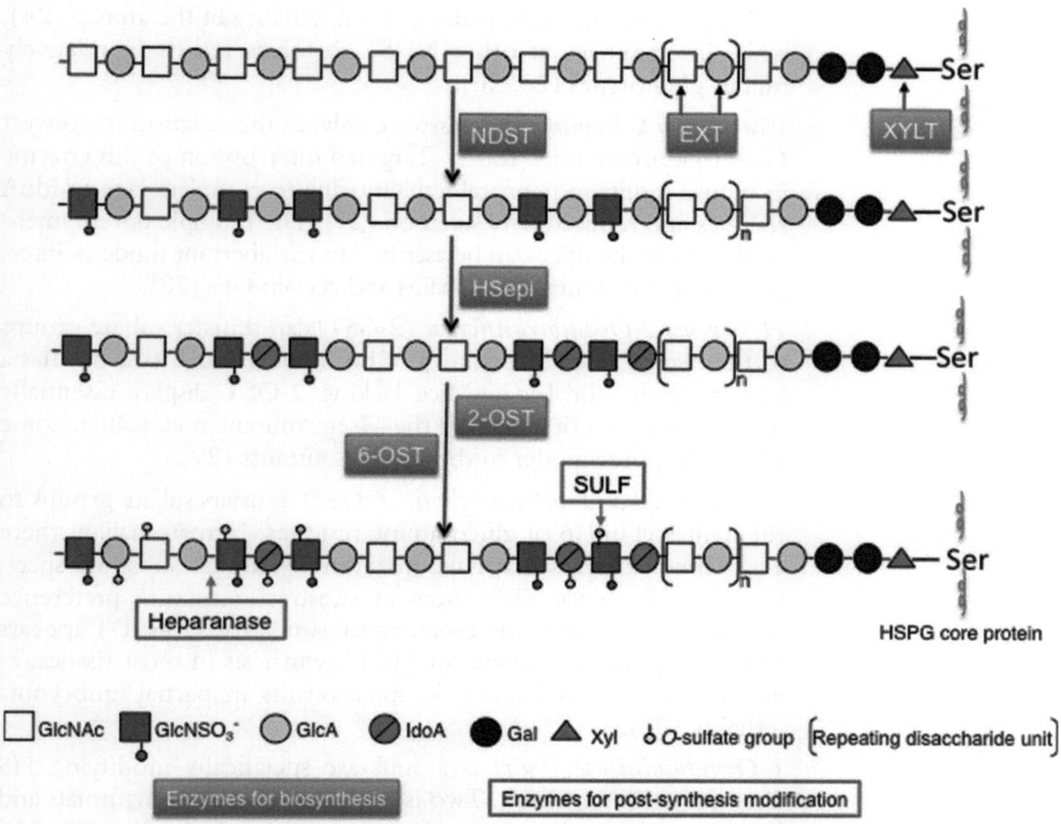

Fig. 1. A simplified illustration of the HS biosynthesis process. The functions of the enzymes described in the text (except UXS) are indicated.

in mammals, xylose is always the first sugar attached to the core protein, leading to synthesis of HS or CS polysaccharides (21). Therefore, inactivation of these two enzymes will abolish synthesis of both HS and CS chains.

HS polymerases (EXT1 and EXT2) are GlcNAc- and GlcA-transferases, responsible for elongation of HS chains by alternate addition of GlcNA and GlcA residues. Although each of the two enzymes has a dual activity, capable of transferring both UDP-GlcNAc and UDP-GlcA to the appropriate substrates as demonstrated by in vitro experiments, the respective functions of the two enzymes in vivo are still elusive. Elimination of either of the enzymes in mice causes a defect in HS production, resulting in early embryonic lethality, indicating that both enzymes are essential (22, 23).

Glucosaminyl N-sulfotransferase/N-deacetylase (NDST) is a family of four members expressed differentially in different tissues. This enzyme catalyzes the first chain modification, providing substrates for the following modification reactions. Among the four isoforms, NDST1 seems to play a dominant role because elimination of the

Ndst1 gene in mice results in neonatal lethality of the animal (24), while interruption of other NDST isoforms results in relatively milder phenotypes (25, 26).

Glucuronyl C5-epimerase (Hsepi) catalyzes the reaction to convert GlcA to iduronic acid (IdoA). Targeted interruption of this enzyme in mouse results in neonatal lethality, due to complete lack of IdoA residues and reduced 2-*O*-sulfation (27). The multiple developmental defects of the mice can be ascribed to the aberrant mode of interaction between mutant HS chains and certain GFs (28).

Hexuronyl 2-O-sulfotransferase (2-OST) can transfer sulfate groups to the hydroxyl in C2 position of both GlcA and IdoA, but has a higher affinity for IdoA. Mice lacking 2-OST display essentially overlapping phenotypes with the Hsepi mutant mice, albeit some of which appear milder in the 2-OST mutants (29).

Glucosaminyl 6-O-sulfotransferase (6-OST) transfers sulfate groups to the hydroxyl in C6 of glucosamine residues. In mammalian, there exist three isoforms that have qualitatively similar substrate specificities, with minor differences in substrate structure preference in vitro (30). Despite the biochemical similarity, HS6ST-1 appears to be the primary enzyme for HS biosynthesis in most tissues, as elimination of this enzyme in mice results in partial embryonic lethality (31).

6-O-endosulfatase (Sulf) is a sulfatase specifically modifying HS chains post-biosynthesis. Two isoforms are found in mammals and knockout of genes in mice results in a variety of phenotypes (32, 33). Expression of Sulf 2 is correlated with tumors, making the enzyme a potential target for cancer treatment (34).

Heparanase (Hpa) is an endoglucuronidase degrading HS. The enzyme activity is mostly detected in lysosomes and involved in HS catabolism; however, it is also associated with the cell membrane and secreted into the ECM where it degrades HS thus releasing bound ligands. Transgenic overexpression of heparanase in mouse causes extensive degradation of HS in organs (35, 36), while elimination of the heparanase gene results in nondegraded HS chains (37). Nonetheless, neither the Hpa-tg nor the Hpa-KO mice show any spontaneous overt defect, making these animals ideal models for different applications.

In the following sections we describe the use of cellular model systems with altered HS biosynthesis to address questions related to establishment of the role of cellular HS structures as co-receptor. We focus on the example of FGF activities that can stand as model for many other molecular interactions of this type. FGF-HS interactions can thus be considered prototypic for other types of cytokine-HS interactions, but also for related studies focusing on e.g., pathologic proteins such as amyloidotic peptides or microbial adhesion proteins from viruses, bacteria, or parasites.

2. Materials

2.1. Cellular Model Systems

In this section, we provide a short description of the diverse cell lines available with modifications in HS biosynthesis that are excellent tools to not only study both the role of HS as receptors for candidate ligands but also to establish consequences of structural alterations in HS for physiological and pathological functions.

1. *Mouse embryonic stem (ES)* cells are derived from the inner cell mass of the preimplantation blastocyst with the pluripotent property to differentiate into numerous cell types. This characteristic renders ES cells favorable to study early embryonic development. As HS is a mandatory co-receptor for many GFs and morphogens involved in embryonic development (38), existing mutant ES cells derived from mouse embryos with deficiency in HS biosynthesis enzymes are used to illustrate the functions of HS in cell differentiation.

   – *EXT1-ES* cells have been generated from embryos with a defect in *Ext1*. The mutant cell lacking HS retains the self-renewal potential but fails to transit from self-renewal to differentiation upon removal of leukemia inhibitory factor. This aberrant cell fate commitment is caused by defects in FGF signaling (39).

   – *NDST-ES* cells are derived from *Ndst1*, *Ndst2*, and the double mutant embryos. These cells, like the corresponding mice, produce HS with a reduced sulfation degree and are defective in in vitro differentiation. Using these cells it has been demonstrated that HS chains are critical co-receptors for signals-inducing ES cell differentiation (40).

   – *Hsepi-ES* cells have been derived from the *Glce* mutant mice and produce HS structures without IdoA resembling HS in the mutated mouse embryos (27).

2. *Mouse embryonic fibroblasts (MEFs)* are prepared from mouse embryos at day 12.5–14.5. Several MEF cell lines have been established from mice that are defect in the genes coding for enzymes involved in HS biosynthesis. Although the primary MEF cells can be used, like many other primary mammalian cells, the HS deficient MEFs have limited life span in cell culture. Therefore, most of the HS mutant MEF cells have been immortalized (41).

   – The *EXT-MEF* is defect in one of the HS polymerases (EXT1). This cell produces shortened HS chains (42) and accordingly shows a decreased FGF2-induced signaling and reduced ability to attach to collagen I (43). The cells are also resistant to infection by certain strains of alpha viruses, presumably due to hampered interaction capacity of the short cell surface HS chains with the virus (44).

- *NDST1-MEFs* are defect in NDST1 which is the prime isoform involved in HS synthesis. NDST1-MEF expresses defective HS with reduced *N*-sulfation that limits PDGF-BB-receptor interaction, resulting in impaired pericyte recruitment (45).
- The *Hsepi-MEF* is defect in hexuronyl C5-epimerase. The HS expressed in this cell line completely lacks IdoA residues which results in a reduced 2-*O*-sulfation, but upregulated *N*- and 6-*O*-sulfation. The cells display defects in response to FGF2, but not to FGF10 (28).
- The *2-OST-MEF* is defect in the only 2-*O*-sulfotransferase gene. The cells produce HS devoid of 2-*O*-sulfate, but with a normal content of IdoA. Like the Hsepi-MEF, 2-OST-MEF has an upregulated *N*- and 6-*O*-sulfation to maintain the overall charge density. Even though the apparent affinity of the HS for FGF1 and FGF2 is reduced, 2-OST-MEFs, differently from the Hsepi-MEFs, are able to mount an apparently normal signaling response to these GFs (46).
- The *6-OST-MEF* is defect in either 6-*O*-sulfotransferase 1 or 2. The double 6-OST deficient MEF produces HS with little 6-*O*-sulfate, but with increased 2-*O*-sulfation. The cells show reduced FGF1-, FGF2-, and FGF4-induced signaling (47).
- The *SULF-MEF* is defect in SULFs. The SULF-MEF cells display a dynamic influence of Sulf on the enzymes of the HS biosynthetic machinery, suggesting that the enzyme activity is involved in a potential cellular feedback mechanism, in which the SULFs regulate cellular signaling through editing HS sulfation (48).
- The *Hpa-KO-MEF* is defect in heparanase. The cells do not show an obvious phenotype, although the HS chains isolated from the cells are substantially longer than in the corresponding wild-type MEF (unpublished).

3. *CHO cells* have been isolated from the ovary of Chinese hamsters and form adherent monolayers. These cells have been broadly used in experiments relating to gene expression, genetics, toxicity screening, and nutritional studies. By a chemical mutagenesis approach with ethyl methane sulfonate treatment, clones defect in synthesis of HS or both HS and CS have been selected by screening of ^{35}S incorporation (49, 50). The mutant clones were then identified through analysis of their GAGs, leading to well-characterized cell lines with specified defects in GAG biosynthesis. These cells have been extensively used in studies to define the functions of cell surface HS with regard to interaction with diverse protein ligands. Selected cell lines that are defect in the HS biosynthetic enzymes are briefly described in the following.

- *CHO-pgsI-208* is deficient in the UDP-xylose synthase (UXS) (51). Cells express neither HS nor CS as the lack of UXS prevents initiation of HS/CS synthesis.
- *CHO-pgsA-745* is deficient in the xylosyltransferase (XylT2) that transfers UDP-xylose to the serine residue in core proteins (49). As the other mammalian xylosyltransferase (XylT1) seems to be silenced in these cells (20), this cell line, like the CHO-pgsI-208, fails to produce HS and CS.
- *CHO-pgsD-677* is deficient in EXT1, one of the polymerases responsible for elongation of HS chains. The cell fails to produce HS, but compensates with an increased production of CS (52).
- *CHO-pgsE-606* is partially deficient in NDST, resulting in reduced *N*-sulfation of HS chains (53).
- *CHO-pgsF-17* is deficient in hexuronyl 2-*O*-sulfotransferase (2-OST). This cell produces HS that lacks 2-*O*-sulfate, but contains a higher proportion of 6-*O*-sulfation. The cells fail to interact with FGF2 (54).

4. For cell culture, prepare basic medium corresponding to the cell line that shall be applied: For ES cultivation prepare Dulbecco's or Glasgow α-minimal essential medium (MEM) supplemented with 10–20% serum (see Note 1), 1,000 IU/mL leukemia inhibitory factor (LIF) (Chemicon), L-glutamine, 100 units/mL penicillin, 100 μg/mL streptomycin (see Note 2), and 0.1 mM β-mercaptoethanol. For MEF cells prepare Dulbecco's MEM supplemented with 10% fetal bovine serum, 2 mM glutamine, 100 units/mL penicillin, and 100 μg/mL streptomycin (see Note 2). For CHO cell culture prepare Ham's F-12 medium supplemented with 2 mM glutamine and 7–10% fetal bovine serum. Antibiotics may be used as above (see Note 2). For cell splitting a 10× Trypsin/EDTA-solution may be purchased. Cell culture reagents are supplied by Invitrogen. Culture dishes (use 10-cm dishes or 75-mL flasks for routine culture and 3-cm, 6-cm dishes and 6-well plates for different binding and cell stimulation assays) are purchased from Nunc.

2.2. Ligand Binding

1. The ligand, e.g., FGF2, is purchased from Preprotech. FGF may be labeled in different ways (see Note 3).
2. For testing ligand binding to cell surface prepare following medium: A serum-free variant of the medium suitable for the chosen cell type.
3. Heparin lyases can be purchased from different companies (see Note 4). Prepare heparin lyase buffer (10 mM Hepes pH 7.2, 150 mM NaCl, 2 mM $CaCl_2$, 0.1% BSA) to use in cell culture and 0.22-μm sterile filter before use.

2.3. Cell Stimulation

1. The ligand source (e.g., FGF2) is chosen considering the points described in Subheading 2.2. For detection of cell stimulation purchase suitable antibodies, e.g., polyclonal rabbit anti-pERK1/2 (Thr202/Thr204) primary antibody, rabbit anti-total ERK1/2 antibody, and horseradish peroxidase-conjugated secondary goat anti-rabbit antibody (Cell Signaling Technology, Inc.).
2. For cell stimulation prepare following buffers and medium: Phosphate buffered saline (PBS) according to your lab's formulation (Ca-Mg-free). As starvation medium, prepare a serum-free variant of the cells' normal medium supplemented with 1% bovine serum albumin (BSA). Stock solution of BSA is sterile filtered before addition to the medium.
3. For extraction of cells prepare a sample buffer (0.2 M Tris, pH 8, 1% SDS, 0.5 M sucrose, 5 mM EDTA, 0.2 mM Na_3VO_4, 10 μg/mL pepstatin). Add protease and phosphatase inhibitors just before use from aliquoted stock solution kept at −20°C.
4. Prepare transfer buffer (25 mM Tris, pH 8.3, 192 mM glycine, 20% methanol) and following Western blot buffers: ECL-compatible blocking buffer is prepared of 50 mM Tris, pH 7.4, 150 mM NaCl, 0.1% Tween 20 containing 5% (w/v) "ECL Advance blocking agent" (GE Healthcare Biosciences). A light version of above buffer with only 2% blocking agent is prepared for antibody incubations. Use the same buffer without any blocking agent for washing purpose between incubations with different antibodies.
5. Purchase or prepare a 12% SDS-polyacrylamide gel, purchase Immobilon-P (Millipore) or equivalent membrane (see Note 5). Purchase BCA protein assay solution (Pierce), ECL Plus Western blotting detection kit (GE Healthcare Biosciences), and Fuji Super RX X-ray film.
6. Additionally you need following items: rubber policeman, ultrasonication bath, and gel electrophoresis system (e.g., Minigel system from BioRad).

2.4. Cell Proliferation

1. Purchase [^{3}H-Methyl]thymidine, specific activity ~25 Ci/mmol, and Optiphase "Hisafe 3" scintillation cocktail (both from PerkinElmer).
2. Purchase scintillation vials suitable for β-scintillation counter (Beckman).
3. Prepare freshly a cell solubilization solution of final 0.5 M NaOH, 0.5% SDS, and 5% trichloroacetic acid.

3. Methods

In this chapter we describe how an existing or newly produced cell line may be used to establish, whether HS chains serve as co-receptors for a ligand of interest, and explain with the example of FGF how binding to HS as co-receptors may affect cellular signaling events.

3.1. Cell Culture

ES and MEF cells defect in HS production are routinely cultured under the same conditions as the corresponding wild-type cells. The mutant MEF cells are cultured in the same medium, but often grow slower at the initial phase after thawing (see Note 6). CHO cells are commonly grown in Ham's F-12 medium (55). Cell culture is performed at 37°C in 5% carbon dioxide. Medium is changed adequately (daily to every second to third day) and cells are passaged before confluence by using 1× Trypsin/EDTA diluted in sterile PBS to detach them from culture dishes.

3.2. Testing Ligand Binding to Cell Surface HS

1. Seed wild-type and HS mutant cells at convenience in a 3-cm dish or a 6-well plate and let grow to confluence in normal medium (see Note 6).
2. Wash confluent cell layer carefully (see Note 7) with serum-free medium to remove floating cells and cell debris.
3. Add ligand of interest, e.g., labeled FGF (see Note 3), in a minimal volume to cover the cell monolayer and let incubate for 30 min at room temperature (see Note 8). A parallel sample is treated as control either with an excess of unlabeled ligand protein or free tag (see Note 9).
4. Following the incubation period rinse the cell layer again carefully with serum-free medium to remove unbound ligand and collect cell layer for measuring bound ligand in a suitable system (see Note 3).
5. A positive outcome, i.e., a difference in binding between the wild-type cell and an HS mutant cell may be confirmed by treatment of the wild-type cell with heparin lyases to remove surface exposed HS chains on the wild-type cell. Treat cells with 5–10 mIU/mL heparin lyase mixture (see Note 10) in heparin lyase buffer in a minimal volume to cover the cell layer for 1–2 h at 37°C (see Note 11).
6. Carefully wash cell layer with serum-free medium to remove digestion products before addition of the ligand and analysis as described in step 3.

3.3. FGF2-Stimulated Cell Signaling

1. Seed wild-type and HS mutant cells (250×10^3) in 3-cm dishes or 6-well plates (see Note 12) in complete medium as defined for the cell type. After 24 h, remove the medium and wash the cells once with plain medium or PBS before changing to

starvation medium. Let cells incubate in the starvation medium for 3–24 h (see Note 13).

2. Add carefully FGF2 to a final concentration of 10 ng/mL and stimulate the cells for suitable time periods (see Note 14).
3. Carefully remove the medium, wash cell layer twice with PBS, add 130 μL of boiling sample buffer (see Note 15) and scrap off cells with a rubber policeman.
4. Collect the sample into an Eppendorf tube, heat for 4 min at 96°C, followed by ultrasonication for 6 s before centrifugation for 15 min at 13,000 × g at 4°C. Collect supernatant and determine the protein concentration in an aliquot of the supernatant.
5. Equal amounts of total protein (20 μg convenient for minigel system) are then separated by SDS-PAGE and subsequently transferred to an Immobilon-P membrane by overnight transfer at 4°C in transfer buffer.
6. Block transfer membrane with blocking buffer and add primary rabbit anti-p-ERK antibody (see Note 16) diluted 1:1,000 in above buffer containing 2% blocking agent and incubate for 1–2 h at room temperature with gentle shaking.
7. Wash the membrane 4 times with above buffer without blocking agent, before adding 1:2,000 diluted horseradish peroxidase-conjugated secondary antibody and incubate for 1 h at room temperature (see Note 17).
8. Wash the membrane as in step 3 before development by an ECL detection kit as described by the manufacturer and expose the membrane to X-ray film (see Note 18).

3.4. FGF2-Stimulated Cell Proliferation

1. Seed wild-type and HS mutant cells on a 24-well plate at a density of 1.5×10^4 cells/well in complete medium and culture for 24 h. Change to serum-free medium and culture for another 24–48 h (consider also Note 6).
2. Change the cells to fresh serum-free medium and add FGF2 at suitable dose (see Note 19). As positive and negative control, respectively, keep cells in medium with or without 10% FCS complement. Culture cells for 20 h (see Note 20).
3. Add [^{3}H]thymidine (see Note 21) to the cell (final concentration 1 μCi/mL) and culture for 4 h.
4. Aspirate medium from each well (see Note 22). Wash the cells three times with PBS and discard washing buffer. Add 2 mL of cold, freshly made 5% trichloroacetic acid and incubate for 30 min on ice. Aspirate carefully TCA solution and discard. Wash once with PBS at room temperature and add 2 mL of a freshly prepared 0.5 M NaOH, 0.5% SDS solution. Incubate 5 min at room temperature. Pipette cell suspension up and down and add mixture directly to scintillation vial.

5. Incorporation of [^{3}H]thymidine is determined by addition of two volumes of scintillation cocktail to the lysate and scintillation counting in a β-counter (Beckman).

3.5. Effect of Exogenous Soluble GAGs on Cell Function

1. Rescue of HS function in HS-deficient cells. To confirm that an absent response in an HS-deficient cell upon stimulation with a ligand is indeed caused by the lack of HS, a rescuing experiment may be performed. Exogenously added HS chains or oligosaccharides may thus replace missing cell surface HS chains and provide assistance/facilitated interaction between a soluble ligand and its cognate surface receptor in the mutant cell line with rescued cellular signaling. To perform a rescue experiment, in principle, the same types of cellular assays as described in Chapter 3, Subheadings 3.1 and 3.2 are applied with mutant cell lines by supplying ligand (e.g., FGF) in combination with purified HS chains or defined oligosaccharide pools (see Note 23).
2. Screening of GAG competitors. In analogous experimental setup different types of HS, heparin, or other GAGs may be tested for interaction and potential competition of ligand binding to endogenous HS-based receptors. In this case reduced binding of the ligand of interest is measured by adding increasing concentrations of GAGs. On wild-type cells a potential inhibitor shall compete out ligand binding to cell surface HS, whereas on an HS mutant cell line the same GAG should be able to rescue ligand binding at a lower concentration as described in step 1 while competing at a higher concentration.

3.6. Conclusions

Assays described above could easily be adapted to different types of ligand molecules that may be dependent on HS (or other types of GAGs) for their activities. Identification of a potential interaction could thus be envisaged as starting point to elucidate molecular mechanisms of physiological phenomena or for the development of a treatment strategy, e.g., for microbial intruders.

4. Notes

1. The serum concentration is dependent on the cell type. The serum shall be of high quality preferably "ES cell qualified" (Invitrogen) or completely replaced by a "knockout serum replacement" (Invitrogen). Depending on the degree of serum replacement, add nonessential amino acids (Invitrogen).
2. Consider the use of antibiotics carefully. Antibiotic-free solutions are preferable whenever cells can easily be retrieved or cured in case a contamination problem arises.

3. Depending on the molecule of interest a number of different tagging methods are available (also commercially) and will determine the exact mode of detection. Classically, radioiodination with ^{125}I (detection by γ-counting) has been used to label protein ligands, but may also be replaced with small biomolecular tags (such as biotin—detection by a streptavidin or anti-biotin antibody-based detection system) or fluorescent molecules (e.g., FITC or fluorescent protein domains—detection by fluorescence microscopy or spectrometry) (see also (56)). No matter which tagging method is used, one has to bear in mind not to modify potential HS binding domains in the ligand of interest (often domains/clusters of basic amino acids). These domains may be protected during the labeling process by e.g., heparin or by choosing linkage methods not making use of amino functions in the protein. If heparin is used to protect the sample it is very important that the protein shall be purified from the protective chain by dissociation in high salt and removal of the carbohydrate.
4. Heparin lyases (also described as heparinases and heparitinases) are bacterial endoglycosidases from *Flavobacterium heparinum* that cleave HS chains with different substrate specificity into disaccharides containing an unsaturated uronic acid ring at the nonreducing end of the cleavage site. These enzymes can be purchased from several suppliers (IBEX, Iduron, or Sigma-Aldrich). Most products are sold defined with the international unit that describes the release of 1 μmol of disaccharides per minute at standard conditions, while Sigma-Aldrich uses a Sigma unit as release of 0.1 μmol of disaccharides per hour. The conversion between these two types of units is thus 1 IU = 600 ΣU.
5. Make sure the membrane you apply is compatible with the molecule and detection method chosen.
6. Many cells deficient in HS biosynthesis grow slower and adhere worse to culture dishes, especially in initial phase. Bear therefore in mind to adapt culture conditions (especially timing) in experiments when different cell lines are to be compared. Mutant cells may need to be seeded at higher concentration and allowed to settle longer before an experiment is initiated. To gain comparable values for resulting data it is highly recommended to relate findings to cell number or protein level at the end point of each experiment.
7. As the HS mutants have a tendency to adhere worse to culture dishes due to the engagement of HS in matrix interactions the washing process must be performed extremely smoothly avoiding any shear forces during rinsing.

8. Depending on the ligand of interest live cells may take up the molecule. The incubation period as well as incubation temperature (to slow down cellular metabolism) will have to be established carefully in each occasion. Alternatively, fixation of the cell layer is an option. In this case, avoid cross-linking agents (especially containing aldehydes) that make use of hydroxylfunctions or carboxy-groups in the carbohydrates to avoid cross-linking and blocking of GAG structures.

9. Unspecific binding of a tagged ligand needs to be controlled by either use of an excess of non-labeled ligand or saturation of tag-binding sites on the cell surface by free tag molecules.

10. Use a heparin lyase mixture of isoform II and III (heparitinases) for removal of HS exposed at the cell surface (see also Note 4).

11. The success of removal of HS from the cell surface may be followed by either following the disappearance of HS-related antibody epitopes or appearance of the stub of HS chains with the terminal unsaturate uronic acid ring structure remaining after endoglycolytic digestion and recognized by monoclonal antibody 3G10 (Seikagaku).

12. In case multiwell dishes are used, it is advisable to combine parallel samples with the same incubation period in one multiwell dish for simultaneous harvest.

13. To get a good signal, starvation time and ligand doses need to be tested for individual ligand and cell systems. Most importantly, conditions must be consistent between treatment series that shall be compared.

14. FGF2 is kept as aliquots at a concentration of 0.1–1 mg/mL in 10 mM Tris pH 8.5 at −80°C. Do not freeze and thaw! It is of crucial importance that the ligand is free of heparin contamination when studying the role of HS for FGF-induced signaling, i.e., the recombinant protein must not have been purified by heparin affinity chromatography (which may bleed heparin into the protein eluate) or stabilized by the addition of heparin. The exact GF concentration required depends on quality of GF and cell type. Perform a pilot experiment titrating GF addition vs. effect and use the same concentration for following experiments.

15. As some of the intracellular effector molecules stimulated by binding of ligand to cells may translocate into the nucleus it is important to use a powerful lysis buffer to release also nuclear contents. Boiling buffer will improve the lysis and help in inactivating proteases.

16. When same membranes are tested with different antibodies, always use the antibody producing weakest signals first. As equal

loading control detect either an independent housekeeping protein such as actin or glyceraldehyde-3-phosphate dehydrogenase using a corresponding antibody or strip membranes with 0.1 M glycine/HCl, pH 2.5 and reprobe with anti-total ERK1/2 antibody (1:1000). Test success of stripping conditions by performing a "dummy" development!

17. Keep membranes and antibody solutions protected from light all through steps 7 and 8.
18. Expose membrane for different time-points (timing dependent on signal intensity) to catch a non-saturated signal on the X-ray film.
19. The optimal dose of FGF is governed by the same aspects as discussed in Notes 13 and 14. A convenient dose to start with for many cell systems is 5–10 ng/mL of FGF2.
20. Medium including FCS serves as positive control for maximal stimulation of the cells, while medium without FCS and FGF provides a control for baseline level of cell activation.
21. For use of radioactive isotope consult local routines. Tritium is a weak β-emitter and the sample solution must be complemented with scintillation cocktail in order to detect decay in a β-counter.
22. Remember that all fractions of these cultures are radioactively contaminated. Handle them therefore according to local routines for radioactive waste.
23. It is important to titrate the dose of exogenous GAGs added carefully. As a rule of thumb, test a range of 1:10–10:1 molar ratios between ligand and exogenous saccharide. It should be pinpointed that several protein binding sites can occur per single HS chain that may result in an underestimation of HS molarity. Size-defined heparin oligosaccharides (commercially available e.g., from Iduron) can therefore be recommended as prime alternative for rescuing studies also allowing for the possibility to identify the minimal size of oligosaccharide permitting functional interaction.

Acknowledgment

The authors are supported by grants from the Swedish Research Council (K2009-67X-21128-01-3), the Swedish Cancer Foundation (09 0717 and 09 0762), and Polysackaridforskning AB (Uppsala).

References

1. Bernfield, M., Götte, M., Park, P. W., Reizes, O., Fitzgerald, M. L., and Lincecum, J., and Zako, M. (1999) Functions of cell surface heparan sulfate proteoglycans. *Annu Rev Biochem* **68**, 729–777.
2. Lindahl, U. and Li, J.-P. (2009) Interactions between heparan sulfate and proteins-design and functional implications. *Int Rev Cell Mol Biol* **276**, 105–159.
3. Jastrebova, N., Vanwildemeersch, M., Lindahl, U., and Spillmann, D. (2010) Heparan sulfate domain organization and sulfation modulate FGF2 induced cell signaling. *J Biol Chem* **27**, 26842–26851.
4. Rapraeger, A. C. (1995) In the clutches of proteoglycans: how does heparan sulfate regulate FGF binding? *Chem Biol.* **2**, 645–649.
5. Yayon, A., Klagsbrun, M., Esko, J. D., et al (1991) Cell surface, heparin-like molecules are required for binding of basic fibroblast growth factor to its high affinity receptor. *Cell* **64**, 841–848.
6. Mohammadi, M., Olsen, S. K., and Ibrahimi, O. A. (2005) Structural basis for fibroblast growth factor receptor activation. *Cytokine Growth Factor Rev* **16**, 107–137.
7. Schlessinger, J., Plotnikov, A. N., Ibrahimi, O. A., Eliseenkova, A. V., Yeh, B. K., Yayon, A., et al (2000) Crystal structure of a ternary FGF-FGFR-heparin complex reveals a dual role for heparin in FGFR binding and dimerization. *Mol Cell* **6**, 743–750.
8. Pellegrini, L., Burke, D. F., Von Delft, F., Mulloy, B., and Blundell, T. L. (2000) Crystal structure of fibroblast growth factor receptor ectodomain bound to ligand and heparin. *Nature* **407**, 1029–1034.
9. Takazaki, R., Shishido, Y., Iwamoto, R., and Mekada, E. (2004) Suppression of the biological activities of the epidermal growth factor (EGF)-like domain by the heparin-binding domain of heparin-binding EGF-like Growth Factor. *J Biol Chem.* **279**, 47335–47343.
10. Barnett, M. W., Fisher, C. E., Perona-Wright, G., and Davies, J. A. (2002) Signalling by glial cell line-derived neurotrophic factor (GDNF) requires heparan sulphate glycosaminoglycan. *J Cell Sci.* **115**, 4495–4503.
11. Davies, J. A., Yates, E. A., and Turnbull, J. E. (2003) Structural determinants of heparan sulphate modulation of GDNF signalling. *Growth Factors* **21**, 109–119.
12. Kemp, L. E., Mulloy, B., and Gherardi, E. (2006) Signalling by HGF/SF and Met: the role of heparan sulphate co-receptors. *Biochem Soc Trans.* **34**, 414–417.
13. Lustig, F., Hoebeke, J., Östergren-Lunden, G., Velge-Roussel, F., Bondjers, G., Olsson, U., et al (1996) Alternative splicing determines the binding of platelet-derived growth factor (PDGF-AA) to glycosaminoglycans. *Biochemistry* **17**, 12077–12085.
14. Carrasco, H., Olivares, G. H., Faunes, F., Oliva, C., and Larraín, J. (2005) Heparan sulfate proteoglycans exert positive and negative effects in Shh activity. *J Cell Biochem* **96**, 831–838.
15. Ito, N. and Claesson-Welsh, L. (1999) Dual effects of heparin on VEGF binding to VEGF receptor-1 and transduction of biological responses. *Angiogenesis* **3**, 159–166.
16. Selleck, S. B. (2006) Signaling from across the way: transactivation of VEGF receptors by HSPGs. *Mol Cell* **22**, 431–432.
17. Rider, C. C. (2006) Heparin/heparan sulphate binding in the TGF-beta cytokine superfamily. *Biochem Soc Trans* **34**, 458–460.
18. Tsuda, M., Kamimura, K., Nakato, H., Archer, M., Staatz, W., Fox, B., Humphrey, M., et al (1999) The cell-surface proteoglycan Dally regulates Wingless signalling in Drosophila. *Nature* **400**, 276–280.
19. Kikuchi, A., Yamamoto, H., and Sato, A. (2009) Selective activation mechanisms of Wnt signaling pathways. *Trends Cell Biol* **19**, 119–129.
20. Cuellar, K., Chuong, H., Hubbell, S. M., and Hinsdale, M. E. (2007) Biosynthesis of chondroitin and heparan sulfate in Chinese hamster ovary cells depends on xylosyltransferase II. *J Biol Chem.* **282**, 5195–5200.
21. Fransson, L-Å., Silverberg, I., and Carlstedt, I. (1985) Structure of the heparan sulfate-protein linkage region. Demonstration of the sequence galactosyl-galactosyl-xylose-2-phosphate. *J Biol Chem.* **260**, 14722–14726.
22. Lin, X., Wei, G., Shi, Z., Dryer, L., Esko, J. D., Wells, D. E., and Matzuk, M. M. (2000) Disruption of gastrulation and heparan sulfate biosynthesis in EXT1- deficient mice. *Dev Biol.* **224**, 299–311.
23. Stickens, D., Zak, B. M., Rougier, N., Esko, J. D., and Werb, Z. (2005) Mice deficient in Ext2 lack heparan sulfate and develop exostoses. *Development* **132**, 5055–5068.

24. Ringvall, M., Ledin, J., Holmborn, K., van Kuppevelt, T., Ellin, F., Eriksson, I., et al (2000) Defective heparan sulfate biosynthesis and neonatal lethality in mice lacking *N*-deacetylase/*N*-sulfotransferase-1. *J Biol Chem.* **275**, 25926–25930.
25. Forsberg, E., Pejler, G., Ringvall, M., Lunderius, C., Tomasini-Johansson, B., Kusche-Gullberg, M., et al (1999) Abnormal mast cells in mice deficient in a heparin-synthesizing enzyme. *Nature* **400**, 773–776.
26. Pallerla, S. R., Lawrence, R., Lewejohann, L., Pan, Y., Fischer, T., Schlomann, U., et al (2008) Altered heparan sulfate structure in mice with deleted NDST3 gene function. *J Biol Chem* **283**, 16885–16894.
27. Li, J. P., Gong, F., Hagner-Mcwhirter, Å., Forsberg, E., Abrink, M., Kisilevsky, R., et al (2003) Targeted disruption of murine glucuronyl C5-epimerase gene results in heparan sulfate lacking *L*-iduronic acid and in neonatal lethality. *J Biol Chem.* **278**, 28363–28366.
28. Jia, J., Maccarana, M., Zhang, X., Bespaļov, M., Lindahl, U., and Li, J. P. (2009) Lack of *L*-iduronic acid in heparan sulfate affects interaction with growth factors and cell signaling. *J Biol Chem.* **284**, 15942–15950.
29. Bullock, S. L., Fletcher, J. M., Beddington, R. S., and Wilson, V. A. (1998) Renal agenesis in mice homozygous for a gene trap mutation in the gene encoding heparan sulfate 2-sulfotransferase. *Genes Dev.* **12**, 1894–1906.
30. Smeds, E., Habuchi, H., Do, A. T., Habuchi, H., Kimata, K., Lindahl, U., and Kusche-Gullberg, M. (2003) Substrate specificities of mouse heparan sulphate glucosaminyl 6-*O*-sulphotransferases. *Biochem J.* **372**: 371–380.
31. Habuchi, H., Nagai, N., Sugaya, N., Atsumi, F., Stevens, R. L., and Kimata, K. (2007) Mice deficient in heparan sulfate 6-*O*-sulfotransferase-1 exhibit defective heparan sulfate biosynthesis, abnormal placentation, and late embryonic lethality. *J Biol Chem.* **282**, 15578–15588.
32. Ai, X., Kitazawa, T., Do, A. T., Kusche-Gullberg, M., Labosky, P. A., and Emerson, C. P. Jr. (2007) SULF1 and SULF2 regulate heparan sulfate-mediated GDNF signaling for esophageal innervation. *Development* **134**, 3327–3338.
33. Lum, D. H., Tan, J., Rosen, S. D., and Werb, Z. (2007) Gene trap disruption of the mouse heparan sulfate 6-*O*-endosulfatase gene, Sulf2. *Mol Cell Biol.* **27**, 678–688.
34. Rosen, S. D. and Lemjabbar-Alaoui, H. (2010) Sulf-2: an extracellular modulator of cell signaling and a cancer target candidate. *Expert Opin Ther Targets* **14**, 935–949.
35. Zcharia, E., Metzger, S., Chajek-Shaul, T., Aingorn, H., Elkin, M., Friedmann, Y., et al (2004) Transgenic expression of mammalian heparanase uncovers physiological functions of heparan sulfate in tissue morphogenesis, vascularization, and feeding behavior. *FASEB J* **18**, 252–263.
36. Escobar Galvis, M. L., Jia, J., Zhang, X., Jastrebova, N., Spillmann, D., Gottfridsson, E., et al (2007) Transgenic or tumor-induced expression of heparanase upregulates sulfation of heparan sulfate. *Nat Chem Biol* **3**, 773–778.
37. Zcharia, E., Jia, J., Zhang, X., Baraz, L., Lindahl, U., Peretz, T., et al (2009) Newly generated heparanase knock-out mice unravel co-regulation of heparanase and matrix metalloproteinases. *PLoS ONE* **4**, e5181.
38. Baldwin, R. J., Ten Dam, G. B., Van Kuppevelt, T. H., et al (2008) A developmentally regulated heparan sulfate epitope defines a subpopulation with increased blood potential during mesodermal differentiation. *Stem Cells* **26**, 3108–3118.
39. Kraushaar, D. C., Yamaguchi, Y., and Wang, L. (2010) Heparan sulfate is required for embryonic stem cells to exit from self-renewal. *J Biol Chem.* **285**: 5907–5916.
40. Lanner, F., Lee, K. L., Sohl, M., Holmborn, K., Yang, H., Wilbertz, J., et al (2010) Heparan sulfation-dependent fibroblast growth factor signaling maintains embryonic stem cells primed for differentiation in a heterogeneous state. *Stem Cells* **28**, 191–200.
41. Sun, H. and Taneja, R. (2007) Analysis of transformation and tumorigenicity using mouse embryonic fibroblast cells. *Methods Mol Biol.* **383**, 303–310.
42. Yamada, S., Busse, M., Ueno, M., Kelly, O. G., Skarnes, W. C., Sugahara, K., and Kusche-Gullberg, M. (2004) Embryonic fibroblasts with a gene trap mutation in Ext1 produce short heparan sulfate chains. *J Biol Chem.* **279**, 32134–32141.
43. Osterholm, C., Barczyk, M. M., Busse, M., Grønning, M., Reed, R. K., and Kusche-Gullberg, M. (2009) Mutation in the heparan sulfate biosynthesis enzyme EXT1 influences growth factor signaling and fibroblast interactions with the extracellular matrix. *J Biol Chem.* **284**, 34935–34943.
44. Zhu, W., Wang, L., Yang, Y., Jia, J., Fu, S., Feng, Y., et al (2010) Interaction of E2 glycoprotein with heparan sulfate is crucial for cellular infection of Sindbis virus. *PLoS ONE* **5**, e9656.

45. Abramsson, A., Kurup, S., Busse, M., Yamada, S., Lindblom, P., Schallmeiner, E., et al (2007) Defective *N*-sulfation of heparan sulfate proteoglycans limits PDGF-BB binding and pericyte recruitment in vascular development. *Genes Dev.* **21**, 316–331.

46. Merry, C. L. R., Bullock, S. L., Swan, D. C., Backen, A. C., Lyon, M., Beddington, R. S., et al (2001) The molecular phenotype of heparan sulfate in the Hs2st–/– mutant mouse. *J Biol Chem.* **276**, 35429–35434.

47. Sugaya, N., Habuchi, H., Nagai, N., Ashikari-Hada, S., and Kimata, K. (2008) 6-*O*-sulfation of heparan sulfate differentially regulates various fibroblast growth factor-dependent signalings in culture. *J Biol Chem.* **283**, 10366–10376.

48. Lamanna, W. C., Frese, M. A., Balleininger, M., and Dierks, T. (2008) Sulf loss influences *N*-, 2-*O*-, and 6-*O*-sulfation of multiple heparan sulfate proteoglycans and modulates fibroblast growth factor signaling. *J Biol Chem.* **283**, 27724–27735.

49. Esko, J. D., Stewart, T. E., and Taylor, W. H. (1985) Animal cell mutants defective in glycosaminoglycan biosynthesis. *Proc Natl Acad Sci USA* **82**, 3197–3201.

50. Bai, X., Crawford, B., and Esko, J. D. (2001) Selection of glycosaminoglycan-deficient mutants. *Methods Mol Biol.* **171**, 309–316.

51. Bakker, H., Oka, T., Ashikov, A., Yadav, A., Berger, M., Rana, N. A., et al (2009) Functional UDP-xylose transport across the endoplasmic reticulum/Golgi membrane in a Chinese hamster ovary cell mutant defective in UDP-xylose synthase. *J Biol Chem.* **284**, 2576–2583.

52. Lidholt, K., Weinke, J. L., Kiser, C. S., Lugemwa, F. N., Bame, K. J., Cheifetz, S., et al (1992) A single mutation affects both *N*-acetylglucosaminyltransferase and glucuronosyltransferase activities in a Chinese hamster ovary cell mutant defective in heparan sulfate biosynthesis. *Proc Natl Acad Sci USA* **89**, 2267–2271.

53. Bame, K. J. and Esko, J. D. (1989) Undersulfated heparan sulfate in a Chinese hamster ovary cell mutant defective in heparan sulfate *N*-sulfotransferase. *J Biol Chem.* **264**, 8059–8065.

54. Bai, X. and Esko, J. D. (1996) An animal cell mutant defective in heparan sulfate hexuronic acid 2-*O*-sulfation. *J Biol Chem.* **271**, 17711–17717.

55. Bai, X., Crawford, B., and Esko, J. D. (2001) Selection of glycosaminoglycan-deficient mutants. *Methods Mol Biol.* **171**, 309–316.

56. Friedl, A., Filla, M., and Rapraeger, A. C. (2001) Tissue-specific binding by FGF and FGF receptors to endogenous heparan sulfates. *Methods Mol Biol.* **171**, 535–546.

Part IV

Proteoglycans Involvement in Pathophysiology

Chapter 17

Models for Studies of Proteoglycans in Kidney Pathophysiology

Scott J. Harvey

Abstract

Proteoglycans (PGs) impact many aspects of kidney health and disease. Models that permit genetic dissection of PG core protein and glycosaminoglycan (GAG) function have been instrumental to understanding their roles in the kidney. Matrix-associated PGs do not serve critical structural roles in the organ, nor do they contribute significantly to the glomerular barrier under normal conditions, but their abnormal expression influences fibrosis, inflammation, and progression of kidney disease. Most core proteins are dispensable for nephrogenesis (glypican-3 being an exception) and for maintenance of function in adult life, but their loss alters susceptibility to experimental kidney injury. In contrast, kidney development is exquisitely sensitive to GAG expression and fine structure as evidenced by the severe phenotypes of mutants for genes involved in GAG biosynthesis. This article reviews PG expression in normal kidney and the abnormalities caused by their disruption in mice and man.

Key words: Heparan sulfate, Nephrogenesis, Glomerulus, Podocyte

1. Introduction

Proteoglycans (PGs) consist of a core protein with covalently bound glycosaminoglycan (GAG) side chains formed of heparan sulfate (HS), chondroitin sulfate (CS), dermatan sulfate (DS), or keratan sulfate (KS). Whether secreted or decorating the cell surface, PGs serve important structural and functional roles in the kidney by influencing cell–matrix and cell–cell interactions and by modulating growth factor activity. PGs are required for kidney development, as revealed by the defects arising from their chemical or enzymatic disruption, and in genetic knockout models lacking specific core proteins or biosynthetic enzymes. In adult kidney, PGs on the surface of glomerular endothelial and epithelial cells or incorporated within the glomerular basement membrane (GBM) are thought to

Françoise Rédini (ed.), *Proteoglycans: Methods and Protocols*, Methods in Molecular Biology, vol. 836,
DOI 10.1007/978-1-61779-498-8_17,

Table 1
Proteoglycan core proteins in kidney disease

Core protein	Expression pattern	Mouse kidney phenotype	Human kidney disease
Agrin	GBM, CD, SMC (C-terminus); all BMs (N-terminus)	GBM charge defect in podocyte-specific KO mice, kidney function normal (13)	Diminished in DN (14), alloantibodies in transplant glomerulopathy (17)
Perlecan	Ubiquitous, GBM very weak	No kidney defect reported in *Hspg2*$^{-/-}$ mutants, but *Hspg2*$^{\Delta 3/\Delta 3}$ mutants have a GBM charge defect (27) and Perlecan decreased in HIVAN models (29, 30)	None
Collagen XVIII	BC, TBM, SMC, MM	KO more susceptible to anti-GBM nephritis (41), aged mutants have TBM thickening, MM expansion, elevated serum creatinine (40). *Col18a1*$^{-/-}$ x *Hspg2*$^{\Delta 3/\Delta 3}$ double KOs less susceptible to IRI (44)	Rare cases of nephropathy associated with Knoblach syndrome (49, 50)
Syndecan-1	Podocyte, mesangial	KO more sensitive to anti-GBM nephritis (56)	None
Syndecan-4	Podocyte, mesangial	KO more sensitive to acute kidney injury (55), gender-dependent effect on disease progression after unilateral nephrectomy (52)	Increased in IgA nephropathy (58)
Glypican-3	Embryonic progenitors (mesenchyme and UB), levels diminish by birth	KO has enhanced UB branching and proliferation leading to cystic and dysplastic kidneys (67)	Simpson-Golabi-Behemel syndrome; large dysplastic kidneys, pelvic duplication, hydronephrosis (70)
Decorin	Interstitial fibroblasts, vessels, mesangium	KO more susceptible to DN (87, 88) and nephropathy after UUO (86, 89)	Upregulated in CKD (84, 85)
Biglycan	Interstitial and glomerular endothelial cells, fibroblasts, SMC, DT, CD	KO has striking glomerulocystic changes after UUO (86)	Upregulated in CKD (84, 85)

Many PGs in the kidney show a temporally or spatially restricted expression profile. Genetic disruption of PG core proteins in mice does not generally affect development or baseline structure or function in adult life, but mutants show altered sensitivity to experimental kidney injury. PGs have been linked, directly or indirectly, to the pathogenesis of several human kidney diseases

GBM glomerular basement membrane; *CD* collecting duct; *SMC* smooth muscle cell; *BM* basement membrane; *KO* knockout; *DN* diabetic nephropathy; *HIVAN* human immunodeficiency virus-associated nephropathy; *TBM* tubular BM; *MM* mesangial matrix; *IRI* ischemia-reperfusion injury; *IgA* Immunoglobulin-A; *UB* ureteric bud; *UUO* unilateral ureteral obstruction; *CKD* chronic kidney disease; *DT* distal tubule

be important for the permselective properties of the capillary wall and its role as an ultrafiltration barrier. Various experimental and human kidney diseases are associated with a reduction in glomerular HS, and binding of HSPG antibodies to the GBM causes proteinuria, supporting the notion that HSPGs are important for barrier function (1). In recent years, our understanding of PGs in glomerular function has advanced considerably. The concept that HS in the GBM contributes significantly to filtration has been challenged and attention has shifted from podocyte-derived HSPGs to those on the surface of glomerular endothelial cells.

This article reviews PGs in the context of kidney pathophysiology with emphasis on their roles in development and glomerular function. It begins with an overview of nephrogenesis and the glomerular barrier, then summarizes the renal phenotypes of mice lacking PG core proteins (see Table 1) or mutated for various GAG-modifying/degrading enzymes (see Table 2). Human kidney diseases associated with these molecules are highlighted.

2. Generalized Roles for PGs During Nephrogenesis

The basic excretory unit of the kidney is the nephron. Distinct components along its length are responsible for various physiologic functions including ultrafiltration of blood, reabsorption of nutrients, and secretion of wastes. The nephron is essentially a specialized tortuous tubule that comes to a blind end in the renal cortex where it envelops a compact vascularized tuft (the glomerulus), which is the site of ultrafiltration. At its opposite end, each nephron is continuous with a system of collecting tubules that convey the filtrate to the renal pelvis and ureter. Each human kidney is endowed with ~300,000 to 1 million nephrons by birth, while in mice the mature complement of ~20,000 nephrons is established by 2 weeks postnatally.

Kidney development (nephrogenesis) involves a series of iterative reciprocal inductive events between an epithelial outgrowth of the mesonephric duct (the ureteric bud; UB) and the metanephric mesenchyme (MM), a process under exquisite molecular control (2). The UB branches as it invades the MM, and at the tips of these branches, mesenchymal cells are induced to form a condensate (the nephrogenic vesicle) and undergo epithelialization. The UB continues to branch and grow, sending ramifications deeper into the uninduced blastema where these initiate further nephrogenesis. The vesicle passes through several developmental stages, ultimately giving rise to the glomerulus, proximal tubule, loop of Henle, and distal tubule.

These processes are associated with dynamic changes in the expression of PG core proteins and GAGs. Their specific roles in the kidney will be addressed in the sections that follow, but it is

Table 2
PG-modifying enzymes in kidney disease

Gene (function)	Effect of disruption	Mouse kidney phenotype	Human kidney disease
Ext1 (glycosyltransferase)	Block in HS-GAG polymerization	Podocyte-specific KO has mild glomerular structural changes, GBM charge defect, microalbuminuria (96)	Single case report of nephropathy associated with osteochondroma (98)
Xylt2 (xylosyltransferase)	Block in initiation step of GAG synthesis	KO develops late-onset cystic kidney disease. Normal renal HS/CS levels may reflect compensation by Xylt1 (99)	Genetic modifier of DN (100)
Glce (epimerase)	HS-GAG lacks IdoA and 2-O sulfation	KO shows bilateral renal agenesis. Impaired growth factor binding/signaling of mutant HS (101, 102)	None
Hs2st (sulfotransferase)	Block in HS 2-O sulfation	KO shows bilateral renal agenesis (103). UB and MM developmentally competent, failure of reciprocal induction (106)	None
Sulf1/Sulf2 (sulfatases)	Increased HS 6-O sulfation	Reduced kidney weight in *Sulf2*$^{-/-}$ and *Sulf1*$^{-/-}$ x *Sulf2*$^{-/-}$ double KOs, attenuation of renal cortex, incomplete penetrance (110). Glomerular structural and functional defects in double KOs (114)	Endothelial Sulf-1 expression attenuated in transplant rejection (44)
Hpse (heparanase)	Up-regulation causes loss of HS	No deleterious effect of transgenic overexpression or exogenous administration (116)	Upregulated in various glomerular diseases (115)

Genetic knockout models demonstrate that the synthesis and fine structure of HS are essential for normal kidney development and function. Despite the striking phenotypes of mutant mice, there is very little evidence implicating these enzymes in human disease. Heparanase was once thought to be a key contributor to human glomerular disease, but this concept has been refuted by studies in animal models

HS heparan sulfate; *CS* chondroitin sulfate; *GAG* glycosaminoglycan; *GBM* glomerular basement membrane; *KO* knockout; *DN* diabetic nephropathy

worthwhile first noting a study that highlights their general requirement in nephrogenesis (3). Steer and colleagues showed that the levels of some PGs, including most glypicans and syndecan-1, peak between embryonic days 13–15 (E13-15) during the period of early UB branching and MM induction, but that their levels later taper off. Other PGs such as agrin, syndecan-2 and -4, and decorin show the opposite pattern, their levels rising as development proceeds. Biochemical analysis of E13 kidneys revealed synthesis of both HS and CS, with the latter thought to derive from the mesenchyme and accounting for ~75% of GAGs. Nonselective inhibition of sulfation with chlorate impaired proliferation and branching morphogenesis and could be rescued by sulfate. Branching was also disrupted with naphthalene xyloside that inhibits HS and CS attachment but not with the CS-selective decalin xyloside. Heparitinase led to strong inhibition of UB branching and blocked nephrogenesis, an effect not seen with chondroitinase. Thus, despite being less abundant, HS may play a more important role in modulating UB branching. The defects in whole kidneys were reproduced in isolated UB, suggesting a direct effect on this tissue rather than a failure of the MM to induce branching.

3. The Glomerular Filtration Barrier

The filtration barrier separating the lumen of the glomerular capillaries from Bowman's space is classically viewed as having three layers. The first is formed by glomerular endothelial cells that have fenestrations 70–100 nm in diameter allowing access of plasma proteins to the underlying GBM. The endothelial surface is coated by a glycocalyx of remarkable thickness (~300 nm) rich in PGs, glycoproteins, and glycolipids. This layer, which also overlies the fenestral openings, is thought to contribute to hydraulic resistance and form an electrostatic barrier to the passage of anionic serum proteins. The meshwork-like GBM forms the second layer and it may act to sieve molecules based on size and charge. The third layer is formed by specialized epithelial cells, termed podocytes that extend slender foot processes that interdigitate and organize into an intricate network embracing the outer surface of the capillary loops. Neighboring foot processes are connected by a slit diaphragm ~40 nm in width with a zipper-like structure. Disruptions of proteins forming the slit diaphragm complex (e.g., nephrin or podocin) cause glomerular disease (4). Each of the layers contributes in an important way to the functional integrity of the system, and together they ensure that proteins ~40 Å or larger, including notably albumin, are effectively retained within the circulation.

4. Basement Membrane HSPGs in Glomerular Function

The side chains of HSPGs are formed of unbranched repeats of the disaccharide glucuronic acid (GlcA) and *N*-acetylglucosamine (GlcNAc) that are extensively modified through epimerization of GlcA to iduronic acid (IdoA) and deacetylation and sulfation at various positions. HSPGs may be expressed as membrane-spanning or glycophosphatidylinositol (GPI)-linked forms at the cell surface or secreted and incorporated into extracellular matrices or BMs. The latter is mediated by binding of the core proteins or HS chains to various BM-associated molecules including nidogen, collagen IV, and laminin, and also with dystroglycan and integrin receptors. While HSPGs serve less of a structural role in BMs than the major networks of collagen IV and laminin, they act to bind and regulate the bioactivity of many growth factors relevant to kidney health and disease, including members of the GDNF, PDGF, FGF, TGF, BMP, and VEGF families.

The GBM carries significant anionic charge imparted by sulfated GAGs, and by carboxyl and sialyl groups of glycoproteins. The former were identified as HSPGs based on their susceptibility to enzymatic digestion (5), and other sulfated GAGs are not present in the GBM at significant levels. HSPGs were long thought to be the main determinant of charge-selective filtration. Indeed, disruptions in HSPGs have been documented in many human and experimental glomerular diseases (1). Within the GBM, HSPGs are concentrated along the lamina rara externa (LRE) underlying podocytes, and to a lesser extent, along the subendothelial lamina rara interna (LRI). Here they are arranged in quasiregular arrays with a ~60 nm periodicity (5). Although the basis for this is unknown, it presumably reflects an intrinsic molecular organization determined by interaction of HSPGs with their receptors and other matrix components. Three genetically distinct BM-HSPGs are recognized; agrin, perlecan, and collagen XVIII (6, 7) that are reviewed below.

4.1. Agrin

Agrin is characterized by a ~2,000 residue core protein (~225 kDa) complexed to 3 HS-GAG chains, with the mass of the glycanated form being ~400 kDa. Its N-terminus consists of a globular domain (NtA) that binds laminin γ1, followed by multiple follistatin-like domains. The central portion of the molecule bears EGF-like repeats and serine/threonine-rich and sperm protein/enterokinase/agrin (SEA) domains, and at its C-terminus are laminin globular (LG) domains that bind integrin and dystroglycan (7). Agrin exists as multiple tissue-specific alternatively spliced isoforms that are BM- or cell-associated, with the latter specific to neurons (8). C-terminal antibodies stain the GBM and the BM of collecting ducts and smooth muscle in kidney, whereas N-terminal antibodies stain all renal BMs (9). The basis for this pattern is not entirely clear and may reflect masking of epitopes or proteolytic processing, as neurotrypsin-dependent cleavage of the agrin C-terminus occurs in brain and kidney (10). Agrin is the predominant GBM-HSPG in all species (11) which prompted speculation that it was especially important for podocyte adhesion and glomerular filtration.

Agrin is required for proper assembly of nerve-muscle synapses, and knockout of the gene in mice leads to neuromuscular defects and death shortly after birth due to paralysis (12). To investigate the role of agrin in the kidney, we studied newborn agrin-deficient mice and generated conditional mutants lacking agrin in podocytes (13). Newborn agrin-deficient mice showed no renal histologic abnormalities indicating agrin is dispensable for kidney development, but their severe phenotype precluded functional studies. Podocyte-specific knockout mice are viable and had no renal histopathology up to 9 months of age. Their GBM labeled weakly for agrin at 1 week of age, but staining diminished to nearly undetectable levels by 3 weeks. Thus, agrin in mature GBM is derived principally from podocytes. Other GBM components were normally expressed in mutant glomeruli. Anionic sites in the GBM were assessed by electron microscopy after labeling with the cationic probe polyethyleneimine (PEI). The number of anionic sites along the subepithelial LRE was significantly reduced in agrin-deficient embryos, whereas those in the LRI were unchanged. By 3 weeks of age, PEI labeling in normal mice becomes prominent along the LRE, while labeling of the LRI is weaker and less discrete. In conditional knockouts 3 weeks and older, PEI labeling was nearly undetectable throughout the GBM indicating that agrin contributes the bulk of negative charge in this site. Despite the severe charge disruption, albumin excretion was comparable in mutant and control animals suggesting the glomerular barrier was intact. An anionic Ficoll tracer was used to test whether mutants might have a filtration defect that was compensated by tubular resorption, but no differences between mutants and controls were found. Thus, although agrin contributes significantly to the negative charge of the GBM, both it and this charge are dispensable for glomerular function.

To date, there is no evidence of a causative role for agrin in human kidney disease. Glomerular labeling for the core protein was reported to be diminished in diabetic nephropathy by some (14), but not others (15, 16), with the differences likely reflecting severity or duration of disease. A humoral immune response to agrin is present in many patients with transplant glomerulopathy, a form of chronic kidney transplant rejection characterized by GBM duplications (17), although the pathologic significance of this finding is unclear.

4.2. Perlecan

Perlecan is a ubiquitous BM-HSPG that appears as a linear molecule of ~80 nm with a series of bead-like globular domains. The core protein of ~400 kDa is linked to 3 or 4 HS-GAG chains at its

N-terminus that can be variably substituted by CS, and which brings its mass to ~700 kDa. Other structural domains starting from its N-terminus include a SEA module, a low-density-lipid receptor ligand-binding domain, a series of globular domains homologous to laminin domain IV that are separated by multiple laminin-EGF-like repeats, and multiple N-CAM-like Ig repeats. The C-terminus of the molecule is comprised of three laminin G globular domains separated by EGF-like repeats and this portion of the molecule can be cleaved to liberate a ~85 kDa angiostatic fragment termed endorepellin. Perlecan is a component of all renal BMs, and its distribution within the glomerulus changes during development. It is detected along with agrin in the immature GBM where it is thought to derive primarily from endothelial cells. Perlecan expression in this site diminishes with glomerular maturation, and ultimately it is only prominent in the mesangium and Bowman's capsule (18–20).

Perlecan is essential for normal development. Mice lacking perlecan (*Hspg2*$^{-/-}$) have an embryonic lethal phenotype with brain, skeletal, and heart abnormalities and BM disruptions in regions of mechanical stress (21, 22). Roughly half of the mutants die at midgestation (E10.5), while the remainder survive until birth with severe skeletal defects including abnormal endochondral ossification and disorganized poorly proliferative chondrocytes. Specific kidney defects were not reported. To investigate the importance of the GAG chains of perlecan relative to the core protein, knockout mice (*Hspg2*$^{\Delta3/\Delta3}$) were generated that lack 3 of 4 potential attachment sites (23). Mutants are viable, but at birth they have small eyes and their lenses show increased apoptosis and structural defects that progress by 3 weeks of age. Mutants also have increased smooth muscle proliferation and intimal hyperplasia following experimental vascular injury, and impaired angiogenesis, wound healing, and tumor growth (24, 25). Morita and colleagues studied the kidneys of *Hspg2*$^{\Delta3/\Delta3}$ mice in detail (26). Mutants showed no glomerular ultrastructural changes and a normal number of anionic sites along the subepithelial aspect of the GBM. These were identified as HS based on their sensitivity to heparitinase digestion. The lack of a charge defect is consistent with the fact that agrin is the principal HSPG in this site. Levels of agrin, perlecan, and collagen XVIII mRNAs in mutant glomeruli were normal. Although glomerular HS levels were not assessed, immunostaining revealed an upregulation of CS in mutant GBM, perhaps reflecting its attachment to alternative sites in the mutant core protein. Baseline kidney function was normal in mutants, but their urinary protein excretion was significantly higher than controls in a model of protein-overload nephropathy induced by BSA administration. Based on this finding following experimental challenge, it was proposed that perlecan-HS contributes to glomerular filtration.

Mutants lacking either podocyte-derived agrin or perlecan-HS would not manifest a kidney phenotype if only one of these PGs were sufficient for normal function. To test this hypothesis, we generated mice lacking both components (27). PEI labeling revealed a significant reduction in anionic sites only along the subendothelial GBM in *Hspg2*$^{\Delta3/\Delta3}$ mice. Therefore, perlecan contributes to some extent to GBM negative charge. Immunostaining revealed a loss of GBM-HS in agrin mutants, although some mesangial labeling remained. In contrast, mesangial labeling was attenuated in *Hspg2*$^{\Delta3/\Delta3}$ mice while some GBM staining remained. As expected, HS levels were significantly reduced in all glomerular BMs of double mutants. These animals showed no histologic or ultrastructural abnormalities up to 15 months of age, similar to their counterparts lacking either single component. Importantly, despite a GBM charge defect, baseline kidney function and excretion of an anionic Ficoll tracer in double mutants were not different from controls.

Mutations of *HSPG2* in man cause the allelic disorders Schwartz-Jampel syndrome and the Silverman-Handmaker type of dyssegmental dysplasia, but neither has kidney involvement. Glomerular perlecan levels are reduced in human and experimental diabetic nephropathy (28) and in two mouse models of human HIV-associated nephropathy (HIVAN) (29, 30). Infection of cultured human podocytes with HIV-1 also reduces perlecan mRNA levels. The authors proposed that down-regulation of perlecan promotes mesangial proliferation in these models and in human HIVAN.

4.3. Collagen XVIII/Endostatin

Collagen XVIII is a member of the multiplexin collagen subfamily that is formally classified as a HSPG. The core protein of 180 kDa is linked to 3 HS-GAG chains at its N-terminus and central noncollagenous interruptions (31, 32), and it assembles into a triple helical homotrimer of ~1 MDa. Three isoforms differing by their signal peptides and N-terminal noncollagenous domains arise from use of an internal promoter and alternative splicing. The longest isoform is widely expressed, the intermediate form is restricted to liver, and the short form is expressed in many tissues but is enriched in kidney (33, 34). Staining with a C-terminal antibody labels mouse glomerular and tubular BMs from E14 onwards and colocalizes with perlecan. In adult mouse kidney, it is detected in BMs of Bowman's capsule, tubules, interstitial capillaries and vessels, and weakly in the mesangium, but it is not present in the GBM (35). Human fetal and adult kidney have a comparable staining pattern (33). Cleavage of the noncollagenous C-terminus within the hinge region separating it from the triple helical domain of the molecule liberates a ~20 kDa fragment that was named endostatin due to its anti-angiogenic properties (36). Endostatin binds glypicans via HS on the surface of endothelial and renal tubular epithelial cells, and it inhibits renal branching morphogenesis (37, 38).

Endostatin also interacts with the BM components perlecan, laminin, and nidogen, as well as with integrins.

Knockout of *Col18a1* in mice causes an eye phenotype characterized by delayed hyaloid vessel regression, abnormalities of the retinal vasculature and retinal pigment epithelial cells, and defects of the iris and ciliary body (39). Mutants on a different genetic background develop hydrocephalus and have a mild kidney phenotype (40). Although renal histology by light microscopy was normal, these mice show tubular BM thickening and mesangial expansion, and as adults (6–16 months) they have a small increase in serum creatinine levels suggesting a filtration defect. Collagen XVIII is upregulated in the GBM and in Bowman's capsule in a mouse model of anti-GBM nephritis. In this model, glomerular and tubulointerstitial injury, endothelial damage, and matrix remodeling were more severe in *Col18a1*$^{-/-}$ mice than controls, and disease progression was not influenced by exogenous endostatin (41). In contrast, endostatin has a protective effect when administered in the early stages to mice with streptozocin-induced diabetic nephropathy (42).

Whereas the studies above point to a positive effect of collagen XVIII/endostatin on kidney structure and function, others suggest that in a different setting it can be injurious. Renal ischemia-reperfusion injury, whether in experimental models or in man (as can occur during transplantation), causes endothelial damage and promotes the interaction of collagen XVIII with L-selectin and monocyte chemoattractant protein-1 (MCP-1) (43). This influences adhesion and activation of inflammatory cells and promotes their infiltration to the renal interstitium. Although leukocyte influx following ischemia-reperfusion in *Col18a1*$^{-/-}$ or *Hspg2*$^{\Delta3/\Delta3}$ mutant mice is comparable to controls, double mutants have a significant reduction in early monocyte/macrophage influx, in support of a role for BM-HSPGs as mediators of renal inflammation (44). The absolute level of collagen XVIII/endostatin increases in response to renal injury and may further exacerbate the disease process. For example, endostatin expression by tubular epithelial cells is increased at both the protein and mRNA in mouse models of ischemia-reperfusion injury (45) and ureteral obstruction (46). In a model of acute renal failure induced by lipopolysaccharide, there is increased endostatin staining in tubular BMs, but mRNA levels are reduced suggesting it derives from increased metabolism of collagen XVIII (47).

Mutations in *COL18A1* cause Knoblach syndrome (48), an autosomal recessive disorder characterized by myopia, vitreoretinal degeneration, retinal detachment, and encephalocele. Nephropathy has been described in association with this condition, but only in two cases with documented mutations. One patient had bilateral duplex kidneys and vesicoureteral reflux (49), while the second showed unilateral duplication of the collecting system and bifid ureter (50). Whether these defects are a rare feature of this disease or are unrelated is unclear.

5. Cell Surface HSPGs in Kidney Function

5.1. The Syndecan Family

The syndecans are a family of four type I transmembrane HSPGs expressed by lymphocytes, endothelial, and epithelial cells (51). Its members can be cell surface-associated or shed via proteolysis. By virtue of their cytoplasmic tails that couple indirectly to the cytoskeleton, they transmit extracellular signals that influence processes such as cell adhesion, proliferation, and migration. Syndecans also serve a barrier function and play roles in tumorigenesis, inflammation, and viral infection. Syndecans have multiple N-terminal HS attachment sites, although some can carry CS. In the kidney, expression of syndecan-1 is detected by immunostaining in podocytes and mesangial cells in normal mice (52), but at the RNA level it is additionally detected in epithelial cells of some tubules and Bowman's capsule. Syndecan-2 (fibroglycan) is expressed in the mesenchyme of embryonic mouse kidney, but is absent in the adult organ (52, 53). Syndecan-3 (N-syndecan) is restricted to neural tissues. Finally, syndecan-4 (ryodocan) is enriched in kidney and its distribution is the same as that of syndecan-1 (54, 55).

Knockout mice have been described for all but syndecan-2. Mutants show defects in wound healing and angiogenesis (for syndecan-1 and -4), learning, feeding behavior, and muscle development (syndecan-3), but none have an overt kidney phenotype. Experimental injury models, however, have uncovered roles for some family members in kidney function. Glomerular expression of syndecan-1 and -4 is transiently upregulated in mice with anti-GBM nephritis. This response appears to be protective, as syndecan-1 knockout mice (*Sdc1*$^{-/-}$) are more sensitive to injury and have increased glomerular lymphocyte influx during the early heterologous phase of disease (56). Labeling of glomerular endothelial cells with antibodies defining inflammatory *N*- and 6-*O*-sulfated HS moieties is upregulated in all treated mice, but staining is significantly greater in *Sdc1*$^{-/-}$ mutants. Syndecan-1 deficiency also impacts the later autologous phase of disease in this model associated with the host response to the GBM-bound foreign antibodies. Mutant glomeruli show increased T cell influx and matrix deposition, and the animals have elevated albuminuria and serum creatinine levels. The immune response is also qualitatively different in mutants which have higher IgG_1 and lower IgG_{2a} levels of anti-rabbit antibodies indicating a shift from a Th1 to a Th2 response. In both disease phases, circulating levels of Th2-promoting cytokines (interleukins-1β, -4, -6, -10, -13, and MCP-1) are higher in *Sdc1*$^{-/-}$ mice. This model is typically associated with a Th1 response, and the induced antibodies are more effective at complement activation than Th2 class, so the consequence of this shift and how it aggravates the mutant phenotype is unclear. Syndecan-1 and -4, glypican-1, and collagen XVIII are upregulated by cultured glomerular endothelial cells upon activation with TNF-α or interleukin-1β (57). The proinflammatory HS epitopes induced upon glomerular endothelial injury that promote leukocyte adhesion may be associated with one or more of these upregulated PGs and could represent novel therapeutic targets.

Mice lacking syndecan-4 (*Sdc4*$^{-/-}$) show elevated blood urea nitrogen and mortality in response to κ-carrageenan which deposits in the collecting system causing obstructive nephropathy (55). This susceptibility was not seen with other nephrotoxic agents (cisplatin, bromoethylamine hydrobromide, or glycerol), and the relevance of this particular model to human kidney disease is unclear. Others studying a different line of knockout mice reported a gender-dependent protective effect of syndecan-4 following unilateral nephrectomy (52). This study showed firstly that kidney-to-body weight ratios of *Sdc4*$^{-/-}$ mice were higher in mutants under baseline conditions. Two months after nephrectomy, male mutants showed mesangial expansion, GBM thickening, and increased deposition of fibronectin and collagen. These changes were not seen in nephrectomized wild-type animals or female mutants. There was a corresponding up-regulation of syndecan-2 and TGF-β in diseased glomeruli. The authors proposed that syndecan-2 might be induced as a form of molecular compensation that is ultimately injurious as it promotes the prosclerotic activity of TGF-β. Syndecans have not been directly implicated in any human kidney disease, although increased mesangial and tubulointerstitial expression of syndecan-4 was described in IgA nephropathy (58).

5.2. The Glypican Family

The glypicans are a family of GPI-anchored cell surface PGs of 60–70 kDa that bear 3 or 4 HS-GAG chains (59). By virtue of their GPI anchor, they are localized (although not restricted) to the apical aspect of polarized cells within lipid rafts, distinct microdomains enriched with receptors that serve as signaling platforms. Humans and mice express six glypican isoforms, each encoded by a different gene (*GPC1-6*). Some of these are clustered in pairs in the genome (*GPC3/GPC4* and *GCP5/GPC6*) reflecting duplication events. Glypicans are important modulators of the WNT, Hedgehog, FGF, and BMP signaling pathways.

All except glypican-2 which is neural-specific are present in the kidney and show developmentally regulated expression. The brain-enriched isoform glypican-1 is detected in fetal mouse glomeruli, but not tubular epithelial cells (60). Glypican-3 is expressed in the mesenchyme and UB in midgestation mouse embryos, but its levels diminish significantly by birth (61, 62). Glypican-4 is expressed later during nephrogenesis and is prominent in differentiated tubules but absent from glomeruli (63). Glypican-5 is expressed early in the mesonephric duct and later in condensing mesenchyme but not in mature tubules or glomeruli, and it is present in fetal and adult human kidney (64). Finally, glypican-6 is expressed by condensing mesenchyme of embryonic mouse kidney, and also in adult human kidney (65).

Targeted or gene-trapped alleles for all glypican genes in mice exist, but phenotypes have only been reported for *Gpc1* and *Gpc3* in the literature. The former has a brain phenotype consistent with prominent expression of glypican-1 in this site (66). The same study reports that *Gpc2*$^{-/-}$ mice are normal, perhaps reflecting functional redundancy. *Gpc3*$^{-/-}$ mice have a perinatal lethal phenotype with cardiac malformations with delayed formation of the coronary vascular plexus, defects in lung development, and cystic and dysplastic kidneys (67, 68). Mutants have enhanced UB branching during early nephrogenesis and increased proliferation of UB epithelium. Hartwig and colleagues showed that glypican-3 stimulates BMP2-dependent signaling via SMADs in a pathway that inhibits UB proliferation and branching (69). *Gpc3* loss-of-function and even subtle changes brought about by reducing gene dosage relieves this repression and enhances branching morphogenesis. *GPC3* mutations cause Simpson-Golabi-Behmel syndrome, a disorder characterized by pre- and postnatal overgrowth, cardiac and other malformations, and cancer predisposition (70). Renal involvement is a feature of the disease, and findings include large and dysplastic kidneys, duplication of the renal pelvis, and hydronephrosis.

6. CSPGs (NG2, Collagen XV, Leprecan, and "Bamacan")

Several CSPGs have been characterized in kidney, including neural/glial cell 2 (NG2) which is membrane-bound, and the matrix-associated CSPGs collagen XV, leprecan and bamacan. NG2, encoded by the *CSPG4* gene, is an integral membrane PG with a core protein of ~300 kDa. It is expressed by oligodendrocyte progenitors, chondroblasts, pericytes, and smooth muscle, as well as several tumor cell types. In the kidney, it is expressed by mesangial cells and pericytes of interstitial capillaries (71). *Cspg4*$^{-/-}$ mice have brain, skin, and smooth muscle defects, but no kidney phenotype was reported. NG2 is upregulated in the kidneys of diabetic rats and was shown to influence mesangial proliferation and matrix synthesis in vitro suggesting a possible role in diabetic nephropathy.

Collagen XV is a component of the BM zone of many tissues that serves an important role in anchoring the BM to the underlying stroma. It is assembled as a homotrimer of ~400 kDa bearing CS-GAGs within its N-terminal half, but in the kidney it can also carry significant amounts of HS (72). In human fetal kidney, it is localized to the GBM, mesangium, interstitial capillaries, and some tubules, whereas in adults it is detected in vessels and some tubules, but glomerular labeling is significant only along Bowman's capsule (73). The distribution of collagen XV in mouse kidney is comparable to that in man, although there is prominent mesangial and GBM staining at maturity (74). *Col15a1*$^{-/-}$ mice have normal BMs and no kidney phenotype, but they show exercise-induced muscle and vascular defects. In chronic human kidney disease, there is accumulation of collagen XV in areas of interstitial and periglomerular fibrosis and within thickened GBM that could be relevant to disease progression.

Leprecan, named such as it is a "leucine and proline-enriched proteoglycan," is a 80–100 kDa CSPG encoded by the *LEPRE1* gene. An integrin-binding motif and four CXXXC tetratricopeptide repeats are present at its N-terminus, while at the C-terminus is a prolyl hydroxylase domain and ER retention motif that point to a role in the posttranslational modification of collagen. Leprecan expression in rat kidney is highest during development, where there is diffuse cytoplasmic staining of condensing mesenchyme, mesangial cells, Bowman's capsule, and some tubules (75). In mature glomeruli, there is also weak labeling of podocytes. Mesangial staining is slightly upregulated in hypertrophic glomeruli of rats following unilateral nephrectomy. This pattern appears species-specific, as leprecan is restricted to the renal pelvis in chick and absent from human kidney. The discrepancy may reflect expression of alternatively spliced transcripts, other leprecan family members, or the use of different antibodies or probes. Mutations in the human *LEPRE1* gene interfere with collagen 3-hydroxylation and cause osteogenesis imperfecta type VIII. Kidney disease has not been reported in these patients, nor in *Lepre1*$^{-/-}$ mice (76), arguing against a role for leprecan in kidney function.

Another CSPG that was characterized based on its reactivity with monoclonal antibodies and later termed bamacan is localized to the mesangium, Bowman's capsule, and tubular and vascular BMs, but is not present in the GBM except transiently during fetal and early postnatal life (77–79). Its expression is reduced in a model of polycystic kidney disease (80), whereas it is upregulated in the GBM in diabetic nephropathy (81). Cloning of the murine gene forced a reevaluation of bamacan, as it proved homologous to the SMC3 subclass of the "structural maintenance of chromosomes" family of intracellular proteins that regulate chromosomal dynamics (82). The CSPG recognized by the various antibodies and that which was subsequently cloned as bamacan appear to be unrelated, and thus, the identity of the BM CSPG disrupted in these studies is unclear.

7. Small Leucine-Rich Proteoglycans

Members of this class of mesenchymally derived PGs are characterized by an N-terminal domain of leucine-rich repeats flanked by cysteine clusters (83). They bind collagen I and modulate fibrillogenesis, and also to TGF-β to inhibit its signaling. Two members of the family that carry mixed CS/DS-GAG chains, decorin and biglycan, have been implicated in the pathogenesis of fibrotic kidney disease but appear to have opposing effects.

7.1. Decorin

Decorin is detected in human kidney at very low levels in mesangial cells and podocytes, but is prominent in interstitial fibroblasts and vessels (84, 85). It is upregulated with biglycan in the interstitium and scarred glomeruli in advanced human and experimental kidney diseases (85, 86). Decorin-deficient mice (*Dcn*$^{-/-}$) are more susceptible to streptozocin-induced diabetic nephropathy (87). Mutants have elevated albuminuria and serum creatinine, more advanced glomerular lesions and macrophage infiltration, and earlier mortality than diabetic wild-type controls, and they show increased glomerular levels of TGF-β and collagen I. In the same model, others showed mutants had enhanced apoptosis and overexpression of TGF-β, CTGF, and IGF-I receptor by tubular epithelial cells, and more biglycan-positive infiltrating mononuclear cells. Treatment of human tubular epithelial cells with decorin in vitro inhibits the apoptotic response to hyperglycemia by binding the IGF-I receptor (88). *Dcn*$^{-/-}$ mice also have more severe renal histopathology, inflammation, and apoptosis than controls in a model of obstructive nephropathy (89). Gain-of-function approaches further support the protective role of decorin in the kidney. Transgenic or exogenous decorin abrogates up-regulation of TGF-β and improves kidney function in rat models of anti-Thy-1 glomerulonephritis (90, 91) and diabetic nephropathy (92). Mesangial accumulation of decorin in diseased glomeruli was proposed to reflect its reduced turnover or increased retention rather than enhanced local synthesis. Stable transfection of decorin in cultured rat mesangial cells leads to up-regulation of caspase-3 promoting apoptosis, and it inhibits TGF-β expression (93).

7.2. Biglycan

Biglycan is expressed by endothelial cells of interstitial and glomerular capillaries, fibroblasts, distal and collecting duct epithelium, and throughout the width of large vessels of normal human kidney (85). Although its prominent interstitial accumulation in chronic kidney disease derives principally from infiltrating inflammatory cells, particular attention has focused on biglycan regulation in mesangial cells. Cytokine stimulation of both cell types activates iNOS and enhances NO production in glomerular disease. This can influence the bioactivity of PGs by depolymerizing their GAG chains and by altering their expression levels. Biglycan synthesis is repressed in cultured rat mesangial cells by NO, and at early time points in the anti-Thy1 model of glomerulonephritis in vivo (94). Biglycan levels later increase in response to TGF-β stimulation, which may promote mesangial expansion through its antiadhesive effects, by counteracting PDGF signaling to promote proliferation, and by inhibiting apoptosis via caspase-3. Knockout mice (*Bgn*$^{-/-}$) subjected to unilateral ureteral obstruction have reduced expression of fibrillin-1 compared to controls, and they develop striking cystic dilatation of Bowman's capsule (86). *Dcn*$^{-/-}$ mice have the same phenotype, although the lesions develop much later following injury.

8. GAG Modifying/ Degrading Enzymes

8.1. EXT1

Exostosin-1 and exostosin-2 are type II transmembrane glycosyltransferases and subunits of the widely expressed ER-resident copolymerase that catalyzes the elongation of HS-GAG chains. Knockout of *Ext1* in mice leads to synthesis of PG core proteins lacking HS, and mutants have an embryonic lethal phenotype in keeping with its critical role in many developmental processes (95).

To study the role of HS in glomerular filtration, Chen and colleagues generated podocyte-specific *Ext1*$^{-/-}$ mice (96). Mutants had a significant decrease in glomerular HS without a compensatory up-regulation of CS, but did not develop overt kidney disease. Mutant podocytes showed microvillous transformation and foot process effacement by 1 month of age, and these changes persisted into adulthood. By 8 months, they display glomerular hypertrophy and vacuolation of proximal tubule epithelial cells. Mutants have a significant reduction in anionic charge along the subepithelial GBM, comparable to that in podocyte-specific agrin mutants. These defects were accompanied by microalbuminuria, as detected by ELISA at 2 months, and although mutant urinary albumin/creatinine ratios were elevated compared to controls at 2 and 8 months of age, this increase was not statistically significant. The same group generated immortalized Ext1-deficient podocytes for in vitro studies and showed HS-GAGs influence podocyte adhesion, spreading, and migration (97). Mutant podocytes were negative with an antibody recognizing a 3-*O*-sulfated HS epitope enriched in rat GBM. Syndecan-4 was localized to focal contacts of control cells cultured on a fibronectin matrix, but it was disrupted in mutant cells. Surface levels of syndecan-4 in mutant podocytes were significantly greater than controls. As a consequence, they had an up-regulation in the amount and activity of membrane-localized PKCα that is known to regulate syndecan-4-mediated stress fiber formation. The authors noted the in vivo correlate that glomeruli of podocyte-specific *Ext1*$^{-/-}$ mice lack syndecan-4.

Mutations in the human *EXT1* or *EXT2* genes cause hereditary multiple exostoses (osteochondroma). This rare autosomal dominant disorder is characterized by the formation of bony lesions capped by cartilaginous matrix that lacks HS, perlecan, and decorin and shows increased fibrillar collagen. Kidney involvement is not a feature of this disorder; however, a frameshift *EXT1* mutation was described in a patient with osteochondroma and steroid-sensitive nephrotic syndrome (98). There was a history of kidney disease in this patients' family, although its onset was highly variable and only the proband developed exostoses. Renal biopsy revealed mild mesangial expansion, focal GBM thickening, and mesangial and subendothelial deposition of fibrillar collagen. Although kidney disease in this unique case could not be definitively attributed to EXT1, the authors noted that the common features point to a link between abnormal HS biosynthesis, exostoses, and glomerular disease.

8.2. Xylosyltransferase-2

Xylosyltransferase-2 is one of two isozymes that link UDP-xylose to serine residues of PG core proteins, the initiating step in the assembly of GAG chains. Xylt2 is expressed in many mouse and human tissues and is the principal form in liver. Its counterpart, Xylt1, is only prominent in brain and kidney. *Xylt2*$^{-/-}$ knockout mice are born in the expected Mendelian ratios, and their livers lack both isoforms and have significantly reduced HS, while core protein expression (assessed for decorin) is normal. In the kidney, Xylt1 levels were only slightly decreased. Mutants develop a phenotype characteristic of human autosomal dominant polycystic kidney disease (ADPKD), with biliary epithelial cysts, dilated renal tubules, organ fibrosis, and BM abnormalities (99). Disease severity varies among animals, but in general mutants show renal hypertrophy and a decline in kidney function by 3–5 months, and hydronephrosis, tubular dilatation, and cysts by 6–10 months. Levels of HS and CS in kidney were normal, which may reflect molecular compensation by Xylt1. PGs are decreased in tissues of ADPKD patients and are thought to play a role in cystogenesis. The function of the human *XYLT2* gene is unclear at present, although it has been reported to be a modifier of diabetic nephropathy (100).

8.3. Glucuronyl C5-Epimerase

This ubiquitous enzyme, encoded by the *GLCE* gene, is responsible for C5-epimerization of GlcA to IdoA residues in HS-GAG chains. Knockout of *Glce* in mice leads to the synthesis of HSPGs lacking IdoA with obligate loss of 2-O sulfation and up-regulation of N- and 6-O sulfation. Mutants die at birth from respiratory failure and have a complex phenotype characterized by bilateral renal agenesis, iris coloboma, poorly developed lungs, and various skeletal abnormalities (101). Despite this, other organs such as brain, heart, liver, and skin are normal indicating specific HS structural requirements in kidney and other tissues. A mechanism that may contribute to the phenotype involving reduced binding of mutant HS to GDNF and FGF2, with downstream defects in proliferation and migration, was proposed based on studies of Glce-deficient fibroblasts (102). No human disease has been associated with this gene.

8.4. HS Sulfotransferases

The enzyme HS2ST1 responsible for 2-O sulfation of IdoA and to a lesser extent GlcA is encoded by a single gene in mammals. It is expressed by both epithelial and mesenchymal progenitors in developing kidney, and when inactivated in mice, results in perinatal lethality with bilateral renal agenesis (103). Mutants also have eye and skeletal abnormalities and developmental brain defects (104, 105). Early events marking the onset of nephrogenesis, specifically outgrowth of the UB from the Wolffian duct and induction of mesenchymal Pax-2 expression, occur normally in mutants. However, the mesenchyme fails to condense around the UB tips, and there is a block in inductive signaling between these tissues necessary for branching morphogenesis. The molecular basis for this block was recently analyzed in detail (106). Mutant UB undergoes branching and induces mesenchymal-to-epithelial transformation (MET) when cocultured with wild-type MM or in the presence of the exogenous probranching factors GDNF and FGF1. Conversely, mutant MM could be induced to undergo MET by wild-type UB tissue. Thus, both mutant tissues are themselves differentiation competent. Culture of isolated wild-type kidney rudiments with native HS in the form of heparin or with heparin depleted for 2-*O* sulfate revealed that MM induction but not UB branching was dependent on this modification. From this, the authors conclude that the mutant phenotype stems from either a failure of reciprocal induction or a primary defect in MM morphogenesis. To date, no diseases have been associated with mutations in the human *HS2ST1* gene.

3-O sulfation of glucosamine is a rare terminal modification in HS. The reaction is carried out by a family of sulfotransferase (HS3ST) isozymes with differing specificities. One isoform, *HS3ST1*, generates an anticoagulant sequence present in heparin and endothelial-derived HS that binds and activates antithrombin. $Hs3st1^{-/-}$ mice have a strain-dependent postnatal lethal phenotype and fertility defects but no coagulation abnormalities. A specific 3-*O*-sulfated containing sequence, IdoA(2S)$_{\alpha 1\text{-}4}$GlcN(3S) lacking anticoagulant activity, is enriched in human and bovine GBM and was reported to be restricted to this site. Its levels in the GBM are significantly decreased in diabetic nephropathy, suggesting a specific sulfation defect (107). This 3-*O*-sulfated residue has been shown to be generated by *HS3ST3A1* and this enzyme is expressed in kidney. It is reasonable that 3-*O*-sulfated anticoagulant HSPGs (aHSPGs) would be enriched in glomerular matrices and cells to maintain an anticoagulant environment in the face of circulating clotting factors and plasma proteins. Labeling of kidney with an antithrombin probe reveals aHPSGs within the glomerular capillary wall attributed to both podocytes and endothelial cells. Glomerular epithelial cell lines express *HS3ST3A1* and aHSPGs recognized by this antithrombin probe. Labeling for tubular and vascular aHSPGs is eliminated in kidneys of *Hs3st1*$^{-/-}$ mice, while glomerular expression is normal suggesting other 3-*O* sulfotransferases, perhaps *Hs3st3a1*, are responsible for their synthesis in glomeruli. Gene trap and targeted mutations of various Hs3st isozymes exist for mice, but phenotypes have not been reported, and no mutations in these genes have been described in man. The contribution of this particular HS modification to glomerular function, if any, may be subtle as even deletion of *Ext1* in podocytes does not cause a striking renal phenotype.

8.5. Sulfatases

Sulf-1 is a secreted 6-*O*-endosulfatase that in the kidney is expressed by podocytes in developing and mature glomeruli, as well as in tubules. It is downregulated in glomeruli of mice with a mutation in *Wt1*, encoding the podocyte transcription factor Wilms' tumor-1, that are a model of the human kidney disease Denys-Drash syndrome (DDS) (108). *Wt1* mutant mice, like human DDS patients, show decreased GBM labeling with an antibody to a low-sulfated HS epitope, a finding consistent with reduction of Sulf-1 activity. *Sulf1* and its counterpart, *Sulf2*, were recently shown to be direct *Wt1* target genes (108, 109), suggesting a role in podocyte function.

Sulf1$^{-/-}$ mice have no overt phenotype, while *Sulf2*$^{-/-}$ mice have a small reduction in body and kidney weight (110). Others using a different allele found that *Sulf2* deficiency impairs lung development and reduces viability and growth, changes that are influenced by genetic background (111). Knockout of both genes causes neonatal lethality indicating they have overlapping yet critical roles during development (110, 112, 113). However, this phenotype is incompletely penetrant, with up to 50% of mice surviving to adulthood with no phenotype except decreased weight and subfertility. Kidney weight and size are reduced in *Sulf1*$^{-/-}$ x *Sulf2*$^{-/-}$ double mutant embryos that show attenuation of the renal cortex and medulla (110). Double mutants show glomerular structural abnormalities involving endothelial cells and podocytes, decreased glomerular labeling for VEGF, and age-dependent proteinuria (114). Disruption of VEGF in this study and of GDNF as reported by others (109, 113) may be particularly relevant as both serve critical roles in the kidney. Sulfatase expression is also altered in other kidney diseases. Sulf-1 and Sulf-2 are increased in mouse models of anti-GBM and lupus nephritis (57), whereas in human transplant rejection Sulf-1 which is normally expressed by endothelial cells of interstitial capillaries is lost (44).

8.6. Heparanase

Glomerular HS levels are reduced in many kidney diseases and this is thought to compromise the integrity of the filtration barrier. Loss of HS could reflect decreased production or functional neutralization (e.g., by binding cationic nucleosomes or Ig complexes). However, the frequent finding that heparanase is upregulated in diseased glomeruli suggested that local degradation of HS might be the culprit (115). In recent years, this theory has been put to the test using various approaches, and collectively this work has called into question the notion that heparanase is injurious to the kidney (116). Transgenic overexpression or systemic administration of heparanase removes glomerular HS, but does not lead to significant alteration of glomerular architecture or function. Heparanase overexpression in the kidney actually may have a beneficial effect, as mice overexpressing the enzyme are protected against experimental renal amyloid deposition (117).

9. Concluding Remarks

The development of models, mostly in mice, has been instrumental to advances in the field of functional glycomics. They have allowed detailed analysis of PG function through targeted knockout of core proteins and biosynthetic enzymes, either constitutive or conditional, and individually or in combination. From these studies, and perhaps contrary to expectations, it was revealed that loss of PG core proteins has only subtle phenotypic effects on the kidney. Experiments to deplete or delete HS from the GBM did not uncover a role for HSPGs in glomerular filtration, forcing a reevaluation of the nature and location of the glomerular charge barrier. Disruption of enzymes that polymerize or modify GAGs disrupts renal branching morphogenesis by altering growth factor signaling. Given the striking phenotypes of these mutants, it is surprising that none of the corresponding genes have been associated with congenital human kidney disease. Additional tools will be required to take this field further, for example, by making it possible to specifically target GAG-modifying enzymes in glomerular endothelial cells to address their role in filtration.

Acknowledgments

The author is supported by grants from the Agence Nationale de la Recherche (ANR) and the Inserm Avenir program.

References

1. Raats, C. J., van den Born, J., and Berden, J. H. (2000) Glomerular heparan sulfate alterations: mechanisms and relevance for proteinuria. *Kidney Int.* **57**, 385–400.
2. Little, M., Georgas, K., Pennisi, D., and Wilkinson, L. (2010) Kidney development: two tales of tubulogenesis. *Curr Top Dev Biol.* **90**, 193–229.
3. Steer, D. L., Shah, M. M., Bush, K. T., Stuart, R. O., Sampogna, R. V., Meyer, T. N., et al (2004) Regulation of ureteric bud branching morphogenesis by sulfated proteoglycans in the developing kidney. *Dev Biol.* **272**, 310–327.
4. Machuca, E., Benoit, G., and Antignac, C. (2009) Genetics of nephrotic syndrome: connecting molecular genetics to podocyte physiology. *Hum Mol Genet.* **18**, R185–194.
5. Kanwar, Y. S., Linker, A., and Farquhar, M. G. (1980) Increased permeability of the glomerular basement membrane to ferritin after removal of glycosaminoglycans (heparan sulfate) by enzyme digestion. *J Cell Biol.* **86**, 688–693.
6. Iozzo, R. V. (2005) Basement membrane proteoglycans: from cellar to ceiling. *Nat Rev Mol Cell Biol.* **6**, 646–6956.
7. Bezakova, G. and Ruegg, M. A. (2003) New insights into the roles of agrin. *Nat Rev Mol Cell Biol.* **4**, 295–308.
8. Burgess, R. W., Skarnes, W. C., and Sanes, J. R. (2000) Agrin isoforms with distinct amino termini: Differential expression, localization, and function. *J Cell Biol.* **151**, 41–52.
9. Raats, C. J., Bakker, M. A., Hoch, W., Tamboer, W. P., Groffen, A. J., van den Heuvel, L. P., et al (1998) Differential expression of agrin in renal basement membranes as revealed by domain-specific antibodies. *J Biol Chem.* **273**, 17832–17838.
10. Reif, R., Sales, S., Hettwer, S., Dreier, B., Gisler, C., Wölfel, J., et al (2007) Specific cleavage of agrin by neurotrypsin, a synaptic protease linked to mental retardation. *FASEB J.* **21**, 3468–3478.
11. Groffen, A. J., Ruegg, M. A., Dijkman, H., van de Velden, T. J., Buskens, C. A., van den Born, J., et al (1998) Agrin is a major heparan sulfate proteoglycan in the human glomerular basement membrane. *J Histochem Cytochem.* **46**, 19–27.
12. Lin, W., Burgess, R. W., Dominguez, B., Pfaff, S. L., Sanes, J. R., and Lee, K. F. (2001) Distinct roles of nerve and muscle in postsynaptic differentiation of the neuromuscular synapse. *Nature* **410**, 1057–1064.
13. Harvey, S. J., Jarad, G., Cunningham, J., Rops, A. L., van der Vlag, J., Berden, J. H., et al (2007) Disruption of glomerular basement membrane charge through podocyte-specific mutation of agrin does not alter glomerular permselectivity. *Am J Pathol.* **171**, 139–152.
14. Yard, B. A., Kahlert, S., Engelleiter, R., Resch, S., Waldherr, R., Groffen, A. J., et al (2001) Decreased glomerular expression of agrin in diabetic nephropathy and podocytes, cultured in high glucose medium. *Exp Nephrol.* **9**, 214–222.
15. van den Hoven, M. J., Rops, A. L., Bakker, M ., Aten, J., Rutjes, N., Roestenberg, P., et al (2006) Increased expression of heparanase in overt diabetic nephropathy. *Kidney Int.* **70**, 2100–2108.
16. Wijnhoven, T. J., van den Hoven, M. J., Ding, H., van Kuppevelt, T. H., van der Vlag, J., Berden, J. H., et al (2008) Heparanase induces a differential loss of heparan sulfate domains in overt diabetic nephropathy. *Diabetologia* **51**, 372–382.
17. Joosten, S. A., Siipkens, Y. W., van Ham, V., Trouw, L. A., van der Vlag, J., van den Heuvel, B., et al (2005) Antibody response against the glomerular basement membrane protein agrin in patients with transplant glomerulopathy. *Am J Transplant.* **5**, 383–393.
18. Groffen, A. J., Hop, F. W., Tryggvason, K., Dijkman, H., Assmann, K. J., Veerkamp, J. H., et al (1997) Evidence for the existence of multiple heparan sulfate proteoglycans in the human glomerular basement membrane and mesangial matrix. *Eur J Biochem.* **247**, 175–182.
19. Groffen, A. J. A., Veerkamp, J. H., Monnens, L. A. H., and van den Heuvel, L. P. (1999) Recent insights into the structure and functions of heparan sulfate proteoglycans in the human glomerular basement membrane. *Nephrol Dial Transplant.* **14**, 2119–2129.
20. Handler, M., Yurchenco, P. D., and Iozzo, R. V. (1997) Developmental expression of perlecan during murine embryogenesis. *Dev Dyn.* **210**, 130–145.
21. Arikawa-Hirasawa, E., Watanabe, H., Takami, H., Hassell, J. R., and Yamada, Y. (1999) Perlecan is essential for cartilage and cephalic development. *Nature Gen.* **23**, 354–358.
22. Costell, M., Gustafsson, E., Aszodi, A., Mörgelin, M., Bloch, W., Hunziker, E., et al (1999) Perlecan maintains the integrity of cartilage and some basement membranes. *J Cell Biol.* **147**, 1109–1122.

23. Rossi, M., Morita, H., Sormunen, R., Airenne, S., Kreivi, M., Wang, L., et al (2003) Heparan sulfate chains of perlecan are indispensable in the lens capsule but not in the kidney. *EMBO J.* **22**, 236–245.
24. Zhou, Z., Wang, J., Cao, R., Morita, H., Soininen, R., Chan, K. M., et al (2004) Impaired angiogenesis, delayed wound healing and retarded tumor growth in perlecan heparan sulfate-deficient mice. *Cancer Res.* **64**, 4699–4702.
25. Tran, P. K., Tran-Lundmark, K., Soininen, R., Tryggvason, K., Thyberg, J., and Hedin, U. (2004) Increased intimal hyperplasia and smooth muscle cell proliferation in transgenic mice with heparan sulfate-deficient perlecan. *Circ Res.* **94**, 550–55.8
26. Morita, H., Yoshimura, A., Inui, K., Ideura, T., Watanabe, H., Wang, L., et al (2005) Heparan sulfate of perlecan is involved in glomerular filtration. *J Am Soc Nephrol.* **16,** 1703–1710.
27. Goldberg, S., Harvey, S. J., Cunningham, J., Tryggvason, K., and Miner, J. H. (2009) Glomerular filtration is normal in the absence of both agrin and perlecan-heparan sulfate from the glomerular basement membrane. *Nephrol Dial Transplant.* **24**, 2044–2051.
28. Conde-Knape, K. (2001) Heparan sulfate proteoglycans in experimental models of diabetes: a role for perlecan in diabetes complications. *Diabetes Metab Res Rev.* **17**, 412–421.
29. Husain, M., D'Agati, V. D., He, J. C., Klotman, M. E., and Klotman, P. E. (2005) HIV-1 Nef induces dedifferentiation of podocytes *in vivo*: a characteristic feature of HIVAN. *AIDS* **19**, 1975–1980.
30. Arora, S., Husain, M., Kumar, D., Patni, H., Pathak, S., Mehrotra, D., et al (2009) Human immunodeficiency virus downregulates podocyte apoE expression. *Am J Physiol Renal Physiol.* **297**, F653-F661.
31. Halfter, W., Dong, S., Schurer, B., and Cole, G. J. (1998) Collagen XVIII is a basement membrane heparan sulfate proteoglycan. *J Biol Chem.* **273**, 25404–25412.
32. Dong, S., Cole, G. J., and Halfter, W. (2003) Expression of type XVIII collagen and localization of its glycosaminoglycan attachment sites. *J Biol Chem.* **278**, 1700–1707.
33. Saarela, J., Rehn, M., Oikarinen, A., Autio-Harmainen, H., and Pihlajaniemi, T. (1998) The short and long forms of type XVIII collagen show clear tissue specificities in their expression and location in basement membrane zones in humans. *Am J Pathol.* **153**, 611–626.
34. Suzuki, O. T., Sertie, A. L., Der Kaloustian, V.M., Kok, F., Carpenter, M., Murray, J., et al (2002) Molecular analysis of collagen XVIII reveals novel mutations, presence of a third isoform, and possible genetic heterogeneity in Knobloch syndrome. *Am J Hum Genet.* **71**, 1320–1329.
35. Miosge, N., Simniok, T., Sprysch, P., and Herken, R. (2003) The collagen type XVIII endostatin domain is co-localized with perlecan in basement membranes *in vivo*. *J Histochem Cytochem.* **51**, 285–296.
36. O'Reilly, M. S., Boehm, T., Shing, Y., Fukai, N., Vasios, G., Lane, W. S., et al (1997) Endostatin: an endogenous inhibitor of angiogenesis and tumor growth. *Cell* **88**, 277–285.
37. Karumanchi, S. A., Jha, V., Ramachandran, R., Karihaloo, A., Tsiokas, L., Chan, B., et al (2001) Cell surface glypicans are low-affinity endostatin receptors. *Mol Cell.* **7**, 811–822.
38. Karihaloo, A., Karumanchi, S. A., Barasch, J., Jha, V., Nickel, C. H., Yang, J., et al (2001) Endostatin regulates branching morphogenesis of renal epithelial cells and ureteric bud. *Proc Natl Acad Sci USA* **98**, 12509–12514.
39. Fukai, N., Eklund, L., Marneros, A. G., Oh, S. P, Keene, D. R., Tamarkin, L., et al (2002) Lack of collagen XVIII/endostatin results in eye abnormalities. *EMBO J.* **21**, 1535–1544.
40. Utriainen, A., Sormunen, R., Kettunen, M., Carvalhaes, L. S., Sajanti, E., Eklund, L., et al (2004) Structurally altered basement membranes and hydrocephalus in a type XVIII collagen deficient mouse line. *Hum Mol Genet.* **13**, 2089–2099.
41. Hamano, Y., Okude, T., Shirai, R., Sato, I., Kimura, R., Ogawa, M., et al (2010) Lack of collagen XVIII/endostatin exacerbates immune-mediated glomerulonephritis. *J Am Soc Nephrol.* **21**, 1445–1455.
42. Ichinose, K., Maeshima, Y., Yamamoto, Y., Kitayama, H., Takazawa, Y., Hirokoshi, K., et al (2005) Antiangiogenic endostatin peptide ameliorates renal alterations in the early stage of a type 1 diabetic nephropathy model. *Diabetes* **54**, 2891–2903.
43. Kawashima, H., Watanabe, N., Hirose, M., Sun, X., Atarashi, K., Kimura, T., et al (2003) Collagen XVIII, a basement membrane heparan sulfate proteoglycan, interacts with L-selectin and monocyte chemoattractant protein-1. *J Biol Chem.* **278**, 13069–13076.
44. Celie, J. W., Rutjes, N. W. P., Keuning, E. D., Soininen, R., Heljasvaara, R., Pihlajaniemi, T., et al (2007) Subendothelial heparan sulfate proteoglycans become major L-selectin and monocyte chemoattractant protein-1

ligands upon renal ischemia/reperfusion. *Am J Pathol.* **170**, 1865–1878.

45. Bellini, M. H., Coutinho, E. L., Filgueiras, T. C., Maciel, T. T., and Schor, N (2007) Endostatin expression in the murine model of ischaemia/reperfusion-induced acute renal failure. *Nephrology (Carlton)* **12**, 459–465.
46. Maciel, T. T., Coutinho, E. L., Soares, D., Achar, E., Schor, N., and Bellini, M. H. (2008) Endostatin, an antiangiogenic protein, is expressed in the unilateral ureteral obstruction mice model. *J Nephrol.* **21**, 753–760.
47. Cichy, M. C., Rocha, F. G., Tristão, V. R., Pessoa, E. A., Cenedeze, M. A., Nürmberg Junior, R., et al (2009) Collagen XVIII/endostatin expression in experimental endotoxemic acute renal failure. *Braz J Med Biol Res.* **42,** 1150–1155.
48. Passos-Bueno, M. R., Suzuki, O. T., Armelin-Correa, L. M., Sertié, A. L., Errera, F. I., Bagatini, K., et al (2006) Mutations in collagen 18a1 (COL18A1) and their relevance to the human phenotype. *Anais Acad Bras Cinècmias* **78**, 123–131.
49. Williams, T. A., Kirkby, G. R., Williams, F., and Ainsworth, J. R. (2008) A phenotypic variant of Knobloch syndrome. *Ophthalmic Genet.* **29**, 85–86.
50. Czeizel, A. E., Göblyös, P., Kustos, G., Mester, E., and Paraicz, E. (1992) The second report of Knobloch syndrome. *Am J Med Genet.* **42**, 777–779.
51. Xian, X., Gopal, S., and Couchman, J. R. (2010) Syndecans as receptors and organizers of the extracellular matrix. *Cell Tissue Res.* **339**, 31–46.
52. Cevikbas, F., Schaefer, L., Uhlig, P., Robenek, H., Theilmeier, G., Echtermeyer, F., et al (2008) Unilateral nephrectomy leads to up-regulation of syndecan-2 and TGF-beta-mediated glomerulosclerosis in syndecan-4 deficient male mice. *Matrix Biol.* **27**, 42–52.
53. David, G., Bai, X. M., Van der Schueren, B., Marynen, P., Cassiman, J. J., and Van den Berghe, H. (1993) Spatial and temporal changes in the expression of fibroglycan (syndecan-2) during mouse embryonic development. *Development* **119**, 841–854.
54. Tsuzuki, S., Kojima, T., Katsumi, A., Yamazaki, T., Sugiura, I., and Saito, H. (1997) Molecular cloning, genomic organization, promoter activity, and tissue-specific expression of the mouse ryudocan gene. *J Biochem.* **122**, 17–24.
55. Ishiguro, K., Kadomatsu, K., Kojima, T., Muramatsu, H., Matsuo, S., Kusugami, K., et al (2001) Syndecan-4 deficiency increases susceptibility to k-carrageenan-induced renal damage. *Lab Invest.* **81**, 509–516.
56. Rops, A. L., Götte, M., Baselmans, M. H., van den Hoven, M. J., Steenbergen, E. J., Lensen, J. F., et al (2007) Syndecan-1 deficiency aggravates anti-glomerular basement membrane nephritis. *Kidney Int.* **72**, 1204–1215.
57. Rops, A. L., van den Hoven, M. J., Baselmans, M. M., Lensen, J. F., Wijnhoven, T. J., van den Heuvel, L. P., et al (2008) Heparan sulfate domains on cultured activated glomerular endothelial cells mediate leukocyte trafficking. *Kidney Int.* **73**, 52–62.
58. Yung, S., Woods, A., Chan, T. M., Davies, M., Williams, J. D., and Couchman, J. R. (2010) Syndecan-4 up-regulation in proliferative renal disease is related to microfilament organization. *FASEB J.* **15**, 1631–1633.
59. Filmus, J., Capurro, M., and Rast, J. (2009) Glypicans. *Genome Biol.* **9**, 224.
60. Litwack, E. D., Ivins, J. K., Kumbasar, A., Paine-Saunders, S., Stipp, C. S., and Lander, A. D. (1998) Expression of the heparan sulfate proteoglycan glypican-1 in the developing rodent. *Dev Dyn.* **211**, 72–87.
61. Grisaru, S., Cano-Gauci, D., Tee, J., Filmus, J., and Rosenblum, N. D. (2001) Glypican-3 modulates BMP- and FGF-mediated effects during renal branching morphogenesis. *Dev Biol.* **231**, 31–46.
62. Pellegrini, M., Pilia, G., Pantano, S., Lucchini, F., Uda, M., Fumi, M., et al (1998) Gpc3 expression correlates with the phenotype of the Simpson-Golabi-Behmel syndrome. *Dev Dyn.* **213**, 431–439.
63. Watanabe, K., Yamada, H., and Yamaguchi, Y. (1995) K-glypican: a novel GPI-anchored heparan sulfate proteoglycan that is highly expressed in developing brain and kidney. *J Cell Biol.* **130**, 1207–1218.
64. Saunders, S., Saunders, S. P., and Lander, A. D. (1997) Expression of the cell surface proteoglycan glypican-5 is developmentally regulated in kindey, limb and brain. *Dev Biol.* **190**, 78–93.
65. Veugelers, M., De Cat, B., Ceulemans, H., Bruystens, A. M., Coomans, C., Dürr, J., et al (1999) Glypican-6, a new member of the glypican family of cell surface heparan sulfate proteoglycans. *J Biol Chem.* **274**, 26968–26977.
66. Jen, Y. H., Musacchio, M., and Lander, A. D. (2009) Glypican-1 controls brain size through regulation of fibroblast growth factor signaling in early neurogenesis. *Neural Dev.* **4**, 33.

67. Cano-Gauci, D. F., Song, H. H., Yang, H., McKerlie, C., Choo, B., Shi, W., et al (1999) Glypican-3-deficient mice exhibit developmental overgrowth and some of the abnormalities typical of Simpson-Golabi-Behmel syndrome. *J Cell Biol.* **146**, 255–264.
68. Ng, A., Wong, M., Viviano, B., Erlich, J. M., Alba, G., Pflederer, C., et al (2009) Loss of glypican-3 function causes growth factor-dependent defects in cardiac and coronary vascular development. *Dev Biol.* **335**, 208–215.
69. Hartwig, S., Hu, M. C., Cella, C., Piscione, T., Filmus, J., and Rosenblum, N. D. (2005) Glypican-3 modulates inhibitory Bmp2-Smad signaling to control renal development *in vivo*. *Mech Dev.* **122**, 928–938.
70. Pilia, G., Hughes-Benzie, R. M., MacKenzie, A., Baybayan, P., Chen, E. Y., Huber, R., et al (1996) Mutations in GPC3, a glypican gene, cause the Simpson-Golabi-Behmel overgrowth syndrome. *Nat Genet.* **12**, 241–247.
71. Xiong, J., Wang, Y., Zhu, Z., Liu, J., Wang, Y., Zhang, C., et al (2007) NG2 proteoglycan increases mesangial cell proliferation and extracellular matrix production. *Biochem Biophys Res Commun.* **361**, 960–967.
72. Amenta, P. S., Scivoletti, N. A., Newman, M. D., Sciancalepore, J. P., Li, D., and Myers, J. C. (2005) Proteoglycan-collagen XV in human tissues is seen linking banded collagen fibers subjacent to the basement membrane. *J Histochem Cytochem.* **53**, 165–176.
73. Hägg, P. M., Hägg, P. O., Peltonen, S., Autio-Harmainen, H., and Pihlajaniemi, T. (1997) Location of type XV collagen in human tissues and its accumulation in the interstitial matrix of the fibrotic kidney. *Am J Pathol.* **150**, 2075–2086.
74. Muona, A., Eklund, L., Väisänen, T., and Pihlajaniemi, T. (2002) Developmentally regulated expression of type XV collagen correlates with abnormalities in *Col15a1(−/−)* mice. *Matrix Biol.* **21**, 89–102.
75. Lauer, M., Scruggs, B., Chen, S., Wassenhove-McCarthy, D, and McCarthy, K. J. (2007) Leprecan distribution in the developing and adult kidney. *Kidney Int.* **72**, 82–91.
76. Vranka, J. A., Pokidysheva, E., Hayashi, L., Zientek, K., Mizuno, K., Ishikawa, Y., et al (2010) Prolyl 3-hydroxylase 1 null mice display abnormalities in fibrillar collagen-rich tissues such as tendons, skin, and bones. *J Biol Chem.* **285**, 17253–17262.
77. McCarthy, K. J., Accavitti, M. A., and Couchman, J. R. (1989) Immunological characterization of a basement membrane-specific chondroitin sulfate proteoglycan. *J Cell Biol.* **109**, 3187–3198.
78. McCarthy, K. J., Bynum, K., St John, P. L., Abrahamson, D. R., and Couchman, J. R. (1993) Basement membrane proteoglycans in glomerular morphogenesis: chondroitin sulfate proteoglycan is temporally and spatially restricted during development. *J Histochem Cytochem.* **41**, 401–414.
79. Wu, R. R. and Couchman, J. R. (1997) cDNA cloning of the basement membrane chondroitin sulfate proteoglycan core protein, bamacan: a five domain structure including coiled-coil motifs. *J Cell Biol.* **136**, 433–444.
80. Ehara, T., Carone, F. A., McCarthy, K. J., and Couchman, J. R. (1994) Basement membrane chondroitin sulfate proteoglycan alterations in a rat model of polycystic kidney disease. *Am J Pathol.* **144**, 612–621.
81. McCarthy, K. J., Abrahamson, D. R., Bynum, K. R., St John, P. L., and Couchman, J. R. (1994) Basement membrane-specific chondroitin sulfate proteoglycan is abnormally associated with the glomerular capillary basement membrane of diabetic rats. *J Histochem Cytochem.* **42**, 473–484.
82. Ghiselli, G., Siracusa, L. D., and Iozzo, R. V. (1999) Complete cDNA cloning, genomic organization, chromosomal assignment, functional characterization of the promoter, and expression of the murine Bamacan gene. *J Biol Chem.* **274**, 17384–17393.
83. Schaefer, L. and Iozzo, R. V. (2008) Biological functions of the small leucine-rich proteoglycans: from genetics to signal transduction. *J Biol Chem.* **283**, 21305–21309.
84. Stokes, M. B., Holler, S., Cui, Y., Hudkins, K. L., Eitner, F., Fogo, A., et al (2000) Expression of decorin, biglycan, and collagen type I in human renal fibrosing disease. *Kidney Int.* **57**, 487–498.
85. Schaeffer, L., Grone, H.-J., Raslik, I., Robenek, H., Ugorcakova, J., Budny, S., et al (2000) Small proteoglycans of normal adult human kidney: Distinct expression patterns of decorin, biglycan, fibromodulin and lumican. *Kidney Int.* **58**, 1557–1568.
86. Schaefer, L., Mihalik, D., Babelova, A., Krzyzankova, M., Gröne, H. J., Iozzo, R. V., et al (2004) Regulation of fibrillin-1 by biglycan and decorin is important for tissue preservation in the kidney during pressure-induced injury. *Am J Pathol.* **165**, 383–396.
87. Williams, K. J., Qiu, G., Usui, H. K., Dunn, S. R., McCue, P., Bottinger, E., et al (2007) Decorin deficiency enhances progressive nephropathy in diabetic mice. *Am J Pathol.* **171**, 1441–1450.
88. Merline, R., Lazaroski, S., Babelova, A., Tsalastra-Greul, W., Pfeilschifter, J., Schluter, K. D., et al (2009) Decorin deficiency in

diabetic mice: aggravation of nephropathy due to overexpression of profibrotic factors, enhances apoptosis and mononuclear cell infiltration. *J Physiol Pharmacol.* **60**, 5–13.

89. Schaefer, L., Macakova, K., Raslik, I., Micegova, M., Gröne, H. J., Schönherr, E., et al (2002) Absence of decorin adversely influences tubulointerstitial fibrosis of the obstructed kidney by enhanced apoptosis and increased inflammatory reaction. *Am J Pathol.* **160**, 1181–1191.
90. Isaka, Y., Brees, D. K., Ikegaya, K., Kaneda, Y., Imai, E., Noble, N. A., et al (1996) Gene therapy by skeletal muscle expression of decorin prevents fibrotic disease in rat kidney. *Nat Med.* **2**, 418–423.
91. Border, W. A., Noble, N. A., Yamamoto, T., Harper, J. R., Yamaguchi, Y., Pierschbacher, M. D., et al (1992) Natural inhibitor of transforming growth factor-beta protects against scarring in experimental kidney disease. *Nature* **360**, 361–364.
92. Zhang, Z., Wu, F., Zheng, F., and Li, H. (2010) Adenovirus-mediated decorin gene transfection has therapeutic effects in a streptozocin-induced diabetic rat model. *Nephron Exp Nephrol.* **116**, e11–21.
93. Wu, H., Wang, S., Xue, A., Liu, Y., Liu, Y., Wang, H., et al (2008) Overexpression of decorin induces apoptosis and cell growth arrest in cultured rat mesangial cells *in vitro*. *Nephrology (Carlton)* **13**, 607–615.
94. Schaefer, L., Beck, K. F., Raslik, I., Walpen, S., Mihalik, D., Micegova, M., et al (2003) Biglycan, a nitric oxide-regulated gene, affects adhesion, growth, and survival of mesangial cells. *J Biol Chem.* **278**, 26227–26237.
95. Lin, X., Wei, G., Shi, Z., Dryer, L., Esko, J. D., Wells, D. E., et al (2000) Disruption of gastrulation and heparan sulfate biosynthesis in EXT1-deficient mice. *Dev Biol.* **224**, 299–311.
96. Chen, S., Wassenhove-McCarthy, D. J., Yamaguchi, Y., Holzman, L. B., van Kuppevelt, T. H., Jenniskens, G. J., et al (2008) Loss of heparan sulfate glycosaminoglycan assembly in podocytes does not lead to proteinuria. *Kidney Int.* **74**, 289–299.
97. Chen, S., Wassenhove-McCarthy, D., Yamaguchi, Y., Holzman, L., van Kuppevelt, T. H., Orr, A. W., et al. (2010) Podocytes require the engagement of cell surface heparan sulfate proteoglycans for adhesion to extracellular matrices. *Kidney Int* **78**, 1088–1099.
98. Roberts, I. S. D. and Gleadle, J. M. (2008) Familial nephropathy and multiple exostoses with exostosin-1 (EXT1) gene mutation. *J Am Soc Nephrol.* **19**, 450–453.
99. Condac, E., Silasi-Mansat, R., Kosanke, S., Schoeb, T., Towner, R., Lupu, F., et al (2007) Polycystic disease caused by deficiency in xylosyltransferase 2, an initiating enzyme of glycosaminoglycan biosynthesis. *Proc Natl Acad Sci USA* **104**, 9416–9421.
100. Hendig, D., Tarnow, L., Kuhn, J., Kleesiek, K., and Götting, C. (2008) Identification of a xylosyltransferase II gene haplotype marker for diabetic nephropathy in type 1 diabetes. *Clin Chim Acta* **398**, 90–94.
101. Li, J. P., Gong, F., Hagner-McWhirter, A., Forsberg, E., Abrink, M., Kisilevsky, R., et al (2003) Targeted disruption of a murine glucuronyl C5-epimerase gene results in heparan sulfate lacking L-iduronic acid and in neonatal lethality. *J Biol Chem.* **278**, 28363–28366.
102. Jia, J., Maccarana, M., Zhang, X., Bespalov, M., Lindahl, U., and Li, J. P. (2009) Lack of L-iduronic acid in heparan sulfate affects interaction with growth factors and cell signaling. *J Biol Chem.* **284**, 15942–15950.
103. Bullock, S. L., Fletcher, J. M., Beddington, R. S., and Wilson, V. A. (1998) Renal agenesis in mice homozygous for a gene trap mutation in the gene encoding heparan sulfate 2-sulfotransferase. *Genes and Dev.* **12**, 1894–1906.
104. Merry, C. L., Bullock, S. L., Swan, D. C., Backen, A. C., Lyon, M., Beddington, R. S., et al (2001) The molecular phenotype of heparan sulfate in the *Hs2st*$^{-/-}$ mutant mouse. *J Biol Chem.* **276**, 35429–35434.
105. McLaughlin, D., Karlsson, F., Tian, N., Pratt, T., Bullock, S. L., Wilson, V. A., et al (2003) Specific modification of heparan sulphate is required for normal cerebral cortical development. *Mech Dev.* **120**, 1481–1488.
106. Shah, M. M., Sakurai, H., Sweeney, D. E., Gallegos, T. F., Bush, K. T., Esko, J. D., et al (2010) Hs2st mediated kidney mesenchyme induction regulates early ureteric bud branching. *Dev Biol.* **339**, 354–365.
107. Edge, A. S. B. and Spiro, R. G. (2000) A specific structural alteration in the heparan sulphate of human glomerular basement membrane in diabetes. *Diabetologia* **43**, 1056–1059.
108. Ratelade, J., Arrondel, C., Hamard, G., Garbay, S., Harvey, S., Biebuyck, N., et al (2010) A murine model of Denys-Drash syndrome reveals novel transcriptional targets of WT1 in podocytes. *Hum Mol Genet.* **19**, 1–15.

109. Langsdorf, A., Schumacher, V., Shi, X., Tran, T., Zaia, J., Jain, S., et al. (2011) Expression regulation and function of Sulfs in the spermatogonial stem cell niche. *Glycobiology* **21**, 152–161.

110. Holst, C. R., Bou-Reslan, H., Gore, B. B., Wong, K., Grant, D., Chalasani, S., et al (2007) Secreted sulfatases Sulf1 and Sulf2 have overlapping yet essential roles in mouse neonatal survival. *PLoS One* **2**, e575.

111. Lum, D. H., Tan, J., Rosen, S. D., and Werb, Z. (2007) Gene trap disruption of the mouse heparan sulfate 6-O-endosulfatase gene, Sulf2. *Mol Cell Biol.* **27**, 678–688.

112. Lamanna, W. C., Baldwin, R. J., Padva, M., Kalus, I., Ten Dam, G., van Kuppevelt, T. H., etal(2006)Heparansulfate6-O-endosulfatases: discrete *in vivo* activities and functional cooperativity. *Biochem J.* **400**, 63–73.

113. Ai, X., Kitazawa, T., Do, A. T., Kusche-Gullberg, M., Labosky, P. A., and Emerson, C. P. Jr. (2007) SULF1 and SULF2 regulate heparan sulfate-mediated GDNF signaling for esophageal innervation. *Development* **134**, 3327–3338.

114. Schumacher V.A., Schlötzer-Schrehardt U., Karumanchi S.A., Shi X., Zaia J., Jeruschke S., et al. (2011) WT1-dependent sulfatase expression maintains the normal glomerular filtration barrier. *J Am Soc Nephrol.* **22**, 1286–1296.

115. van den Hoven, M. J., Rops, A. L., Vlodavsky, I., Levidiotis, V., Berden, J. H., and van der Vlag, J. (2007) Heparanase in glomerular diseases. *Kidney Int.* **72**, 543–548.

116. Harvey, S. J. and Miner, J. H. (2008) Revisiting the glomerular charge barrier in the molecular era. *Curr Opin Nephrol Hypertens.* **17**, 393–398.

117. Li, J. P., Galvis, M. L., Gong, F., Zhang, X., Zcharia, E., Metzger, S., et al (2005) *In vivo* fragmentation of heparan sulfate by heparanase overexpression renders mice resistant to amyloid protein A amyloidosis. *Proc Natl Acad Sci.* **102**, 6473–6477.

Chapter 18

Lumican Promotes Corneal Epithelial Wound Healing

Chia-Yang Liu and Winston Whei-Yang Kao

Abstract

Lumican regulates collagenous matrix assembly as a keratan sulfate proteoglycan in the cornea and is also present in the connective tissues of other organs and embryonic corneal stroma as a glycoprotein. In normal unwounded cornea, lumican is expressed by stromal keratocytes. Interestingly, injured mouse corneal epithelium ectopically and transiently expresses lumican during the early phase of wound healing, suggesting a potential lumican functionality unrelated to regulation of collagen fibrillogenesis, e.g., modulation of epithelial cell adhesion or migration. Healing of a corneal epithelial injury in lumican knockout ($Lum^{-/-}$) mice was significantly delayed compared with $Lum^{+/-}$ mice. Addition of purified lumican to cultured medium promoted re-epithelialization and enhanced cell proliferation of wild-type mouse corneal epithelial cells in an organ culture. Therefore, administration of lumican may be beneficial for treating epithelial defects in the cornea and other tissues.

Key words: Lumican, Extracellular matrices, Small leucine-rich proteoglycan, Cornea, Epithelium, Wound healing

1. Introduction

Lumican, a member of small leucine-rich proteoglycan (SLRP) family, is one of the major extracellular components in interstitial collagenous matrices of the corneal stroma, aorta, skin, skeletal muscle, lung, kidney, bone, cartilage, and intervertebral discs (1–8). In the cornea, lumican contains keratan sulfate chains. However, it is present as a low or nonsulfated glycoprotein (50–57 kDa) in most interstitial connective tissues other than corneas (1, 2, 9–11). Its wide distribution implies that it has multiple functions in tissue morphogenesis and maintenance of tissue homeostasis. This is best illustrated by the multiple clinical manifestations observed in $Lum^{-/-}$ mice that exhibit corneal opacity, skin and tendon fragility, delayed wound healing, and low fertility. Indeed, lumican has been shown to play essential roles in corneal transparency by regulating

Françoise Rédini (ed.), *Proteoglycans: Methods and Protocols*, Methods in Molecular Biology, vol. 836,
DOI 10.1007/978-1-61779-498-8_18, © Springer Science+Business Media, LLC 2012

collagen fibrillogenesis (8), in wound healing by modulating epithelial cell migration (12), and in the epithelium-mesenchyme transition of the injured lens (13). These results have led to the speculation that lumican may play an active role in corneal epithelial wound healing.

The mouse corneal epithelium has been an excellent and frequently used model system for the study of wound healing. Commercially available trephine (2 mm in diameter) in conjunction with a corneal rust ring remover Algerbrush can easily generate a consistent epithelial wound for quantitative analysis of epithelial wound closure. Recombinant mouse lumican is commercially available. Bromodeoxyuridine (BrdU) incorporation can be performed to measure cell proliferation index by immunohistochemistry with anti-BrdU antibody.

2. Materials

2.1. Generation of 2-mm Diameter Corneal Epithelial Debridement Wound in Mouse Eye

1. C57BL/6 mice: age 6–8 weeks (Jackson Laboratories, Bar Harbor, ME).
2. Stereomicroscope attached with Axiovision digital camera: We used SV11; (Carl Zeiss Meditec, Dublin, CA) (see Note 1).
3. Trephine (2 mm in diameter).
4. Algerbrush II™—Rust Ring Remover with a 0.5-mm burr (Alger Equipment Co., Inc., Lago Vista, TX).
5. 1× Keratimne/Xylaxin solution: ketamine (20 mg/mL)/xylazin (5 mg/mL). For 10 mL solution, mix 2 mL of ketamine hydrochloride (100 mg/mL) (Bioniche Pharma USA LLC272 Lake Forest, IL) and 0.5 mL of xylazine (100 mg/mL) into 7.5 mL 1× PBS. Store at room temperature (r.t.).
6. Proparacaine hydrochloride 0.5% (Alcaine; Alcon Laboratories, Inc., Fort Worth, TX).
7. Fluorescein strips.
8. 1× Phosphate-buffered saline (PBS): 10 mM Na_2HPO_4, 1.76 mM KH_2PO_4, 2.7 mM KCl, 137 mM NaCl, pH 7.4.

2.2. Ex Vivo Mouse Eyeball Culture

1. Dulbecco's Modified Eagle's Medium (DMEM) (Gibco/BRL, Bethesda, MD) supplemented with 1% fetal bovine serum (FBS, HyClone, Ogden, UT) (see Note 2).
2. Purified lumican (cat# 2745LU, R&D Systems, Inc. Minneapolis, MN) (see Note 3).
3. BD Falcon™ 24-well Multiwell Plate.
4. Image-analysis software (Image Beta 4.02; Scion Corp., Frederick, MD).

2.3. Immuno-histochemistry

1. 4% Paraformaldehyde (PFA): freshly prepare from 16% PFA (Electron Microscopy Sciences, Inc., Hatfield, PA) in 1× PBS.
2. HCl (3 N): freshly prepare from concentrated HCl (12 N) (see Note 4).
3. BrdU (Sigma-Aldrich): 10 mg/mL stock in PBS and store at –20°C.
4. Borate buffer (0.1 M, pH 8.5): 3.8 g Sodium borate (MW 381.4) dissolved in 100 mL distilled water, Mix to dissolve and adjust pH to 8.5.
5. Anti-BrdU antibody: mouse anti-BrdU monoclonal antibody (Lab Vision Corp.) (see Note 5).
6. Fluorophore-tagged secondary antibodies: Alexa Fluor® 488- or Alexa Fluor® 555-labeled rabbit anti-mouse IgG (H+L) (2 mg/mL stock) (Molecular probes, Inc.).
7. 1× DAPI Working Solution: 100 ng/mL or 300 nM in 1× PBS.
8. PBS-T: 1× PBS with 1% Triton-X100.
9. Blocking B: PBS-T with 2.5% (W/V) bovine serum albumin (Sigma).

3. Methods

3.1. Generation of 2-mm Diameter Corneal Epithelial Debridement Wound in Mouse Eye

1. C57BL/6 mice were anesthetized by intraperitoneal injections of ketamine hydrochloride (2 μg/g body weight) and xylazine (0.4 μg/g body weight) (see Note 6).
2. Topical application of one drop of proparacaine to each eye.
3. The central cornea is marked by a trephine (2 mm in diameter).
4. The corneal epithelium is debrided with Algerbrush II™ under a stereomicroscope (SV11; Carl Zeiss Meditec, Dublin, CA).
5. The wounded cornea is flushed with PBS to remove any epithelial debris.
6. Take one fluorescein strip soak into 1 mL PBS, apply 50 μL fluorescein solution to the wounded eye, wait for 30 s, and wash out the fluorescein with PBS.
7. Take photo with blue filter on fluorescein-stained area as a reference of wounded area (see Fig. 1A1, A5).
8. The animal then is sacrificed, and the eyeballs are enucleated and cultured in DMEM supplemented with 1% FBS and 50 μg/mL gentamicin, with or without 10 μg/mL purified lumican, in a humidified atmosphere of 5% CO_2 at 37°C for 12, 24, and 48 h, respectively.

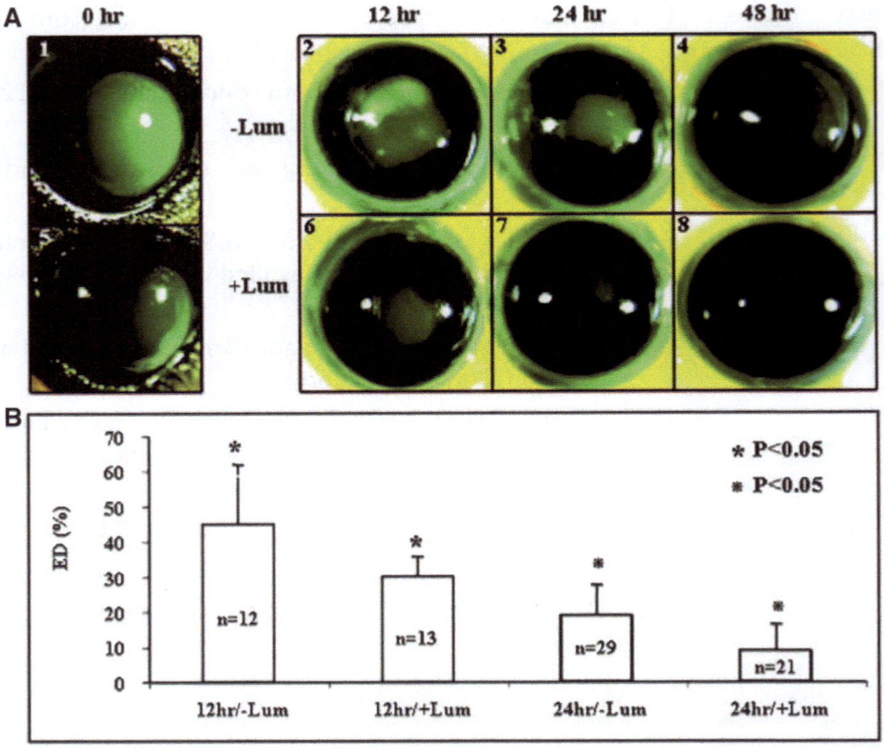

Fig. 1. Purified lumican promoted corneal epithelial wound healing in wild-type eyes. (**a**) Representative photographs of fluorescein-stained eyes (*green*). An epithelial defect (2 mm in diameter) was created at the center of the corneas (*A1*, *A5*). Mouse eyes were enucleated and cultured in DMEM with 1% FBS (*A2–A4*) or in DMEM with 1% FBS plus purified AM lumican (10 μg/mL; *A6–A8*), to allow re-epithelialization for the different periods indicated. (**b**) Percentage of the remaining epithelial defect (ED). Note that the lumican-treated (+Lum) eyes healed better than the nonlumican-treated ones (–Lum) at 12 and 24 h after debridement (courtesy of Yeh et al. (14)).

9. To label the proliferating cells, 40 μg/mL of BrdU is added to the cultured eyeballs 1 h prior to harvest cells after 12, 24, and 48 h of cultivation, respectively.
10. After incubation, the extent of corneal wound closure is examined by fluorescein staining and photographed with a digital camera. 12 h (see Fig. 1A2, A6), 24 h (see Fig. 1A3, A7), and 48 h (see Fig. 1A4, A8), respectively.
11. The circumference of the wound margin of each mouse eye, as projected onto the photograph, is traced on a digitizer, and the remaining defect area is determined by image-analysis software (Image Beta 4.02; Scion Corp., Frederick, MD). All measurements are counted in a masked fashion, and the size of epithelial defect is expressed as a percentage of the total corneal area (see Fig. 1b).
12. The eyeballs are then fixed in 4% PFA in PBS and embedded in paraffin.

3.2. Immunohistochemical Detection of BrdU

BrdU is a synthetic thymidine analog that gets incorporated into a cell's DNA when the cell is dividing (during the S-phase of the cell cycle). Antibodies against BrdU that are conjugated to fluorescent markers can be used to label these cells, thereby providing visual evidence of cell division. In general, cells undergo active proliferation uptake of more BrdU into the nucleus than that in nonproliferative cells.

1. Tissue and cells are fixed with 4% PFA for 16 h or overnight at 4°C.
2. Following fixation, wash in 1× PBS with 1% Triton-X100 (3× 5 min) at room temperature (r.t.).
3. Incubate in HCl (3 N) for 10 min on ice to break open the DNA structure of the BrdU-labeled cells.
4. Immediately after the acid washes, Borate buffer (0.1 M) is added to buffer the cells for 10 min at r.t.
5. Samples are then washed in 1× PBS-T (3× 5 min) at r.t.
6. Incubate in blocking B (1 h) prior to incubating overnight with anti-BrdU (1:100 in blocking B).
7. Following the incubation overnight, wash in PBS-T (3× 5 min) at r.t.
8. Samples can then be treated with AlexaFlour 488- and 555-conjugated anti-mouse IgG secondary antibody (1:250) and DAPI (1:1,000) for 30 min at 37°C, respectively.
9. Following the incubation, wash in PBS-T (3× 5 min) then dd-H_2O (1× 5 min) at r.t.
10. Sample can be mounted on coverslide and visualize the anti-BrdU-labeled and DAPI counterstained cells.

4. Notes

1. Any stereomicroscope should work, too.
2. 1% of FBS is to keep cultured eyeballs' survival and be able to respond to growth promoting agents such as lumican.
3. Lumican can be either purchased from R&Dsystem or prepared from human amniotic membrane (14).
4. For 100 mL, add 25 mL 12 N HCl into 75 mL double-distilled (dd) H_2O.
5. Many sources of anti-BrdU antibodies are commercially available.
6. For 10 g body weight mouse, inject 10 μL of K/X.

Acknowledgment

This work was supported by RO1 EY12486, EY11845, Research to Prevent Blindness, and Ohio Lion Eye Research Foundation.

References

1. Funderburgh, J. L., Funderburgh, M. L., Mann, M. M., and Conrad, G. W. (1991) Arterial lumican: properties of a corneal-type keratan sulfate proteoglycan from bovine aorta. *J Biol Chem.* **266**, 24773–24777.
2. Funderburgh, J. L., Caterson, B., and Conrad, G. W. (1987) Distribution of proteoglycans antigenically related to corneal keratan sulfate proteoglycan. *J Biol Chem.* **262**, 11634–11640.
3. Iozzo, R. V. and Murdoch, A. D. (1996) Proteoglycan of the extracellular environment: clues from the gene and protein side offer novel perspectives in molecular diversity and function. *FASEB J.* **10**, 598–614.
4. Knudson, C. B. and Knudson, W. (2001) Cartilage proteoglycans. *Semin Cell Dev Biol.* **12**, 69–78.
5. Ying, S., Shiraishi, A., Kao, C.W., Converse, R. L., Funderburgh, J. L., Swiergiel, J., et al. (1997) Characterization and expression of the mouse lumican gene. *J Biol Chem.* **272**, 30306–30313.
6. Raouf, A., Ganss, B., McMahon, C., Vary, C., Roughley, P. J., and Seth, A. (2002) Lumican is a major proteoglycan component of the bone matrix. *Matrix Biol.* **21**, 361–367.
7. Iozzo, R. V. and Danielson, K. G. (1999) Transcriptional and posttranscriptional regulation of proteoglycan gene expression. Prog *Nucleic Acids Res Mol Biol.* **62**, 19–53.
8. Chakravarti, S., Magnuson, T., Lass, J. H., Jepsen, K. J., LaMantia, C., and Carroll, H. (1998). Lumican regulates collagen fibril assembly: skin fragility and corneal opacity in the absence of lumican. *J Biol Chem.* **141**, 1277–1286.
9. Grover, J., Chen, X. N., Korenberg, J. R., and Roughley, P. J. (1995) The human lumican gene: organization, chromosomal location, and expression in articular cartilage. *J Biol Chem.* **270**, 21942–21949.
10. Corpuz, L. M., Funderburgh, J. L., Funderburgh, M. L., Bottomley, G. S., Prakash, S., and Conrad, G. W. (1996) Molecular cloning and tissue distribution of keratan. Bovine corneal keratan sulfate proteoglycan 37A. *J Bio Chem.* **271**, 9759–9763.
11. Funderburgh, J. L., Funderburgh, M. L., Mann, M. M., and Conrad, G. W. (1991) Physical and biological properties of keratan sulphate proteoglycan. *Biochem Soc Trans.* **19**, 871–876.
12. Saika, S., Shiraishi, A., and Saika, S. (2000) Role of lumican in the corneal epithelium during wound healing. *J Biol Chem.* **275**, 2607–2612.
13. Saika, S., Miyamoto, T., Tanaka, S. I., Yamanaka, O., Ohnishi, Y., Ooshima, A., et al. (2003) Response of lens epithelial cells to injury: role of lumican in epithelial-mesenchymal transition. *Invest Ophthalmol Vis Sci.* **44**, 2094–2102.
14. Yeh, L. K., Chen, W. L., Li, W., Espana, E. M., Ouyang, J., Kawakita, T., et al (2005) Soluble lumican glycoprotein purified from human amniotic membrane promotes corneal epithelial wound healing. *Invest Ophthalmol Vis Sci.* **46**, 479–486.

Chapter 19

Shedding of Cell Membrane-Bound Proteoglycans

Eon Jeong Nam and Pyong Woo Park

Abstract

Membrane-bound proteoglycans function primarily as coreceptors for many glycosaminoglycan (GAG)-binding ligands at the cell surface. The majority of membrane-bound proteoglycans can also function as soluble autocrine or paracrine effectors as their extracellular domains, replete with all GAG chains, are enzymatically cleaved and released from the cell surface by ectodomain shedding. In particular, the ectodomain shedding of syndecans, a major family of cell surface heparan sulfate proteoglycans, is an important post-translational mechanism that modulates diverse pathophysiological processes. Syndecan shedding is a tightly controlled process that regulates the onset, progression, and resolution of various infectious and noninfectious inflammatory diseases. This review describes methods to induce and measure the shedding of cell membrane-bound proteoglycans, focusing on syndecan shedding as a prototypic example.

Key words: Syndecan, Glypican, Heparan sulfate proteoglycan, Sheddase, Matrix metalloproteinase, ADAM, Inflammation, Infection, Host defense

1. Introduction

All nucleated mammalian cells express at least one membrane-bound proteoglycan on their cell surface. Membrane-bound proteoglycans include the syndecan and glypican families with four and six members in mammals, respectively, NG2 (AN2 in mice), betaglycan, thrombomodulin, α5β1 integrin, neuropilin-1, and CD44 (1–3). Core proteins of syndecans, NG2, CD44, betaglycan, neuropilin-1, α5β1 integrin, and thrombomodulin are type I transmembrane proteins, whereas core proteins of glypicans are attached to the cell surface through a glycosylphosphatidylinositol (GPI) anchor. Core proteins of membrane-bound proteoglycans have discrete glycosaminoglycan (GAG) attachment sites. Heparan sulfate (HS) chains are attached distal to the plasma membrane on all syndecans and chondroitin sulfate (CS) chains are also attached

Françoise Rédini (ed.), *Proteoglycans: Methods and Protocols*, Methods in Molecular Biology, vol. 836, DOI 10.1007/978-1-61779-498-8_19, © Springer Science+Business Media, LLC 2012

proximal to the cell surface on some syndecans (e.g., syndecan-1). GAG attachment sites in NG2 (CS) and CD44 (HS/CS) are located in the middle portion of the core protein, whereas those of glypicans (HS), thrombomodulin (CS), and betaglycan (HS/CS) are proximal to the cell surface.

On the cell surface, membrane-bound proteoglycans function primarily as a coreceptor that catalyzes the encounter between ligands and their respective signaling receptors (1–3). Membrane-bound proteoglycans can also function as soluble proteoglycans since the intact extracellular domain can be released from the cell surface by an enzymatic cleavage mechanism known as ectodomain shedding. Ectodomain shedding is of paramount importance to membrane-bound proteoglycan biology because it both rapidly changes the surface phenotype of affected cells by reducing the amount of ligand-binding GAGs and generating soluble proteoglycan ectodomains replete with all their GAGs that can function as autocrine or paracrine effectors. Syndecans and CD44 are shed by metalloproteinases, whereas glypicans are released from the cell surface by cleavage of their GPI anchor by phospholipases. Other surface proteoglycans are also found in soluble form, but how they are released from the cell surface is not known.

The underlying mechanisms of syndecan-1 shedding have been extensively studied. Syndecan-1 shedding is a highly regulated process that is stimulated in vitro by several inflammatory factors, and in vivo under certain pathological conditions (1, 4). Agonists of syndecan-1 shedding include growth factors, stress-related agonists, heparanase, and bacterial virulence factors, among others (see Table 1). The majority of these agonists enhance metalloproteinase-mediated syndecan-1 shedding at the cell surface in a protein tyrosine kinase-dependent manner. However, several metalloproteinases can shed syndecan-1 ectodomains (see Table 1), suggesting that the shedding agonists and the signaling pathways that they activate tightly control syndecan-1 shedding.

Accumulating evidence suggests that the shedding of membrane-bound proteoglycans is an important posttranslational mechanism that regulates pathophysiological processes. For example, syndecan-1 ectodomains are elevated in blood of patients with sepsis (5, 6), ischemia-reperfusion injury (7), graft-versus-host disease (8), and various cancers (9, 10), among other fluids from injured, inflamed, or cancerous tissues (11–13). Further, mouse studies have shown that the inflammatory response to toxins, chemicals, allergens, and pathogens is dysregulated in the absence of syndecan-1 or when its shedding is inhibited (14–19), suggesting that syndecan-1 shedding is mobilized to assure the adequate and correct functioning of inflammation. Increased levels of CD44 ectodomains are also detected in bronchoalveolar lavage fluids of patients with eosinophilic pneumonia (20), and the inflammatory response in *Escherichia coli*-induced pneumonia is amplified in CD44 null mice compared

Table 1
Partial list of syndecan-1 shedding agonists and sheddases

Shedding agonists	Sheddases
Phorbol esters (PMA, TPA) (24)	MMP7 (Matrilysin) (14, 32)
Sphingomyelinase/Ceramide (24)	MMP9 (gelatinase B) (37)
EGF family growth factors (24, 29)	MMP14 (MT1-MMP) (38)
FGF-2 (32)	ADAM17 (TACE) (39)
Insulin (33)	*Streptococcus pneumoniae* (pneumococcus) ZmpC (23)
TNFα (34, 35)	*Bacillus anthracis* (anthrax) Npr599 and InhA (40)
RANTES (36)	
SDF-1 (37)	
Cellular stress (hyperosmolarity, heat shock) (24)	
Reactive oxygen species (12)	
Heparanase (10)	
Pseudomonas aeruginosa LasA (27)	
Staphylococcus aureus α- and β-toxin (26)	
Bacillus anthracis AnlB, AnlO, and ClnA (28)	

to wild-type mice (21). Together, these data suggest that the shedding of membrane-bound proteoglycans may be a critical mechanism in the pathogenesis of a broad range of infectious and noninfectious diseases, and underscore the importance of developing means to precisely measure and analyze the shedding of membrane-bound proteoglycans. Dot immunoblotting or ELISA is used to measure levels of shed proteoglycan ectodomains. A concomitant decrease in levels of cell surface proteoglycans is measured by dot immunoblotting or FACS. Each of these methods is described below in detail.

2. Materials

2.1. Measurement of Syndecan Ectodomains by Dot Immunoblotting

1. Cells: Normal murine mammary gland (NMuMG) epithelial cell is the standard cell type for syndecan shedding assays because they express syndecan-1, -2, and -4 abundantly and syndecan-3 weakly (22), and NMuMG cells have been

demonstrated to respond to the majority of syndecan shedding agonists (23–28). For shedding assays, culture NMuMG cells to confluency or 1-day post confluency in 96-well plates (see Note 1). Other cell types that have been tested in shedding assays include mouse C127 mammary gland epithelial cells (27), SV40-transformed endothelial cells (24, 29), P3X63 myeloma cells (24), NIH3T3 fibroblasts (27) and LA4 lung epithelial cells (27), and human A549 lung epithelial cells (25), HSAEC small airway epithelial cells (28), A431 epidermal cells (25), CAG myeloma cells (30), and ARH-77 myeloma cells transformed without mouse syndecan-1 (30).

2. NMuMG cell culture medium: Dulbecco's Modified Eagle's Medium (DMEM) with 4.5 g/L glucose, L-glutamine, and sodium pyruvate supplemented with 10% FBS and 10 μg/mL insulin.
3. Trypsin-EDTA: Premade 0.25% trypsin and 2.21 mM ethylenediamine tetraacetic acid (EDTA) in Hank's Balanced Salt Solution (HBSS).
4. Shedding agonists: Phorbol 12-myristate 13-acetate (PMA) is dissolved at 10 mM in dimethyl sulfoxide (DMSO) and stored in aliquots at −80°C. Other commercially available shedding agonists and conditions include EGF (Peptrotech, Rocky Hill, NJ), C2 ceramide (VWR, West Chester, PA), thrombin (EMD Chemicals, Gibbstown, NJ), TNFα (Peprotech), *Staphylococcus aureus* α- and β-toxin (Toxin Technology, Sarasota, FL), hyperosmolarity (e.g., 600 mOsm NaCl for 2 h), and heat shock (e.g., 45°C for 1 h) (see Table 1).
5. Primary antibodies: 281-2 rat anti-mouse syndecan-1 ectodomain monoclonal antibody (BD Biosciences, San Jose, CA) and Ky8.2 rat anti-mouse syndecan-4 ectodomain monoclonal antibody (BD Biosciences). Several polyclonal antibodies for mouse syndecan-1 and -4, and several monoclonal and polyclonal antibodies for other mouse and human syndecans are also available commercially from several sources. Based on our experience, we recommend using 281-2 and Ky8.2 antibodies for mouse syndecan-1 and -4 studies, and B-B4 or BA-38 mouse anti-human syndecan-1 monoclonal antibodies (AbD Serotec, Raleigh, NC; Abcam, Cambridge, MA) and 5G9 mouse anti-human syndecan-4 monoclonal antibodies (Santa Cruz Biotechnology, Santa Cruz, CA) for human syndecan-1 and -4 studies.
6. Secondary antibodies: Horse radish peroxidase (HRP)-conjugated antibodies from Jackson ImmunoResearch Laboratories (West Grove, PA). We recommend using preadsorbed secondary antibodies to minimize background.

7. Purified syndecan standards: Purified syndecan-1 and -4 ectodomains. Native syndecan-1 and -4 ectodomains can be rapidly purified from the culture supernatant of NMuMG cells by DEAE ion exchange chromatography and 281-2 or Ky8.2 affinity chromatography. The typical yield is 5–10 μg from 1 L of NMuMG cell culture supernatant.
8. Buffers and solutions:
   (a) Acetate buffered saline with Tween-20 (ABS-T): 50 mM sodium acetate (NaOAc, pH 4.5), 150 mM NaCl, and 0.1% (v/v) Tween-20 (see Note 2).
   (b) Tris-buffered saline (TBS): 50 mM Tris, pH 7.5, and 150 mM NaCl.
   (c) Blocking buffer 1: TBS containing 10% (w/v) nonfat dry milk from VWR or Fisher Scientific (see Note 3).
   (d) Blocking buffer 2: TBS with 1% (w/v) nonfat dry milk and 0.1% (v/v) Tween-20.
9. Blotting paper: Whatman 3MM paper (VWR or Fisher Scientific). Precut the 3MM paper with a paper cutter to fit the size of the dot blotting apparatus. Handle with gloves to avoid contamination.
10. Cationic nylon membrane: Immobilon Ny+ (Millipore, Billerica, MA). Precut the Immobilon Ny+ membrane with a paper cutter to fit the size of the dot blotting apparatus. Handle with gloves to avoid contamination.
11. Dot blot apparatus: Whatman Minifold I Blotting System (VWR).
12. ECL (enhanced chemiluminescence) development reagent (Pierce Chemical, Rockford, IL).
13. Image analysis system: LAS-4000 Multifunctional Imaging Analysis System (Fujifilm, Stamford, CT).
14. X-ray film: Kodak Biomax MR-1 film (VWR or Fisher Scientific).

2.2. Measurement of Cell Surface Syndecan-1 by Mild Trypsinization and Dot Immunoblotting

1. TBS with 0.5 mM EDTA (TBS/EDTA): Dilute 500 mM stock solution of EDTA, pH 8, in TBS (see Note 4).
2. TPCK (L-1-tosylamide-2-phenylethyl chloromethyl ketone)-treated trypsin: TPCK-treated trypsin (Sigma-Aldrich) is dissolved at 10 μg/mL in TBS/EDTA. Trypsin is autolytic so use freshly dissolved enzyme for best results.
3. Soybean trypsin inhibitor: Soybean trypsin inhibitor (Sigma-Aldrich) is dissolved at 100 μg/mL in TBS/EDTA. Concentrated stock solution of 10 mg/mL can be stored at −80°C.
4. Materials for dot immunoblotting: As described in Subheading 2.1.

2.3. Measurement of Cell Surface Syndecans by FACS

1. Cells: Culture NMuMG cells to confluency or 1 day post-confluency in 6-well plates (BD Falcon Multiwell Flat-Bottom Plates with Lids from VWR).
2. Antibodies: 281-2 anti-mouse syndecan-1 ectodomain antibody; Ky8.2 anti-mouse syndecan-4 ectodomain antibody; and Alexa fluorophore-conjugated secondary antibodies (Invitrogen).
3. Phosphate-buffered saline (PBS).
4. Bovine serum albumin (BSA): BSA powder dissolved fresh to 1% (w/v) in PBS.
5. Paraformaldehyde (PFA): 16% stock ampules (Electron Microscopy Science, Hatfield, PA) diluted to 2% in PBS.
6. Flow cytometer: We use Accuri C6 Flow Cytometer (Accuri Cytometers, Ann Arbor, MI) but most flow cytometers available in individual labs and institutional core programs will suffice.

2.4. Stripping and Reprobing Blots

1. Stripping buffer: Restore Western Blot Stripping Buffer from Pierce Chemical (see Note 5).
2. Buffers and solutions:
   (a) TBS.
   (b) Blocking buffer 1: TBS containing 10% (w/v) nonfat dry milk.
   (c) Blocking buffer 2: TBS containing 1% (w/v) nonfat dry milk and 0.1% Tween-20 as in Subheading 2.1.
3. Materials for reprobing: As described in Subheading 2.1.

3. Methods

Syndecan ectodomains migrate as a broad smear when fractionated by SDS-PAGE due to the heterogeneous GAG chains, which makes precise quantification of syndecan ectodomains by Western immunoblotting difficult. Thus, syndecan ectodomains in cell and tissue samples are measured by dot immunoblotting or ELISA. Human syndecan-1 and -4 ELISA kits are available from several commercial sources (e.g., American Research Products, Belmont, MA; Cell Sciences, Canton, MA, IBL America, Minneapolis, MN; Abcam, Cambridge, MA), but commercial ELISA kits are currently not available for other human syndecans and mouse syndecans. Further, the commercial ELISA kits use recombinant syndecans as standards. In our experience, the anti-syndecan monoclonal antibodies used in the majority of these ELISA kits tend to react less efficiently with recombinant syndecans compared to native purified syndecans. Thus, it is likely that levels of syndecan ectodomains in biological samples are overestimated in the commercial ELISA kits. Another disadvantage of the commercially available ELISA kits is that they are expensive. Regardless, the ELISA kits are suitable for rapid qualitative measurements of syndecan ectodomains and their ease of use is similar to that of the dot immunoblotting method described below.

It is also important to measure cell surface syndecan levels when assessing syndecan shedding. Ectodomain shedding is a posttranslational mechanism that rapidly decreases the amount of surface molecules by proteolytic cleavage. Thus, along with an increase in syndecan ectodomain levels, it is imperative to demonstrate a concomitant decrease in cell surface syndecan levels. When both shed and cell surface syndecan levels are increased, this is most likely due to increased expression and not shedding. This can be confirmed by measuring syndecan mRNA levels. Cell surface syndecan-1 levels can be measured by mild trypsinization and dot immunoblotting of trypsinates. Syndecan-1 ectodomain is apparently cleaved once in the dibasic sequence near the plasma membrane when treated with low levels of trypsin, releasing intact syndecan-1 ectodomains into the culture supernatant. Other syndecans also contain the dibasic sequence, and cell surface levels of mouse syndecan-2 and -3 have been measured by mild trypsinization and dot immunoblotting with polyclonal anti-syndecan-2 and -3 antibodies. However, syndecan-4 ectodomains are extensively hydrolyzed by trypsin and antigenic epitopes for both monoclonal and polyclonal anti-syndecan-4 antibodies are lost even at low doses of trypsin. Thus, flow cytometry is used to measure cell surface syndecan-4 levels (31).

3.1. Measurement of Syndecan Ectodomains by Dot Immunoblotting

1. NMuMG cells are cultured in DMEM with 4.5 g/L glucose, L-glutamine, and sodium pyruvate supplemented with 10% FBS and 10 μg/mL insulin. It is best to perform agonist-induced syndecan shedding assays with confluent or 1 day post-confluent NMuMG cells to maximize the difference between baseline and induced shedding.
2. Wash confluent or 1-day post-confluent NMuMG cells in 96-well plates once with culture medium. Add agonists diluted in culture medium and incubate for the desired time periods at 37°C. In the example shown in Fig. 1, we incubated NMuMG cells with 1 μM PMA for 0, 15, or 30 min.
3. At the end of incubation, collect the culture supernatant, transfer to microfuge tubes, centrifuge at 1,000 × *g* for 15 min to remove detached cells and debris, and transfer the conditioned medium to clean microfuge tubes. Cells remaining on culture vessels are immediately processed for measurement of cell surface syndecan-1 by mild trypsinization and dot immunoblotting as described below in Subheading 3.2.
4. Pre-wet Immobilon Ny+ membranes and 3MM papers in distilled water. Place one sheet of wet 3MM paper on the dot blot apparatus and place one sheet of wet Immobilon

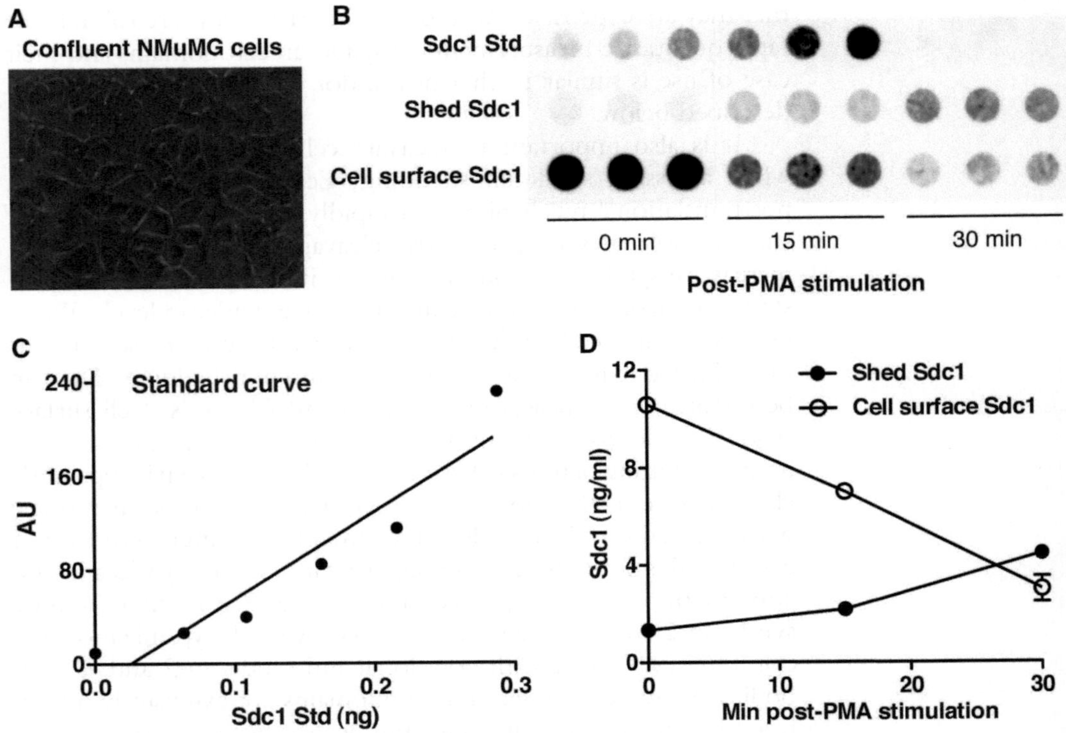

Fig. 1. Time-dependent induction of syndecan-1 shedding by PMA measured by dot immunoblotting. Confluent normal murine mammary gland (NMuMG) cells in 96-well plates were incubated with 1 μM PMA for 0, 15, or 30 min at 37°C in triplicate. For quantification of shed syndecan-1 ectodomains, 35 μL of the conditioned medium was dot blotted onto the Immobilon Ny+ membrane. For quantification of cell surface syndecan-1, following removal of the conditioned medium, cells were washed once with ice-cold TBS/EDTA and incubated with 50 μL of ice-cold 10 μg/mL TPCK-treated trypsin for 10 min at 4°C. Trypsin was inactivated with 50 μL of 100 μg/mL soybean trypsin inhibitor. Trypsinates (35 μL) were dot blotted onto the Immobilon Ny+ membrane. The membrane was immunoblotted with 281-2 antibodies, developed, and quantified as described. (**a**) Picture of confluent NMuMG cells. (**b**) Developed dot blot. (**c**) Standard curve of syndecan-1. (**d**) Quantification of shed and cell surface syndecan-1 at various times post-PMA stimulation.

Ny+ membrane on top of the 3MM paper. Close the dot blot apparatus and connect to a bench-top vacuum line.

5. Dot blot syndecan standards. In Fig. 1, 0.06, 0.11, 0.16, 0.22, 0.29, and 0.38 ng of purified syndecan-1 ectodomains diluted in 100 μL of ABS-T were dot blotted onto the Immobilon Ny+ membrane. Dot immunoblotting is a highly sensitive method that can detect as low as 0.05 ng of native syndecan-1 ectodomains in biological fluids.

6. To obtain measurements within the linear range of standards, dot blot 20–80 μL of the conditioned medium onto Immobilon Ny+ membranes (see Note 6). The volume to be blotted depends on the strength of the syndecan shedding agonist and the level of syndecan expression. For example in Fig. 1, we dot blotted 35 μL of the conditioned medium in triplicate because

NMuMG cells abundantly express syndecan-1 and PMA is a strong and rapid inducer of syndecan-1 shedding.

7. Add ≥1 times volume of ABS-T to each well to acidify samples. Because the isoelectric point (pI) of syndecan ectodomains is considerably lower than pH 4.5 due to the highly anionic HS chains, syndecan ectodomains retain a net negative charge in the ABS-T solution. Further, the pI of syndecan-1 (~4.3), -2 (~4.0), -3 (~4.3), and -4 (~3.8) core proteins is lower than pH 4.5. Thus, by acidifying samples, only highly anionic molecules, such as syndecans, are bound to the cationic Immobilon Ny+ membrane while most proteins pass through the membrane during dot blotting, allowing partial purification of syndecans with this procedure (see Note 7).
8. Apply vacuum to the dot blot apparatus and confirm that all samples completely pass through the membrane. With the vacuum running, add 200 μL of ABS-T to each well and repeat this wash procedure once (see Note 8).
9. Turn the vacuum off, disassemble the dot blot apparatus, and remove the membrane. Label the membrane (e.g., experiment number, date, name) (see Note 9). Also, immediately wash the dot blot apparatus thoroughly to eliminate cross contamination in future experiments.
10. Block the membrane with Blocking buffer 1: TBS containing 10% (w/v) nonfat dry milk for 1 h at room temperature with agitation on a bench-top orbital shaker.
11. Discard Blocking buffer 1 and incubate the membrane twice with Blocking buffer 2: TBS with 1% (w/v) nonfat dry milk and 0.1% (v/v) Tween-20 for 30 min at room temperature.
12. Incubate with 0.2 μg/mL of 281-2 antibody diluted in Blocking buffer 2 for 2–4 h at room temperature or overnight at 4°C with agitation on an orbital shaker (see Note 10).
13. Wash the membrane twice with Blocking buffer 2 for 20 min at room temperature (see Note 11).
14. Incubate with HRP-conjugated secondary antibodies at the dilution recommended by the vendor (1:5,000 for the HRP-conjugated anti-rat antibody in Fig. 1) in Blocking buffer 2 for 2 h at room temperature or overnight at 4°C with agitation on an orbital shaker.
15. Wash the membrane twice with TBS (without milk or Tween-20) for 30 min at room temperature with agitation.
16. Develop membrane with ECL reagent for 1–2 min at room temperature. Remove membrane from ECL solution, remove remnants of ECL solution by blotting with 3MM paper, place membrane in a clear plastic wrap, and press out bubbles from the surface of the membrane.

17. Place the membrane in an image reader and acquire image of the developed blot in 1–5 min intervals. If using film, place the membrane in a film cassette, expose to X-ray films for the desired time periods (e.g., 10 s to 15 min), and develop films with a developer. Scan the acquired image or developed film with commercially available software (e.g., Multi Gauge, Fujifilm) or public domain software (e.g., Image J, NIH Image).
18. Plot the standard graph and quantify dots within the linear range of the standard curve using graph software (e.g., Prism, Kaleidagraph, Excel).

3.2. Measurement of Cell Surface Syndecan-1 by Mild Trypsinization and Dot Immunoblotting

1. Immediately after collecting the conditioned medium, wash NMuMG cells remaining on 96-well plates twice with ice-cold TBS/EDTA.
2. Incubate cells with 10 μg/mL TPCK-treated trypsin in TBS/EDTA for 10 min at 4°C. The volume depends on the culture vessel used. For example, when using 96-well plates, add 50 μL of the TPCK-trypsin solution to each well.
3. Add an equal volume (e.g., 50 μL for 96-well plates) of 100 μg/mL soybean trypsin inhibitor to each well and incubate for 5 min on ice to inactivate trypsin.
4. Transfer trypsinates to microfuge tubes, centrifuge at 1,000 × *g* for 15 min to remove cells and debris, and transfer trypsinates to clean microfuge tubes.
5. Perform dot immunoblotting as described above in Subheading 3.1, steps 4–18. The volume of trypsinates to be dot blotted depends on the expression level of cell surface syndecans and strength of the shedding agonist. For example in Fig. 1, we dot blotted 35 μL of the conditioned medium in triplicate because NMuMG cells abundantly express syndecan-1 and PMA is a strong inducer of syndecan-1 shedding.

3.3. Measurement of Cell Surface Syndecans by FACS

1. Stimulate confluent or 1 day post-confluent NMuMG cells in 6-well plates with agonist. In the example shown in Fig. 2, NMuMG cells were stimulated with 1 μM PMA for 15 or 30 min at 37°C.
2. Collect culture supernatants and proceed to measurement of shed syndecan ectodomains by dot immunoblotting (see Subheading 3.1 above). Wash remaining NMuMG cells once with PBS.
3. Incubate NMuMG cells with PBS containing 5 mM EDTA (PBS/EDTA) for 30 min at 4°C.
4. Remove cells from each well by scraping on ice and transferring the cell suspension to microfuge tubes.
5. Pellet cells by centrifugation at 1,000 × *g* for 10 min at 4°C and wash 3 times with ice-cold PBS.

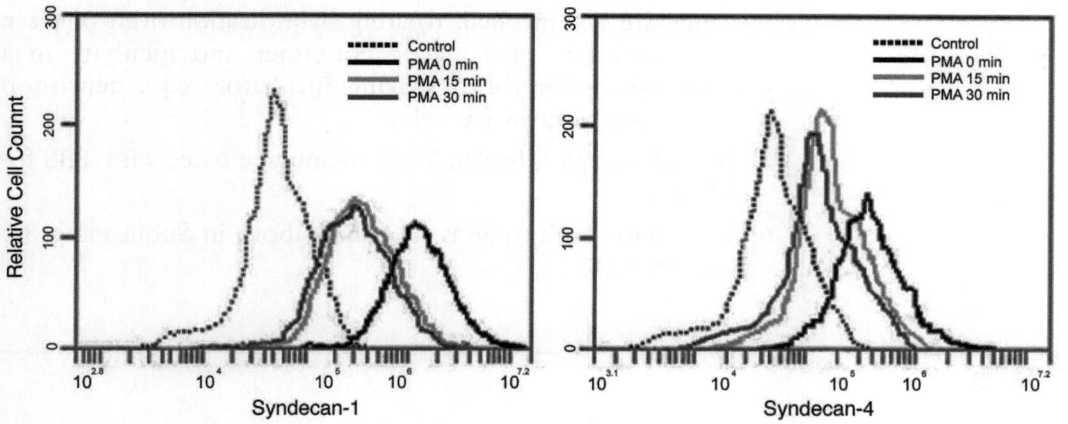

Fig. 2. Time-dependent reduction of cell surface syndecan-1 and -4 by PMA measured by FACS. Confluent NMuMG cells in 6-well plates were incubated with 1 μM PMA for 15 or 30 min at 37°C. Cells were detached by incubating with PBS/EDTA and treated with primary antibodies (281-2 for syndecan-1; Ky8.2 for syndecan-4) and Alexa 488 anti-rat secondary antibodies. After fixing cells with 2% paraformaldehyde (PFA), the fluorescence associated with cell surfaces was measured by FACS. *Black line*, cell surface syndecans without PMA treatment; *light gray line*, cell surface syndecans 15 min after PMA treatment; *dark gray line*, cell surface syndecans 30 min after PMA treatment; and *dotted line*, background fluorescence measured with secondary antibody only.

6. Incubate cells with 10 μg/mL of 281-2 anti-syndecan-1 or 20 μg/mL of Ky8.2 anti-syndecan-4 antibodies in PBS with 1% (w/v) BSA (BSA/PBS) for 1 h at 4°C. Centrifuge at 1,000 × *g* for 10 min at 4°C and wash 3 times with ice-cold BSA/PBS.
7. Incubate cells with Alexa 488 donkey anti-rat antibodies for 1 h at 4°C (see Note 12). Wash cells 3 times with ice-cold BSA/PBS.
8. Fix cells with 2% PFA in PBS for 10 min on ice. Wash fixed cells 3 times with ice-cold PBS.
9. Determine the level and number of cells expressing membrane-bound syndecan-1 and -4 by measuring the fluorescence associated with the cell surface at 519 nm using a flow cytometer (see Fig. 2 for examples).

3.4. Stripping and Reprobing Blots for Shed and Cell Surface Syndecans

1. Multiple syndecans can be measured from the same samples by stripping and reprobing blotted membranes with different anti-syndecan antibodies (see Note 13).
2. Wash previously immunoblotted membrane twice with TBS for 30 min at room temperature.
3. Warm the Restore Western Blot Stripping Buffer to room temperature. Add sufficient volume of stripping buffer to cover the whole surface of the membrane.
4. Incubate at 37°C for 30 min with agitation. We either insert the membrane in a hybridization bottle and incubate in a

temperature-controlled, rotating hybridization oven or place the membrane in a plastic container and incubate in a temperature-controlled shaking incubator (e.g., bench-top shaker used to grow bacteria).

5. Discard stripping buffer. Wash membrane twice with TBS for 30 min.
6. Start immunoblotting as described above in Subheading 3.1 from step 10.

4. Notes

1. Use larger culture vessels when using cell types expressing lower levels of syndecans relative to NMuMG cells. For example, LA4 and NIH3T3 cells should be cultured in 24-well plates and A549 cells should be cultured in 48-well plates to confluency and tested in shedding assays.
2. Stock solutions of 1 M sodium acetate, pH 4.5, and 5 M NaCl can be used to make ABS-T. Use acetic acid to adjust pH when making 1 M sodium acetate. Tween-20 is viscous so cut the tip of the pipet tip with scissors or use a large orifice tip (VWR Signature Genomic Pipet Tips, VWR) to aliquot this detergent from stock bottles.
3. Milk-containing products are easily contaminated and should be made fresh on the day of use.
4. EDTA stock solution can be prepared in-house. Dissolve EDTA disodium salt to 500 mM in deionized water. Must adjust to pH 8 while stirring to dissolve EDTA.
5. Stripping buffer can also be prepared in-house. Prepare a stock solution of 62.5 mM Tris, pH 6.7, 2% (w/v) SDS and store at room temperature. Warm to 60°C and add β-mercaptoethanol to 100 mM. Incubate the membrane in the home-made stripping buffer for 30 min at 60°C to strip antibodies. Follow procedures of the commercial stripping buffer in Subheading 3.4 starting at step 5 for re-blotting with a different antibody.
6. Air can get trapped in the membrane while applying samples. If the wells appear clogged, turn the vacuum off and tap the side of the dot blot apparatus using caution not to spill the samples in wells. If the wells are still clogged, mix the samples in wells by gentle pipetting using a 20–200 μL micropipet.
7. Some samples with high protein concentrations, such as serum and some tissue homogenates, precipitate when acidified. To measure

syndecan levels in these samples, dot immunoblotting should be performed with neutral buffer, such as TBS, pH 7.5, instead of ABS-T.

8. Attach a filtering flask trap between the bench-top vacuum source and dot blot apparatus to prevent clogging the vacuum source. Verify that the material of the filtering flask (glass or polypropylene) can withstand the pressure applied by the vacuum source before use.

9. We recommend using VWR lab markers (fine tip or extra fine tip) because other markers tend to wash away while immunoblotting.

10. For best results, we recommend incubating blotted membranes in primary and secondary antibodies overnight at 4°C. This reduces background.

11. Use caution not to dry the Immobilon Ny+ membrane while immunoblotting. After the membrane is developed, it can be stored dry in a plastic wrap.

12. Other fluorophore-conjugated secondary antibodies (e.g., FITC-, PE-, GFP-conjugated antibodies) can also be used. Alternatively, some pre-labeled anti-syndecan antibodies are available commercially (e.g., Cell Sciences, Abcam) and anti-syndecan antibodies can be directly labeled using labeling kits available from commercial sources (e.g., Alexa fluor labeling kits from Invitrogen).

13. When stripping and reprobing, we recommend determining shedding of syndecans that are expressed at lower levels first because the antigenicity of the blotted samples decreases every time the blot is stripped and reprobed. For example, if examining both syndecan-1 and -4 shedding from the same sample, it is better to determine syndecan-4 levels before those of syndecan-1. For the same reason, we do not recommend stripping and reprobing a membrane more than 2 times.

Acknowledgments

The authors would like to thank past and present members of the Park laboratory for developing essential reagents and constantly improving the described procedures. This work was supported by NIH grants R01 HL094613 and R01 HL107472.

References

1. Bernfield, M., Götte, M., Park, P. W., Reizes, O., Fitzgerald, M. L., Lincecum, J., and Zako, M. (1999) Functions of cell surface heparan sulfate proteoglycans, *Annu. Rev. Biochem.* **68,** 729–777.
2. Couchman, J. R. (2010) Transmembrane Signaling Proteoglycans, *Annu Rev Cell Dev Biol.*, **26**, 89–114.
3. Park, P. W., Reizes, O., and Bernfield, M. (2000) Cell surface heparan sulfate proteoglycans: selective regulators of ligand-receptor encounters, *J. Biol. Chem.* **275**, 29923–29926.
4. Hayashida, K., Bartlett, A. H., Chen, Y., and Park, P. W. (2010) Molecular and cellular mechanisms of ectodomain shedding, *Anat. Rec.* **293**, 925–937.
5. Steppan, J., Hofer, S., Funke, B., Brenner, T., Henrich, M., Martin, E., et al. (2011) Sepsis and Major Abdominal Surgery Lead to Flaking of the Endothelial Glycocalix, *J Surg Res*, **165**, 136–141.
6. Nelson, A., Berkestedt, I., Schmidtchen, A., Ljunggren, L., and Bodelsson, M. (2008) Increased levels of glycosaminoglycans during septic shock: relation to mortality and the antibacterial actions of plasma, *Shock* **30**, 623–627.
7. Rehm, M., Bruegger, D., Christ, F., Conzen, P., Thiel, M., Jacob, M., et al. (2007) Shedding of the endothelial glycocalyx in patients undergoing major vascular surgery with global and regional ischemia, *Circulation* **116**, 1896–1906.
8. Seidel, C., Ringdén, O., and Remberger, M. (2003) Increased levels of syndecan-1 in serum during acute graft-versus-host disease, *Transplantation* **76**, 423–426.
9. Joensuu, H., Anttonen, A., Eriksson, M., Makitaro, R., Alfthan, H., Kinnula, V., and Leppa, S. (2002) Soluble syndecan-1 and serum basic fibroblast growth factor are new prognostic factors in lung cancer, *Cancer Res.* **62**, 5210–5217.
10. Yang, Y., Yaccoby, S., Liu, W., Langford, J. K., Pumphrey, C. Y., Theus, A., et al. (2002) Soluble syndecan-1 promotes growth of myeloma tumors *in vivo*, *Blood* **100**, 610–617.
11. Kainulainen, V., Wang, H., Schick, C., and Bernfield, M. (1998) Syndecans, heparan sulfate proteoglycans, maintain the proteolytic balance of acute wound fluids, *J. Biol. Chem.* **273**, 11563–11569.
12. Kliment, C. R., Englert, J. M., Gochuico, B. R., Yu, G., Kaminski, N., Rosas, I., and Oury, T. D. (2009) Oxidative stress alters syndecan-1 distribution in lungs with pulmonary fibrosis, *J Biol Chem* **284**, 3537–3545.
13. Kato, M., Wang, H., Kainulainen, V., Fitzgerald, M. L., Ledbetter, S., Ornitz, D. M., and Bernfield, M. (1998) Physiological degradation converts the soluble syndecan-1 ectodomain from an inhibitor to a potent activator of FGF-2, *Nat. Med.* **4**, 691–697.
14. Li, Q., Park, P. W., Wilson, C. L., and Parks, W. C. (2002) Matrilysin shedding of syndecan-1 regulates chemokine mobilization and transepithelial efflux of neutrophils in acute lung injury, *Cell* **111**, 635–646.
15. Xu, J., Park, P. W., Kheradmand, F., and Corry, D. B. (2005) Endogenous attenuation of allergic lung inflammation by syndecan-1, *J. Immunol.* **174**, 5758–5765.
16. Park, P. W., Pier, G. B., Hinkes, M. T., and Bernfield, M. (2001) Exploitation of syndecan-1 shedding by *Pseudomonas aeruginosa* enhances virulence, *Nature* **411**, 98–102.
17. Hayashida, A., Bartlett, A. H., Foster, T. J., and Park, P. W. (2009) *Staphylococcus aureus* beta-toxin induces acute lung injury through syndecan-1, *Am. J. Pathol.* **174**, 509–518.
18. Hayashida, K., Parks, W. C., and Park, P. W. (2009) Syndecan-1 shedding facilitates the resolution of neutrophilic inflammation by removing sequestered CXC chemokines, *Blood* **114**, 3033–3043.
19. Hayashida, K., Chen, Y., Bartlett, A. H., and Park, P. W. (2008) Syndecan-1 is an *in vivo* suppressor of Gram-positive toxic shock, *J. Biol. Chem.* **283**, 19895–19903.
20. Katoh, S., Taniguchi, H., Matsubara, Y., Matsumoto, N., Fukushima, K., Kadota, J., et al. (1999) Overexpression of CD44 on alveolar eosinophils with high concentrations of soluble CD44 in bronchoalveolar lavage fluid in patients with eosinophilic pneumonia, *Allergy* **54**, 1286–1292.
21. Wang, Q., Teder, P., Judd, N. P., Noble, P. W., and Doerschuk, C. M. (2002) CD44 deficiency leads to enhanced neutrophil migration and lung injury in *Escherichia coli* pneumonia in mice, *Am. J. Pathol.* **161**, 2219–2228.
22. Kim, C. W., Goldberger, O. A., Gallo, R. L., and Bernfield, M. (1994) Members of the syndecan family of heparan sulfate proteoglycans are expressed in distinct cell-, tissue-, and development-specific patterns, *Mol. Biol. Cell* **5**, 797–805.
23. Chen, Y., Bennett, A., Hayashida, A., Hollingshead, S., and Park, P. W. (2007) *Streptococcus pneumoniae* sheds syndecan-1 ectodomains via ZmpC, a metalloproteinase virulence factor, *J. Biol. Chem.* **282**, 159–167.

24. Fitzgerald, M. L., Wang, Z., Park, P. W., Murphy, G., and Bernfield, M. (2000) Shedding of syndecan-1 and -4 ectodomains is regulated by multiple signaling pathways and mediated by a TIMP-3 sensitive metalloproteinase, *J. Cell Biol.* **148**, 811–824.
25. Hayashida, K., Stahl, P. D., and Park, P. W. (2008) Syndecan-1 ectodomain shedding is regulated by the small GTPase Rab5, *J. Biol. Chem.* **283**, 35435–35444.
26. Park, P. W., Foster, T. J., Nishi, E., Duncan, S. J., Klagsbrun, M., and Chen, Y. (2004) Activation of syndecan-1 ectodomain shedding by *Staphylococcus aureus* alpha-toxin and beta-toxin, *J. Biol. Chem.* **279**, 251–258.
27. Park, P. W., Pier, G. B., Preston, M. J., Goldberger, O., Fitzgerald, M. L., and Bernfield, M. (2000) Syndecan-1 shedding is enhanced by LasA, a secreted virulence factor of *Pseudomonas aeruginosa*, *J. Biol. Chem.* **275**, 3057–3064.
28. Popova, T. G., Millis, B., Bradburne, C., Nazarenko, S., Bailey, C., Chandhoke, V., and Popov, S. G. (2006) Acceleration of epithelial cell syndecan-1 shedding by anthrax hemolytic virulence factors, *BMC Microbiol.* **6**, 8–24.
29. Subramanian, S. V., Fitzgerald, M. L., and Bernfield, M. (1997) Regulated shedding of syndecan-1 and -4 ectodomains by thrombin and growth factor activation, *J. Biol. Chem.* **272**, 14713-14720.
30. Yang, Y., Macleod, V., Miao, H. Q., Theus, A., Zhan, F., Shaughnessy, J. D. Jr., et al. (2007) Heparanase enhances syndecan-1 shedding: a novel mechanism for stimulation of tumor growth and metastasis, *J. Biol. Chem.* **282**, 13326–13333.
31. Hayashida, K., Johnston, D. R., Goldberger, O., and Park, P. W. (2006) Syndecan-1 expression in epithelial cells is induced by TGF-beta through a PKA-dependent pathway, *J. Biol. Chem.* **281**, 24365–24374.
32. Ding, K., Lopez-Burks, M., Sanchez-Duran, J. A., Korc, M., and Lander, A. D. (2005) Growth factor-induced shedding of syndecan-1 confers glypican-1 dependence on mitogenic responses of cancer cells, *J. Cell Biol.* **171**, 729–738.
33. Reizes, O., Goldberger, O., Smith, A. C., Xu, Z., Bernfield, M., and Bickel, P. E. (2006) Insulin promotes shedding of syndecan ectodomains from 3T3-L1 adipocytes: a proposed mechanism for stabilization of extracellular lipoprotein lipase, *Biochemistry* **45**, 5703–5711.
34. Day, R. M., Mitchell, T. J., Knight, S. C., and Forbes, A. (2003) Regulation of epithelial syndecan-1 expression by inflammatory cytokines, *Cytokine* **21**, 224–233.
35. Henry-Stanley, M. J., Zhang, B., Erlandsen, S. L., and Wells, C. L. (2006) Synergistic effect of tumor necrosis factor-alpha and interferon-gamma on enterocyte shedding of syndecan-1 and associated decreases in internalization of *Listeria monocytogenes* and *Staphylococcus aureus*, *Cytokine* **34**, 252–259.
36. Charnaux, N., Brule, S., Chaigneau, T., Saffar, L., Sutton, A., Hamon, M., et al. (2005) RANTES (CCL5) induces a CCR5-dependent accelerated shedding of syndecan-1 (CD138) and syndecan-4 from HeLa cells and forms complexes with the shed ectodomains of these proteoglycans as well as with those of CD44, *Glycobiology* **15**, 119–130.
37. Brule, S., Charnaux, N., Sutton, A., Ledoux, D., Chaigneau, T., Saffar, L., and Gattegno, L. (2006) The shedding of syndecan-4 and syndecan-1 from HeLa cells and human primary macrophages is accelerated by SDF-1/CXCL12 and mediated by the matrix metalloproteinase-9, *Glycobiology* **16**, 488–501.
38. Endo, K., Takino, T., Miyamori, H., Kinsen, H., Yoshizaki, T., Furukawa, M., and Sato, H. (2003) Cleavage of syndecan-1 by membrane type matrix metalloproteinase-1 stimulates cell migration, *J. Biol. Chem.* **278**, 40764–40770.
39. Pruessmeyer, J., Martin, C., Hess, F. M., Schwarz, N., Schmidt, S., Kogel, T., et al. (2010) A disintegrin and metalloproteinase 17 (ADAM17) mediates inflammation-induced shedding of syndecan-1 and -4 by lung epithelial cells, *J Biol Chem* **285**, 555–564.
40. Chung, M. C., Popova, T. G., Millis, B. A., Mukherjee, D. V., Zhou, W., Liotta, L. A., et al. (2006) Secreted neutral metalloproteases of *Bacillus anthracis* as candidate pathogenic factors, *J. Biol. Chem.* **281**, 31408–31418.

Chapter 20

Modulatory Effects of Proteoglycans on Proteinase Activities

Steven Georges, Dominique Heymann, and Marc Padrines

Abstract

Proteoglycans (PGs), composed of a core protein and one or more covalently attached sulfated glycosaminoglycan (GAG) chains, interact with a wide range of bioactive molecules, such as growth factors and chemokines, to regulate cell behaviors in normal and pathological processes. Additionally, PGs, through their compositional diversity, play a broad variety of roles as modulators of proteinase activities. Interactions of proteinases with other molecules on the plasma membrane anchor and activate them at a specific location on the cell surface. These interactions with macromolecules other than their own protein substrates or inhibitors result in changes in their activity and/or may have important biological effects. Thus, GAG chains induce conformational changes upon their binding to peptides or proteins. This behavior may be related to the ability of GAGs to act as modulators for some proteins (1) by acting as crucial structural elements by the control of proteinase activities, (2) by increasing the protein stability, (3) by permitting some binding to occur, exposing binding regions on the target protein, or (4) by acting as coreceptors for some inhibitors, playing important roles for the acceleration of proteinase inhibition. Understanding the modulatory effects exerted by PGs on proteinase activities is expected to lead to new insights in the understanding of some molecular systems present in pathological states, providing new targets for drug therapy.

Key words: Proteinase activities, Proteoglycans, Glycosaminoglycans

1. Introduction

Proteoglycans (PGs) are a family of biomolecules that are composed of a core protein and one or more covalently attached sulfated glycosaminoglycan (GAG) chains. Synthesis and sulfation of GAGs occur on genetically distinct acceptor core proteins within the Golgi, followed by rapid translocation to the cell surface (1). GAGs are linear polymers of repeated disaccharidic units of hexosamine and hexuronic acid, except for keratan sulfate in which hexuronic acid is replaced by galactose. The presence of either two hexosamine isomers, D-glucosamine or D-galactosamine, divides the GAGs into two groups: glucosaminoglycans (heparin [HP]/heparan sulfate [HS]

Françoise Rédini (ed.), *Proteoglycans: Methods and Protocols*, Methods in Molecular Biology, vol. 836,
DOI 10.1007/978-1-61779-498-8_20, © Springer Science+Business Media, LLC 2012

and keratan sulfate [KS]) and galactosaminoglycans (chondroitin [CS]/dermatan sulfate [DS]) (2). Hexuronic acid is also present as two epimers: D-glucuronic acid and L-iduronic acid. Hyaluronic acid, which is not attached to a protein core, is a nonsulfated GAG composed of D-glucuronic acid and D-glucosamine. The degree and position of the sulfate group/moiety as well as the degree and position of 50 epimerization are extremely variable in sulfated GAGs depending on the tissular/cellular/metabolic context, ensuring structural variability of these polysaccharides (3).

PGs are ubiquitous, being present as cell surface molecules anchored in the plasma membrane, as components of the extracellular matrix (ECM) or as soluble molecules present in ECM and serum. Soluble PGs as well as those bound within the ECM are derived from cell secretions or by shedding from the cell surface. PGs play a role in both cell–cell and cell–ECM adhesion and can also act to promote assembly of ECM molecules (4). Additionally, PGs interact with a wide range of bioactive molecules such as growth factors and chemokines via their GAG chains to regulate cell behaviors in normal and pathological processes. Thus, these molecules are increasingly thought to participate in regulating a wide variety of biological processes, including the inflammatory response and tumor cell metastasis. Their ability to control proteinase activities is established. Thus, GAGs induce conformational changes upon their binding to peptides or proteins. This behavior may be related to the ability of GAGs to act as modulators for some proteins by changing their catalytic activity, by increasing protein stability, or by permitting some binding to occur by exposing binding regions on the target protein.

2. Proteinases

Proteinases comprise a group of enzymes that catalyze the cleavage of a peptide bond in a protein or a peptide by nucleophilic attack on the carbonyl carbon. This is mediated by an amino acid (cysteine, serine, threonine) or by water molecules either fitted between two aspartate amino acids and/or complexed by an ionized metal on cysteine, serine, aspartate, and matrix metalloproteinases (MMPs), respectively, or by some other mechanism (5).

Proteinases are also classified according to their substrate specificity, which implies the recognition of peptide bonds or residues at the amino or carboxyl terminus of the molecule as well as side chains of the surrounding amino acids at the bond to be cleaved. These endoproteinases play an important role in cell migration and invasion, the remodeling of the ECM, and the liberation and modification of growth factors. These proteinase activities are ultimately regulated by the balance between activation of inactive proforms

Table 1
Catalytic class and subcellular location of proteinases

Proteinases	Catalytic class	Extracellular	Cytosolic	Plasma membrane associated	Lysosomal
Cathepsin D	Aspartate	+	–	–	+++
BACE1	Aspartate	+	–	–	+++
MMP (1–13, 19–21, 26–28)	Metallo	+++	–	+	+
MT-MMP (MMP14–17, 23–25)	Metallo	+	–	+++	–
ADAM	Metallo	+	–	–	–
ADAMTS	Metallo	+	–	+++	–
Cysteine cathepsins (B, L, S, K)	Cysteine	–	–	++	+++
Neurophil elastase	Serine	–	–	+	+++
Cathepsin G	Serine	–	–	+	+++
Thrombin	Serine	+++	–	–	–
Plasminogen activator	Serine	–	+	+++	–

and preproforms and the levels of their endogenous inhibitors in cellular or tissue compartments, of which tissue inhibitors of metalloproteinases (TIMPs), serine proteinase inhibitors (Serpin), and cystatins represent major classes (6). Many of these proteinases act on the cell surface either because they contain a membrane spanning or binding domain or as the result of their interaction with specific receptors on the cell surface (see Table 1).

3. MMP Docking

Proteinases can be either free in the cytosol or bound to the surface of a cell. The cell surface-bound forms are thought to enhance inflammatory cell functions. It remains unclear how these enzymes make it to the correct location at the cell surface and how the proteolytic activity is controlled at the pericellular space. However, it is becoming increasingly clear that extracellular proteolysis is a cell-regulated process. After all, cells do not release proteinases indiscriminately. Rather they rely on precise interactions to accurately degrade, cleave, or process specific substrates in the pericellular space. Indeed, an emerging concept is that MMPs, as for some serine and cysteine proteinases (7), are anchored to the cell membrane, thereby targeting their catalytic activity to specific substrates within the pericellular space. However, it has been suggested that proteinase binding to cell surface proteins can affect intracellular signaling, facilitate proenzyme localization and activation, mediate cell motility by disruption of cell contacts with the ECM, and promote internalization of the enzyme. For example, integrins are shown to act as receptors for several proteinases (8). Such interactions have been detected in caveolae, invadopodia, and at the leading edge of migrating cells, where directed proteolytic activity is needed.

Metalloproteinases are secreted proteins belonging to a family of zinc metalloendopeptidases that have the capacity to cleave ECM and a variety of extracellular protein substrates. Using oriented peptide libraries to identify potential metalloproteinase cleavage sites, several integrins, PGs, and chemokines or their receptors have been identified as substrates. Thus, metalloproteinases can release growth factors from the ECM or the cell surface. They can modify both cell–cell and cell–ECM interactions by the proteolysis of cell surface growth factors and adhesion receptors, and they are key regulators of inflammatory responses, which can be pro- or anti-inflammatory (9). This activity makes available active growth factors and cytokines. Because their substrates are diverse, metalloproteinases are involved in a variety of homeostatic functions, such as bone remodeling, wound healing, and several aspects of immunity. However, metalloproteinases are also involved in a number of pathological processes, such as tumor progression, fibrosis, chronic inflammation, and tissue destruction. These metalloproteinases can be subdivided into three subfamilies: MMPs, a disintegrin and metalloproteinase domain (ADAM), and a disintegrin and metalloproteinase with thrombospondin motifs (ADAMTS) (10).

Specific cell-MMP interactions have been reported, such as the binding of MMP-2 to integrin $\alpha v\beta 3$ (11), MMP-9 to integrin $\alpha 4\beta 1$ (12), and MMP-1 to integrin $\alpha 2\beta 1$ (13). The membrane-type MMPs (MMP-14, -15, -16, -17, -24, and -25) are single-pass transmembrane proteins that are fixed and active at the cell surface and, in addition to acting as proteinases, may provide docking sites for other MMPs. Thus, MMP-14 activates pro-MMP-2 at the cell surface (14). Tissue inhibitor of metalloproteinase-2 (TIMP-2) has a specialized role in the activation of pro-MMP-2 by MMP-14. The N-terminal domain of TIMP-2 forms an inhibitory complex with the active site of MMP-14, while the C terminal domain interacts with the hemopexin domain of MMP-2. This trimeric complex is essential for activation of this gelatinase (15). It is likely that other MMPs are also attached to cells via specific interaction to membrane proteins, and determining these anchors will lead to identifying activation mechanisms and relevant substrates.

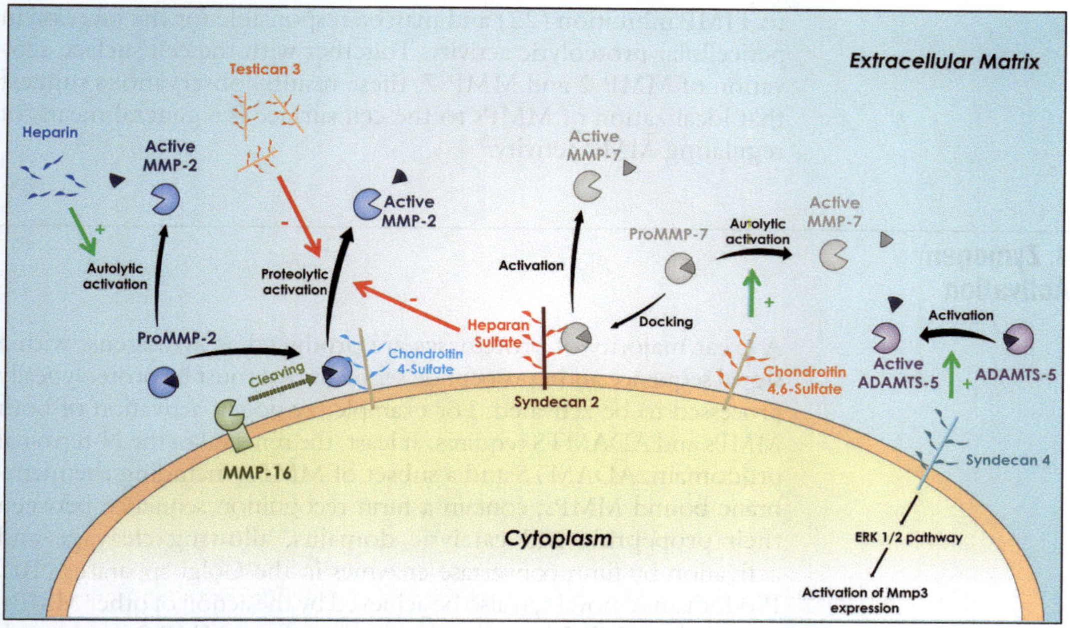

Fig. 1. Regulation of metalloproteinases by GAGs. The activation and activity of MMP-2 is strongly regulated by GAGs. For example, heparin enhanced the autolytic activation of pro-MMP-2. Furthermore, chondroitin-4-sulfate, expressed on cell surface, binds pro-MMP-2 and facilitates its activation by MMP-16, a membranous protease, whereas testican 3 and heparan sulfate inhibit this processing. The activity of MMP-7 is also regulated by GAGs. It's binded by sulfated GAGs like chondroitin-4,6-sulfate which acts as allosteric modulator promoting the autolytic activation. Syndecan-2 interacts directly with pro-MMP7, enhancing its processing into active MMP7. Another syndecan, syndecan-4, is able on the one other to control the ADAMTS-5 activation through direct interaction.

Previous studies have shown that cell surface HSPGs function as docking sites for MMPs at the cell surface (16) (see Fig. 1), and the cell surface localization of MMPs is important for their ability to regulate carcinogenesis (17). Anchoring MMPs to the cell surface or ECM would not only prevent them from rapidly diffusing away but would also enable the cell to keep them under close regulatory control. For example, syndecan-2 acts as a docking receptor for pro-MMP-7. At the cell surface, syndecan-2 interacts directly with pro-MMP-7 at the plasma membrane, enhancing its processing into active MMP-7, which in turn regulates tumorigenic activities of colon cancer cells (18). Another PG, the highly polymorphic facultative cell surface PG CD44, is implicated in a variety of cellular functions, including adhesion, migration, activation, invasion, and cell survival. Several of these functions have been attributed to the ability of CD44 to bind hyaluronan (19), for which CD44 appears to be a major cell surface receptor. CD44 may also dock MMP-7 (20) and MMP-9 (21) to the cell surface. Therefore, the formation of MMP-adhesion receptor complexes appears to be a common pathway through which soluble MMPs are localized to the cell surface. Their localization on the cell surface confers resistance to TIMP inhibition (22) and may be responsible for the increase in pericellular proteolytic activity. Together with the cell surface activation of MMP-2 and MMP-7, these results/observations suggest that localization of MMPs to the cell surface is a general means of regulating MMP activity.

4. Zymogen Activation

A great majority of proteinases are produced as zymogens, with a signal sequence and a propeptide segment that must be proteolytically processed to be activated. For example, zymogen activation of both MMPs and ADAMTS requires, at least, the removal of the N-terminal prodomain. ADAMTS and a subset of MMPs, including the membrane-bound MMPs, contain a furin recognition sequence between their propeptide and catalytic domains, allowing cleavage and activation by furin convertase enzymes in the Golgi apparatus (10). Prodomain removal can also be achieved by the action of other MMPs, such as the MMP-14-mediated activation of pro-MMP-2, or through activation cascades involving coactivators such as plasmin (23).

Sulfated GAGs could play important roles in controlling the activation and thereby the activity of MMP (see Fig. 1). The autolytic activation of proMMP2 is enhanced by HP (24), suggesting that sulfated GAGs may have wide roles in controlling MMP proteolysis. HS regulates ADAM-12 through a molecular switch mechanism. The noncovalently associated prodomain in concert with the catalytic domain of ADAM-12 forms a novel molecular switch critical for the regulation of the ADAM-12 proteolytic activity by HSPGs (25). Via direct interaction with pro-MMP-7, sulfated GAGs such as chondroitin-4,6-sulfate act as allosteric modulators promoting the autolytic activation of the proteinase. Once activated, GAGs may facilitate proteolysis of certain substrates by interacting with the substrate, the enzyme, or both. Activation of pro-MMP-2 by MMP-16 is also significantly enhanced in the presence of excess chondroitin 4-sulfate (C4S), whereas chondroitin 6-sulfate or low-molecular-mass hyaluronan was ineffective. C4S, which is expressed on the tumor cell surface, can bind pro-MMP-2 and facilitate its activation by MMP-16-expressing tumor cells to enhance invasion and metastasis (26).

Among the seven aggrecanases, ADAMTS-4 is mainly expressed in an active form in osteoarthritic cartilage, suggesting that ADAMTS-4 may play an important role in the degradation of aggrecan in human osteoarthritic cartilage (27). However, suppression of ADAMTS-4 and ADAMTS-5, individually or in combination, attenuated the degradation of aggrecan in cytokine-stimulated normal cartilage. The active form of ADAMTS-4 has been reported to colocalize with ADAMTS-mediated aggrecan cleavage in developing long bones in the rat, implying that ADAMTS-4 mediates the developmental turnover of aggrecan during long bone formation (28). However, ADAMTS-5 has been found to be the major aggrecanase in mice (29). In addition, the aggrecanase activity of ADAMTS-5 was at least 1,000-fold greater than that of ADAMTS-4 under physiological conditions (30). The activity of ADAMTS is regulated not only at the transcriptional level but also by posttranslational modifications starting with the processing of the proprotein form by furin or MMP-9. At the cell surface, further activation takes place; glycosylphosphatidyl inositol-anchored MT-MMPs and syndecans may collaborate on the surface of cells to maintain normal ECM homeostasis and also to respond to proinflammatory signals by increased activation of ADAMTS proteinases, which cleave aggregating PGs. ADAMTS-4 activation involves the coordinated activity of both glycosylphosphatidyl inositol-anchored MMP-17 and the PG form of syndecan-1 on the cell surface (31). MMP-17 is the proteinase responsible for ADAMTS4 activation. Moreover, the activated enzyme form can be bound to the cell surface through the GAG chains of membrane-associated syndecan-1. Another syndecan, syndecan-4, controls ADAMTS-5 activation through direct interaction with the proteinase, thereby controlling cartilage breakdown in osteoarthritis (32). Syndecan-4 is crucial in regulating MMP-3 expression by activating ERK1/2 and by targeting ADAMTS-5 to the cell surface of chondrocytes. Consequently, loss of syndecan-4 results in reduced expression of MMP-3 and a marked decrease in aggrecanase activity (see Fig. 1).

In contrast, syndecan-2 acts as a suppressor for MMP-2 activation on the cell surface (33). Syndecans comprise a family of cell surface HSPGs, exhibiting complex biological functions involving the interaction of HS side chains with a variety of soluble and insoluble HP-binding extracellular ligands. Munesue et al. demonstrated an inverse correlation between the expression level of syndecan-2 and the metastatic potential of clones established from Lewis lung carcinoma 3LL. Removal of HS from the cell surface of low metastatic cells by treatment with heparitinase-I promoted MMP-2 activation. In contrast, transfection of syndecan-2 into highly metastatic cells suppressed MMP-2 activation. These results indicate a novel function of syndecan-2, which acts as a suppressor for MMP-2 activation, causing suppression of metastasis in at least the metastatic system used in the present study. Another PG, testican 3, inhibits pro-MMP-2 processing mediated by membrane-type MMPs (34). However, testican 2 abolishes inactivation of membrane-type MMPs by other testican families and permits migration of glioma cells expressing MMP-14 in the presence of other testican family proteins (35). The expression level of testican 2 was the highest among testican family members regardless of histological grade of astrocytic tumors. These results suggest that abundant distribution of testican 2 may contribute to glioma invasion by inactivating other testican family members, which all inhibit membrane-type MMPs.

A proteolytic cascade associated with cancer cell invasion and aimed at the degradation of the ECM, similar to the clotting and fibrinolytic cascades in serum, was proposed (36). This hypothesis is now broadly accepted, although not experimentally proven. The enzymes sitting at the top of the cascade may be cathepsins (lysosomal cysteine proteinases), in particular cathepsin B (37). Cathepsin B activity is often high in various compartments of tumors where invasion by cancer cells, endothelial cells, or inflammatory cells takes place. It can activate prourokinase-type plasminogen activator (pro-uPA), which, by cleaving plasminogen, generates plasmin. The latter is a proteinase of broader specificity that may cleave the propeptide portion of certain MMPs. The cascade is probably more complex than that, as cathepsins may also directly activate a number of MMPs. The propeptide which runs in an extended conformation through the active site cleft forms an α-helical domain on top of the enzyme, which serves to anchor the prodomain to the body of the enzyme (38). Proteolytic removal of the prodomain, which is required for the activation of cysteine cathepsins, occurs in the acidic milieu of the endosomal/lysosomal system (39). Endopeptidases, such as cathepsins B, L, S, and K, can be activated autocatalytically or by other proteinases such as cathepsin D and pepsin (40). Procathepsin B was found to be an active species, suggesting that autocatalytic activation of cysteine cathepsins is a multistep process, starting with a unimolecular conformational change of the zymogen, which unmasks the active site and, in the presence of negatively charged molecules or surfaces, also converts the zymogen into a better substrate. This is followed by the bimolecular proteolytic removal of the propeptide, which can be accomplished in one or more steps (41). In conclusion, GAGs, which are found in the lysosomes, have the potential to interact with and regulate the activities of cysteine cathepsins. Autocatalytic activation of cysteine cathepsins can be substantially accelerated in the presence of GAGs and other negatively charged polysaccharides. GAGs facilitate procathepsin B activation through disruption of propeptide-mature enzyme interactions (42). In the first step, GAGs bind to procathepsin B, thereby inducing a conformational change, which converts the procathepsin B molecule into a better substrate. In the next step, the prodomain is removed by intermolecular proteolytic cleavage. Following propeptide dissociation, the bound GAG probably dissociates and is thus free to bind to another procathepsin B molecule. These observations suggest that the mechanism of insertion of cysteine cathepsins on the plasma membrane and its cellular traffic may depend on HSPGs present on the cell surface. In addition, the activity of cathepsin D, another lysosomal cathepsin, but pertaining to the aspartic proteinase family, which is overexpressed in breast tumors and that activates the cysteine cathepsins, is also regulated by GAG (43). GAGs increase the activity of cathepsin D in vitro. HP increases the activity of the proenzyme form and stimulates the mature (prodomain cleaved) enzyme. In addition, HP increases the limited proteolysis of procathepsin D at acidic pH, concomitant with an increased rate of substrate peptide cleavage. Like cathepsin D, GAGs induce the activation of the β-site amyloid precursor protein (APP) cleaving enzyme (BACE1) (44). BACE1 is a membrane-anchored enzyme that catalyzes the first step in the production of β-amyloid, the protein that accumulates in the brain of Alzheimer's disease patients (45). BACE1 cleavage of APP may possibly occur in the endoplasmic reticulum, endosomes, the Golgi, and the cell surface. HS can colocalize with BACE1 in these subcellular compartments (46). This suggests that HS may be able to regulate BACE1 activity in these compartments. It is also worth noting that APP (a major BACE1 substrate) can bind HP. Thus, endogenous GAGs may not only activate the BACE1 zymogen during early stages of the secretory pathway but they could also help to orientate the enzyme with its substrate, which suggests a mechanism of HP-induced proBACE1 activation in which conformational changes in the monomer may be needed for activation.

In conclusion, interactions of proteinases with other molecules on the plasma membrane anchor and activate them at a specific location on the cell surface. Interactions of proteinases with macromolecules other than their own protein substrates or inhibitors result in changes in their activity and/or may have important biological effects.

5. Proteolytic Activity Regulation

PGs could play important roles in controlling the activation and thereby activity of proteinases. Thus, PGs, via their GAG chains, may facilitate proteolysis of certain substrates by interacting with the substrate, the enzyme, or the proteinase inhibitors.

HP promotes the binding of thrombin to fibrin (47) and stimulates the activity of plasminogen activator (48). HP acts in vitro (49) and in vivo (50) as a tight-binding, hyperbolic, competitive inhibitor of human neutrophil elastase (HNE), and this inhibition is strongly dependent upon polysaccharide chain length and degree of sulfation (51). HNE is an essential component of the phagocytic machinery of polymorphonuclear leukocytes. The concentration of HNE in the azurophil granules of polymorphonuclear leukocytes is thought to be in the millimolar range. An efficient anti-HNE control system must therefore be present at inflammatory sites. Healthy individuals have efficient HNE inhibitors including α1-proteinase inhibitor (α1PI), α2 macroglobulin, secretory leukocyte peptidase inhibitor (SLPI), and elafin (52). In vitro, the sulfated GAG has been shown to significantly increase the rate constant for the inhibition of HNE (53), neutrophil cathepsin G (54), and mast cell chymase (55) by SLPI. In addition, HP in combination with SLPI demonstrated in vivo efficacy reducing early- and late-phase bronchoconstriction.

α1PI is a multifunctional serpin, which circulates in blood and is implicated in a myriad of physiological functions. It is an acute phase protein and protects pulmonary tissues. α1PI depicts a high degree of sequence and structural homology towards other serpins. Serpins adopt a metastable conformation that is required for their inhibitory activity (56). Serpins can inhibit serine or cysteine proteinases by a remarkable conformational change-based mechanism in which the proteinase, having initiated cleavage of the serpin, is translocated 70 Å from one pole of the serpin to the other (57). Translocation results from the existence of a covalent acyl linkage between the proteinase active site residue serine and the P1 residue, and full insertion of the RCL into β-sheet A of the serpin. α1PI, like several other serpins such as antithrombin, proteinase nexin-I, HP cofactor II (HC-II), plasminogen activator inhibitor-I, or protein-C inhibitor, is a HP-binding serpin (58). These serpins possess a positively charged cluster at the protein surface that interacts with various GAGs. Antithrombin becomes fully effective upon activation by HP. The role of inhibition of blood coagulation proteinases by antithrombin is accelerated several thousand fold in the presence of catalytic amounts of HP or related GAGs (59). Binding of a specific HP or HS pentasaccharide to antithrombin induces allosteric activation changes that mitigate the unfavorable interactions and promote template binding of the serpin and proteinase. Like antithrombin, HP brings about significant changes with respect to α1PI structure and function. The intrinsic ability of catalytic quantities of HP accelerates the inhibitory potential of α1PI. Thus, the association rate constants of α1PI in the presence of HP were found to be significantly enhanced compared to the native forms. This activation results from the specific binding of HP to a lysin rich stretch on helix-F (60). α1PI shows a sigmoidal transition upon activation induced by heparin binding indicative of allosteric modulation and appears to be characterized by two steps of binding: a weak followed by a strong binding. Recently, a similar result was described with the squamous cell carcinoma antigens (SCCAs), a serpin of cathepsin L. Higgins et al. (61) showed that the presence of HP accelerated inhibition of cathepsin L by SCCA-1 and SCCA-2 which specifically bind HP and HS but not other GAGs (61). In the case of SCCA-1, HP increased the second-order inhibition rate constant. A templating mechanism was shown, consistent with ternary complex formation. Furthermore, SCCA-1 inhibition of cathepsin L-like proteolytic activity secreted from breast and melanoma cancer cell lines was significantly enhanced by HP.

Acceleration of proteinase inhibition also occurs when HP binds a nonserpin inhibitor such as TIMP-3. TIMP-3, contrary to the other TIMPs, inhibits all known interstitial and membrane-bound MMPs as well as several key ADAMs and ADAMTSs (62). TIMP-3 is unique, because it is the sole TIMP regulator of one of the most versatile ADAM sheddases, ADAM-17 (63). Furthermore, TIMP-3 is also renowned for its ability to induce apoptosis in mammalian cells (64). Another extraordinary quality that distinguishes TIMP-3 from the other TIMPs is its ability to adhere to the ECM. Unlike the other TIMPs, TIMP-3 is not readily soluble (65), but instead is sequestered at the cell surface by association with the GAG chains of PGs, especially HP. TIMP-3 interacts with cell surface and ECM GAGs via the large number of positively charged residues in TIMP-3, and it was shown that this is the basis for its location in the ECM both in vivo and in cell culture. Hence, colocalization of TIMP-3 with proenzymes in the pericellular environment may be a mechanism for increasing the rate of inhibition of MMPs, such as MMP-2, and regulating ECM breakdown during morphogenetic processes (66).

HP and HS were found to increase the activity and stability of the lysosomal cysteine cathepsin. Cathepsin B has been implicated in a variety of diseases involving tissue remodeling states, such as inflammation, and tumor metastasis, by degradation of ECM components (67). The mature form of cathepsin B has been shown to be rapidly inactivated at neutral or alkaline pH and by its endogenous proteins inhibitors, mainly from the cystatin family (68). On the other hand, it has been shown that membrane-bound forms of cathepsin B are very resistant to inactivation at neutral pH (69). The nature of the cathepsin B-GAGs interaction was sensitive to the charge and type of polysaccharide (70). Like papain (71), HP and HS bind cathepsin B specifically, and this interaction induces an increase of α-helix content, which stabilizes the enzyme structure even at alkaline pH (70). This coupling on the cell surface potentiates the endopeptidase activity of the enzyme by increasing fivefold its half-life at physiological pH. The presence of cathepsin B on the plasma membrane results in focal dissolution of ECM proteins and could be enabling the tumor cell to invade the tissue.

Cathepsin K, another cysteine cathepsin, is considered as the principal proteinase responsible for the degradation of most of the bone matrix (10). Controlled degradation of collagen is observed in bone remodeling. Following tight attachment to the bone surface, osteoclasts secrete protons into a closed extracellular compartment enclosed by a sealing zone. This local acidification solubilizes the mineral bone and renders the organic matrix available to proteinases. Cathepsin K is abundantly and predominantly expressed in osteoclasts where it is localized in the lysosomes, in the ruffled border, and in the resorption lacunae on the bone surface (72). GAGs can also participate in bone resorption regulation through the modulation of proteinase activity. The collagenolytic activity of cathepsin K is related to a specific complex constituted of five cathepsin K and five CS molecules. This complex has a triple helical collagen-degrading activity, whereas the monomeric form of cathepsin K can degrade noncollagenous substrates (73). At acidic pH, GAGs predominantly expressed in bone and cartilage, such as CS and KS, enhance the collagenolytic activity of cathepsin K, whereas DS, HS, and HP selectively inhibit its activity (74). The structure of cathepsin K is flexible and converts between multiple conformational states with distinctive characteristics at physiological plasma pH. GAGs increase the activity of the enzyme and promote a conformational change. GAGs act as natural allosteric modifiers of cathepsin K by affecting the distribution of a preexisting equilibrium of conformational states (75). GAGs, which exploit the conformational flexibility of the enzyme, regulate its activity and stability against autoproteolysis. At pH 7.40, the enzyme degrades collagen on its own and CS/DS reduces its collagenolytic activity, whereas HP enhances it. Altogether, this shows that the molecular mechanism behind the unique collagenolytic activity of cathepsin K depends on the environment.

In conclusion, PGs, through their compositional diversity, play a broad variety of roles as modulators of proteinase activities: (1) acting as crucial structural elements by the control of proteinase activities, or (2) acting as coreceptors for some inhibitors, playing important roles for the acceleration of proteinase inhibition.

6. Conclusion

Proteinases mediate vital processes in cells and in the homeostasis of tissues and organs. Proteinases have also been shown to participate in tumor growth and metastasis, with different proteinases exerting different functions at various stages of tumor development. These observations indicate that these enzymes are good targets for the development of anticancer therapies. However, an important question is the extent to which proteinases have structural roles in remodeling tissues vs. a role in regulating access to signaling molecules. This requires a precise understanding of the roles of individual proteinases, with respect to not only ECM degradation, but also modulation of cytokine and growth factor function. Thus, a better understanding of the regulatory mechanisms that control proteinase activities in cancer cells could provide new ways for therapeutic intervention that are more specific and effective. Before inhibitors of proteinases can be applied in therapy, it is important to determine which proteinases are active in which tumor type and the exact role they have. Besides, to elucidate the in vivo roles of proteinases in cellular pathways and thereby identify proteinase drug targets, it will be essential to determine their degradome that illustrates the large substrate specificity of proteinases. Once a proteinase or set of proteinases has been identified as a promising target for anticancer treatment, their production and/or activities can be modulated at each of the various steps that lead to the synthesis and activation of the mature enzyme. The modulatory effect exerted by PGs on proteinase activities is expected to lead to new insights in the understanding of some molecular systems present in pathological states, providing new targets for drug therapy.

Acknowledgments

This work was supported by the Département Loire Atlantique (Program entitled Atlanthèse). Steven GEORGES received a fellowship from the Département Loire Atlantique. Thanks to Verena STRESING for proof reading and helpful discussions.

References

1. Lamoureux, F., Baud'huin, M., Duplomb L, Heymann, D., and Rédini, F. (2007) Proteoglycans: key partners in bone cell biology. *BioEssays.* **29**, 758–771.
2. Ruiz Velasco, C., Colliec-Jouault, S., Rédini, F., Heymann, D., and Padrines, M. (2010) Proteoglycans on bone tumor development. *Drug Discov. Today.* **15**, 553–560.
3. Hardingham, T. E. and Fosang, A. J. (1992). Proteoglycans: many forms and many functions. *FASEB J.* **6**, 861–870.
4. Bernfield, M. (1999) Functions of cell surface heparan sulfate proteoglycans. Annu Rev Biochem. 68, 729–777.
5. Barrett, A. J., Rawlings, N. D., and Woessner, J. F. (2004) The Handbook of Proteolytic Enzymes. 2nd ed. Barrett AJ, Rawlings ND, Woessner JF (Eds), Academic Press, Cambridge, UK.
6. Rawlings, N. D., Tolle, D. P., and Barrett, A. J. (2004) Evolutionary families of peptidase inhibitors. *Biochem. J.* **378**, 705–716.
7. Owen, C. A. (2008) Leukocyte cell surface proteinases : regulation of expression, functions, and mechanisms of surface localization. *Int. J. Biochem. Cell Biol.* **40**, 1246–1272.
8. Stefanidakis, M. and Koivunen, E. (2006) Cell-surface association between matrix metalloproteinases and integrins: role of the complexes in leukocyte migration and cancer progression. *Blood.* **108**, 1441–1450.
9. Mohammed, F. F., Smookler, D. S., and Khokha, R. (2003) Metalloproteinases, inflammation, and rheumatoid arthritis. *Ann Rheum Dis.* **62**, ii43.
10. Georges, S., Ruiz Velasco, C., Trichet, V., Fortun, Y., Heymann, D., and Padrines, M. (2009) Proteases and bone remodeling. *Cytokine Growth Factor Review* **20**, 29–41.
11. Brooks, P. C., Strömblad, S., Sanders, L. C., von Schalscha, T. L., Aimes, R. T., Stetler-Stevenson, W. G., et al. (1996) Localization of matrix metalloproteinase MMP-2 to the surface of invasive cells by interaction with integrin αvβ3. *Cell* **85**, 683–693.
12. Redondo-Muñoz, J., Ugarte-Berzal, E., García-Marco, J. A., del Cerro, M. H., Van den Steen, P. E., Opdenakker, G., et al. (2008) α4β1 integrin and 190-kDa CD44v constitute a cell surface docking complex for gelatinase B/MMP-9 in chronic leukemic but not in normal B cells. *Blood.* **112**, 169–178.
13. Dumin, J. A., Dickeson, S. K., Stricker, T. P., Bhattacharyya-Pakrasi, M., Roby, J. D., Santoro, S. A., et al. (2001) Pro-collagenase-1 (matrix metalloproteinase-1) binds the α2β1 integrin upon release from keratinocytes migrating on type I collagen. *J. Biol. Chem.* **276**, 29368–29374.
14. Knauper, V., Will, H., Lopez-Otin, C., Smith, B., Atkinson, S. J., Stanton, H., et al. (1996) Cellular mechanisms for human procollagenase-3 (MMP-13) activation. Evidence that MT1-MMP (MMP-14) and gelatinase A (MMP-2) are able to generate active enzyme. *J Biol Chem.* **271**, 17124–17131.
15. Nagase, H. (1998) Cell surface activation of progelatinase A (proMMP-2) and cell migration. *Cell Res.* **8**, 179–186.

16. Yu, W. H. and Woessner, J. F. (2000) Heapran sulphate proteoglycans as extracellular docking molecules for matrilysin (matrix metalloproteinase 7). *J. Biol. Chem.* **275**, 4183–4191.
17. Nabeshima, K., Inoue, T., Shimao, Y., and Sameshima, T. (2002) Matrix metalloproteinases in tumor invasion : role for cell migration. *Pathol. Int.* **4**, 255–264.
18. Ryu, H. Y., Lee, J., Yang, S., Par, H., Cho, S., Jung, K. C., et al. (2009) Syndecan-2 Functions as a Docking Receptor for Pro-matrix Metalloproteinase-7 in Human Colon Cancer Cells. *J. Biol. Chem.* **284**, 35692–35701.
19. Bartolazzi, A., Nocks, A., Aruffo, A., Spring, F., and Stamenkovic, I. (1996) Glycosylation of CD44 is implicated in CD44-mediated cell adhesion to hyaluronan. *J. Cell Biol.* **132**, 1199–1208.
20. Yu, W. H., Woessner, J. F., McNeish, J. D., and Stamenkovic, I. (2002) CD44 anchors the assembly of matrilysin/MMP-7 with heparin-binding epidermal growth factor precursor and ErbB4 and regulates female reproductive organ remodelling. *Genes Dev.* **16**, 307–323.
21. Yu, Q. and Stamenkovic, I. (2000) Cell surface-localized matrix metalloproteinase-9 proteolytically activates TGF-β and promotes tumor invasion and angiogenesis. *Genes Dev.* **14**, 163–176.
22. Owen, C. A., Hu, Z., Lopez-Otin, C., and Shapiro, S. D. (2004) Membrane-bound matrix metalloproteinase-8 activated polymorphonuclear cells is a potent, tissue inhibitor of metalloproteinase-resistant collagenase and serpinase. *J. Immunol.* **172**, 7791–7803.
23. Davis, G. E., Pintar Allen, K. A., Salazar, R., and Maxwell, S. A. (2001) Matrix metalloproteinase-1 and -9 activation by plasmin regulates a novel endothelial cell-mediated mechanism of collagen gel contraction and capillary tube regression in three-dimensional collagen matrices. *J. Cell Sci.* **114**, 917–930.
24. Crabbe, T., Ioannou, C., and Docherty, A. J. (1993) Human progelatinase A can be activated by autolysis at a rate that is concentration-dependent and enhanced by heparin bound to the C-terminal domain. *Eur J Biochem.* **218**, 431–438.
25. Sørensen, H. P., Vives, R. R., Manetopoulos, C., Albrechtsen, R., Lydolph, M. C., Jacobsen, J., et al. (2008) Heparan sulfate regulates ADAM12 through a molecular switch mechanism. *J Biol Chem.* **283**, 31920–31932.
26. Iida, J., Wilhelmson, K. L., Ng, J., Lee, P., Morrison, C., Tam, E., et al. (2007) Cell surface chondroitin sulfate glycosaminoglycan in melanoma: role in the activation of pro-MMP-2 (pro-gelatinase A) *Biochem. J.* **403**, 553–563.
27. Naito, S., Shiomi, T., Okada, A., Kimura, T., Chijiiwa, M., Fujita, Y., et al. (2007) Expression of ADAMTS4 (aggrecanase-1) in human osteoarthritic cartilage. *Pathol Int.* **57**, 703–711.
28. Mort, J. S., Flannery, C. R., Makkerh, J., Krupa, J. C., and Lee, E. R. (2003) Use of antineoepitope antibodies for the analysis of degradative events in cartilage and the molecular basis for neoepitope specificity. *Biochem. Soc. Symp.* **70**, 107–114.
29. Stanton, H., Rogerson, F. M., East, C. J., Golub, S. B., Lawlor, K. E., Meeker, C. T., et al. (2005) ADAMTS5 is the major aggrecanase in mouse cartilage in vivo and in vitro. *Nature* **434**, 648–652.
30. Gendron, C., Kashiwagi, M., Lim, N. H., Enghild, J. J., Thøgersen, I. B., Hughes, C., et al. (2007) Proteolytic activities of human ADAMTS-5: comparative studies with ADAMTS-4. *J. Biol.Chem.* **282**, 18294–18306.
31. Gao, G., Plaas, A., Thompson, V. P., Jin, S., Zuo, F., and Sandy, J. D. (2004) ADAMTS4 (Aggrecanase-1) Activation on the Cell Surface Involves C-terminal Cleavage by Glycosylphosphatidyl Inositol-anchored Membrane Type 4-Matrix Metalloproteinase and Binding of the Activated Proteinase to Chondroitin Sulfate and Heparan Sulfate on Syndecan-1. *J. Biol. Chem.* **279**, 10042–10051.
32. Echtermeyer, F., Bertrand, J., Dreier, R., Meinecke, I., Neugebauer, K., Fuerst, M., et al. (2009) Syndecan-4 regulates ADAMTS-5 activation and cartilage breakdown in osteoarthritis. *Nat Med.* **15**, 1072–1076.
33. Munesue, S., Yoshitomi, Y., Kusano, Y., Koyama, Y., Nishiyama, A., Nakanishi, H., et al. (2007) A novel function of syndecan-2, suppression of matrix metalloproteinase-2 activation, which causes suppression of metastasis. *J Biol Chem.* **282**, 28164–28174.
34. Nakada, M., Yamada, A., Takino, T., Miyamori, H., Takahashi, T., Yamashita, J., et al. (2001) Suppression of membrane-type 1 matrix metalloproteinase (MMP)-mediated MMP-2 activation and tumor invasion by testican 3 and its splicing variant gene product, N-Tes. *Cancer Res.* **61**, 8896–8902.
35. Nakada, M., Miyamori, H., Yamashita, J., and Sato, H. (2003) Testican 2 abrogates inhibition of membrane-type matrix metalloproteinases by other testican family proteins. *Cancer Res.* **63**, 3364–3369.
36. Schmitt, M., Jaenicke, F., and Graeff, H. (1992) Tumor-associated proteases. *Fibrinol. Proteol.* **6**, 3–26.

37. Cavallo-Medved, D. and Sloane, B.F. (2003) Cell surface cathepsin B: understanding its functional significance. *Curr. Top. Dev. Biol.* **54**, 313–341.
38. Podobnik, M., Kuhelj, R., Turk, V., and Turk, D. (1997) Crystal structure of the wild-type human procathepsin B at 2.5 A resolution reveals the native active site of a papain-like cysteine protease zymogen. *J. Mol. Biol.* **271**, 774–788.
39. Turk, V., Turk, B., and Turk, D. (2001) Lysosomal cysteine proteases: facts and opportunities. *EMBO J.* **20**, 4629–4633.
40. Turk, B., Turk, D., and Turk, V. (2000) Lysosomal cysteine proteases: more than scavengers. *Biochim Biophys Acta.* **1477**, 98–111.
41. Pungercar, J. R., Caglic, D., Sajid, M., Dolinar, M., Vasiljeva, O., Pozgan, U., et al. (2009) Autocatalytic processing of procathepsin B is triggered by proenzyme activity. *FEBS J.* **276**, 660–668.
42. Caglic, D., Pungercar, J. R., Pejler, G., Turk, V., and Turk, B. (2007) Glycosaminoglycans facilitate procathepsin B activation through disruption of propeptide-mature enzyme interactions. *J. Biol. Chem.* **282**, 33076–33085.
43. Beckman, M., Freeman, C., Parish, C. R., and Small, D. H. (2009) Activation of cathepsin D by glycosaminoglycans. *FEBS J.* **276**, 7343–7352.
44. Klaver, D. W., Wilce, M. C. J., Gasperini, R., Freeman, C., Juliano, J. P., Parish, C., et al. (2010) Glycosaminoglycan-induced activation of the β-secretase (BACE-1) of Alzheimer's disease. *J Neurochem.* **112**, 1552–1561.
45. Sinha, S., Anderson, J. P., Barbour, R., Basi, G. S., Caccavello, R., Davis, D., et al. (1999) Purification and cloning of amyloid precursor protein beta secretase from human brain. *Nature* **402**, 537–540.
46. Scholefield, Z., Yates, E. A., Wayne, G., Amour, A., McDowell, W., and Turnbull, J. E. (2003) Heparan sulfate regulates amyloid precursor protein processing by BACE1, the Alzheimer's beta-secretase. *J. Cell Biol.* **163**, 97–107.
47. Hogg, P. J. and Jackson, C. M. (1990) Heparin promotes the binding of thrombin to fibrin polymer. Quantitative characterization of a thrombin-fibrin polymer-heparin ternary complex. *J. Biol. Chem.* **265**, 241–247.
48. Stein, P. L., van Zonneveld, A. J., Pannekoek, H., and Strickland, S. (1989) Structural domain of tissue-type plasminogen activator that confer stimulation by heparin. *J. Biol. Chem.* **264**, 15441–15444.
49. Frommherz, K. J., Faller, B., and Bieth, J. G. (1991) Heparin strongly decreases the rate of inhibition of neutrophil elastase by α1-proteinase inhibitor. *J. Biol. Chem.* **266**, 15356–15362.
50. Fryer, A., Huang, Y. C., Rao, G., Jacoby, D., Mancilla, E., Whorton, R., et al. (1997) Selective O-desulfation produces nonanticoagulant heparin that retains pharmacological activity in the lung. *J. Pharmacol. Exp.Ther.* **282**, 208–219.
51. Spencer, J. L., Stone, P. J., and Nugent, N. A. (2006) New Insights into the Inhibition of Human Neutrophil Elastase by Heparin. *Biochemistry* **45**, 9104–9120.
52. Moreau, T., Baranger, K., Dadé, S., Dallet-Choisy, S., Guyot, N., and Zani, M. L. (2008) Multifaceted roles of human elafin and secretory leukocyte proteinase inhibitor (SLPI), two serine protease inhibitors of the chelonianin family. *Biochimie* **90**, 284–295.
53. Cadène, M., Boudier, C., Marcillac, G. D., and Bieth, J. G. (1995) Influence of low molecular mass heparin on the kinetics of neutrophil elastase inhibition by mucus proteinase inhibitor. *J. Biol.Chem.* **270**, 13204–13209.
54. Ermolieff, J., Duranton, J., Petitou, M., and Bieth, J. (1998) Heparin accelerates the inhibition of cathepsin G by mucus proteinase inhibitor: potent effect of O-butyrylated heparin. *Biochem. J.* **330**, 1369–1374.
55. Walter, M., Plotnick, M., and Schechter, N. M. (1996) Inhibition of human mast cell chymase by secretory leukocyte proteinase inhibitor: enhancement of the interaction by heparin. *Arch. Biochem. Biophys.* **327**, 81–88
56. Stein, P. E. and Carrell, R. W. (1995) What do dysfunctional serpins tell us about molecular mobility and disease? *Nat. Struct. Biol.* **2**, 96–113.
57. Stratikos, E. and Gettins, P. G. W. (1999) Formation of the covalent serpine proteinase complex involves translocation of the proteinase by more than 70 A and full insertion of the reactive center loop into β-sheet A. *Proc. Natl. Acad. Sci. USA.* **96,** 4808–4813.
58. Gettins, P. G. W. (2002) Serpin structure, mechanism and function. *Chem. Rev.* **102**, 4751–4803.
59. Desai, U. R., Petitou, M., Björk, I., and Olson, S. T. (1998) Mechanism of heparin activation of antithrombin: evidence for an induced-fit model of allosteric activation involving two interaction subsites. *Biochemistry* **37**, 13033–13041.
60. Gupta, V. K. and Gowda, L.R. (2008) Alpha-1-proteinase inhibitor is a heparin binding serpin: molecular interactions with the Lys rich

cluster of helix-F domain. *Biochimie* **90**, 749–761.

61. Higgins, W. J., Fox, D. M., Kowalski, P.S., Nielsen, J. E., and Worrall, D. M. (2010) Heparin enhances serpin inhibition of the cysteine protease cathepsin L. *J. Biol. Chem.* **285**, 3722–3729.
62. Murphy, G., Knäuper, V., Lee, M. H., Amour, A., Worley, J. R., Hutton, M., et al. (2003) Role of TIMPs (tissue inhibitors of metalloproteinases) in pericellular proteolysis: the specificity is in the detail. *Biochem. Soc. Symp.* **70**, 201–212.
63. Amour, A., Slocombe, P. M., Webster, A., Butler, M., Knight, C. G., Smith, B. J., et al. (1998) TNF-alpha converting enzyme (TACE) is inhibited by TIMP-3. *FEBS Lett.* **435**, 39–44.
64. Smith, M. R., Kung, H. F., Durum, S. K., Colburn, N. H., and Sun, Y. (1997) TIMP-3 induces cell death by stabilizing TNF-alpha receptors on the surface of human colon carcinoma cells. *Cytokine* **9**, 770–780.
65. Lee, M. H., Atkinson, S., and Murphy, G. (2007) Identification of the extracellular matrix (ECM) binding motifs of tissue inhibitor of metalloproteinases (TIMP)-3 and effective transfer to TIMP-1. *J. Biol. Chem.* **282**, 6887–6898.
66. Butler, G. S., Apte, S. S., Willenbrock, F., and Murphy, G. (1999) Human tissue inhibitor of metalloproteinases 3 interacts with both the N- and C-terminal domains of gelatinases A and B. Regulation by polyanions. *J. Biol. Chem.* **274**, 10846–10851.
67. Turk, B., Bieth, J. G., Björk, I., Dolenc, I., Turk, D., Cimerman, N., et al. (1995) Regulation of the activity of lysosomal cysteine proteinases by pH-induced inactivation and/or endogenous protein inhibitors, cystatins. *Biol. Chem.* Hoppe Seyler **376**, 225–230.
68. Yan, S. and Sloane, B. F. (2004) Molecular regulation of human cathepsin B: implication in pathologies. *Biol. Chem.* **384**, 845–854.
69. Sloane, B. F., Rozhin, J., Lah, T. T., Day, N.A., Buck, M., Ryan, R. E., et al. (1988) Tumor cathepsin B and its endogenous inhibitors in metastasis. *Adv. Exp. Med. Biol.* **233**, 259–268
70. Almeida, P. C., Nantes, I. L., Chagas, J. R., Rizzi, C. C., Faljoni-Alario, A., Carmona, E., et al. (2001) Cathepsin B activity regulation heparin-like glycosaminoglycans protect human cathepsin B from alkaline ph-induced inactivation. *J. Biol. Chem.* **276**, 944–951.
71. Almeida, P. C., Nantes, I. L., Rizzi, C. C. A., Júdice, W. A., Chagas, J. R., Juliano, L., et al. (1999) Cysteine proteinase activity regulation. A possible role of heparin and heparin-like glycosaminoglycans. *J. Biol. Chem.* **274**, 30433–30438.
72. Littlewood-Evans, A., Kokubo, T., Ishibashi, O., Inaoka, T., Wlodarski, B., Gallagher, J. A., et al. (1997) Localization of cathepsin K in human osteoclasts by in situ hybridization and immunohistochemistry. *Bone* **20**, 81–86.
73. Li, Z., Hou, W. S., Escalante-Torres, C. R., Geld, B. D., and Bromme, D. (2002) Collagenase activity of cathepsin K depends on complex formation with chondroitin sulfate. *J Biol Chem* **277**, 28669–28676.
74. Li, Z., Yasuda, Y., Li, W., Bogyo, M., Katz, N., Gordon, R. E., et al. (2004) Regulation of collagenase activities of human cathepsins by glycosaminoglycans. *J. Biol. Chem.* **279**, 5470–5479.
75. Novinec, M., Kovacic, L., Lenarcic, B., and Baici, A. (2010) Conformational flexibility and allosteric regulation of cathepsin K. *Biochem. J.* **429**, 379–389.

Chapter 21

Proteoglycans and Osteolysis

Marc Baud'Huin, Céline Charrier, Gwenola Bougras, Régis Brion, Frédéric Lezot, Marc Padrines, and Dominique Heymann

Abstract

Osteolysis is a complex mechanism resulting from an exacerbated activity of osteoclasts associated or not with a dysregulation of osteoblast metabolism leading to bone loss. This bone defect is not compensated by bone apposition or by apposition of bone matrix with poor mechanical quality. Osteolytic process is regulated by mechanical constraints, by polypeptides including cytokines and hormones, and by extracellular matrix components such as proteoglycans (PGs) and glycosaminoglycans (GAGs). Several studies revealed that GAGs may influence osteoclastogenesis, but data are very controversial: some studies showed a repressive effect of GAGs on osteoclastic differentiation, whereas others described a stimulatory effect. The controversy also affects osteoblasts which appear sometimes inhibited by polysaccharides and sometimes stimulated by these compounds. Furthermore, long-term treatment with heparin leads to the development of osteoporosis fueling the controversy. After a brief description of the principal osteoclastogenesis assays, the present chapter summarizes the main data published on the effect of PGs/GAGs on bone cells and their functional incidence on osteolysis.

Key words: Bone resorption, Osteoclast, Glycosaminoglycan, RANKL

1. Introduction

Bone metabolism is tightly regulated by a balance between two bone cell types combining catabolic and anabolic activities. Bone catabolism is supported by multinucleated cells specialized in bone catabolism and named osteoclasts. Osteoclasts originate from the monocyte lineage and differentiate by the action of membranous, soluble, and extracellular matrix compounds (1). Among these factors, some are required for proliferation of osteoclast mononuclear progenitors such as macrophage colony-stimulating factor (M-CSF), while other factors such as receptor activator of nuclear factor-κB ligand (RANKL) are more specifically implicated in the commitment of these precursors to their fusion and in the formation

Françoise Rédini (ed.), *Proteoglycans: Methods and Protocols*, Methods in Molecular Biology, vol. 836, DOI 10.1007/978-1-61779-498-8_21, © Springer Science+Business Media, LLC 2012

of multinucleated resorbing osteoclasts (2–4). Bone catabolism depends on the ability of the osteoclast to generate an acid extracellular compartment between itself and the bone surface which is essential for solubilization of the alkaline salts of bone mineral (5). This acidic pH is also necessary for the digestion of the organic bone matrix by lysosomal enzymes secreted by osteoclasts (6). According to their ability to solubilize hydroxyapatite crystals and to digest organic matrix, osteoclasts contribute to the orchestration of the phosphocalcic homeostasis together in concert with the second main bone type cells, the osteoblasts. Osteoblasts originate from mesenchymal stem cells and perform anabolic functions consisting in the formation of extracellular matrix composed by 95% of type I collagen (7). Osteoblast activities are not limited to the formation of bone extracellular matrix, but also extend to the osteolytic process. While osteoblasts produce and secrete gelatinase activities (8) controlling partly the collagenic matrix, they release more particularly cytokines and growth factors regulating osteoclast differentiation and activation (2, 3, 9). Among these polypeptides, RANKL/OPG (osteoprotegerin, OPG) is the main molecular couple involved in the communication between osteoblasts and osteoclasts (2,3). In this system, RANKL expressed by osteoblasts and also by stromal cells binds to its receptor RANK expressed on the surface of osteoclast precursors and consequently activates TRAF6/NF-κB signaling pathway leading to the fusion of osteoclast activity and survival (10, 11). RANK/RANKL interactions are controlled by OPG which is also produced by osteoblasts/stromal cells. OPG acts as a soluble decoy receptor blocking the binding of RANKL to RANK and subsequently the osteoclastogenesis and the osteolytic process (12, 13).

Extracellular matrix components, especially proteoglycans (PGs) and glycosaminoglycans (GAGs), contribute to the bone remodeling and to the maintenance of bone mass (9). Thus, PGs and GAGs are involved in the organization of collagen fibers (9). However, the role of GAGs and PGs in bone metabolism is more complex than initially envisaged and this complexity is mainly related to the structure and the localization of these compounds. PGs exhibit numerous locations and more precisely are associated with intracellular compartments expressed on the cell surface or anchored in the extracellular matrix and basement membrane in almost all tissues in adults (14). The composition of GAGs is very heterogenous and includes linear polymers which are bound to a core protein to form PGs. There is no unifying feature for core protein structures and then PGs display a great diversity of protein forms. Many core proteins have complex modular structures with protein motifs which are of similar in sequence to those found in other protein families. GAGs are composed of repeated disaccharidic units of hexosamine and hexuronic acid, except for keratan sulfate in which hexuronic acid is replaced by galactose. According to the epimerization and sulfation of hexosamine and uronic acid, several families of GAGs have been described. All together, this diversity of composition explains in part their very complex biological activities in all tissues and that GAGs/PGs functions are not limited to the control of fibrillogenesis.

The aim of the present review is to better define the function of GAGs and PGs in bone remodeling and more specifically in osteolysis. The first part of the manuscript will describe the main osteoclastogenesis assays currently used. The review will then focus on the role of PGs in the control of physiological and pathological osteolysis regarding the osteoblastic and osteoclastic components.

2. In Vitro Assays of Osteoclastogenesis

Numerous cell culture systems derived from different species have been established to study the molecular and cellular mechanisms of osteoclastogenesis (3). Recently, osteoclasts have been generated from a single-cell suspension of embryonic stem (ES) cells seeded on a feeder monolayer. Bone-resorbing cells expressing osteoclastic markers such as tartrate-resistant acid phosphatase (TRAP) or RANK were obtained within 11 days (15). However, the main systems used to study the mechanisms of osteoclastogenesis are based on culture of osteoclast progenitors isolated from monocytic cells (peripheral blood monocyte fraction/umbilical cord blood monocytes/spleen cells/monocytic cell lines) in the presence or absence of stromal cells (osteoblastic cells) or in the absence of stromal cells, but after addition of a cytokine cocktail including M-CSF and RANKL. The following paragraphs will describe the most effective methods to study osteoclastogenesis in vitro.

2.1. Differentiation Assay from the Murine RAW 264.7 Monocytic Cell Line

1. Materials and reagents
   - Murine RAW 264.7 monocytic cells (ATCC, Promochem) (16).
   - Phenol red-free α-Minimal Essential Medium (α-MEM) (Invitrogen).
   - Fetal calf serum (FCS) (Perbio), batch specifically selected for osteoclast differentiation.
   - Non-essential amino acids (Invitrogen).
   - Solution of trypsin (0.25%) and ethylenediamine tetraacetic (EDTA) (1 mM) (Invitrogen).
   - Human or mouse RANKL (R&D Systems) is dissolved in phosphate buffer/0.1% BSA at 1 mg/mL and stored in single use aliquots at –80°C until use. Final concentration used is 100 μg/mL (dilution in α-MEM supplemented with 10% FCS).
   - Leukocyte (TRAP) staining kit no. 387A (Sigma).

2. Cell culture

RAW 264.7 cells were routinely cultured in α-MEM supplemented with 10% FCS and 1% non-essential amino acids. Fresh medium is replaced twice a week and cell culture amplifications are performed after cell detachment by scraping. RAW 264.7 are frozen in DMSO solution diluted at 20% in FCS and frozen at 5×10^6 cells/mL in liquid nitrogen until use.

3. Osteoclast differentiation

To induce osteoclast formation, RAW 264.7 cells are scraped then incubated at 37°C for 2 min to allow adherence of the most differentiated cells. Non-adherent cells are then collected and seeded in fresh medium, at a density of 3×10^3 cells/well in a 96-well plate. After 2 h, medium is changed for a fresh one containing 100 ng/mL hRANKL. After 5 days of culture, multinucleated cells (>3 nuclei) are counted under a light microscope (contrast phase) (see Fig. 1a) or after TRAP staining according to the manufacturer's recommendation (16, 17). Osteoclasts can be observed from 30 ng/mL of RANKL.

2.2. Differentiation from Murine CD11b+ Monocytes

1. Materials and reagents
   - 4-week-old C57BL6 male mice.
   - CD11b microbeads and MACS technology (Miltenyi Biotec).
   - Phenol red-free α-MEM (Invitrogen).
   - FCS (Perbio), batch specifically selected for osteoclast differentiation.
   - Mouse M-CSF (R&D Systems) is dissolved in phosphate buffer/0.1% BSA at 25 μg/mL and stored in single use aliquots at −80°C until use. Final concentration used is 25 ng/mL (dilution in α-MEM supplemented with 10% FCS).
   - Human or mouse RANKL (R&D Systems) is dissolved in phosphate buffer/0.1% BSA at 100 μg/mL and stored in single use aliquots at −80°C until use. Final concentration used is 100 ng/mL (dilution in α-MEM supplemented with 10% FCS).
   - Leukocyte (TRAP) staining kit (Sigma).
2. Cell preparation and osteoclast differentiation

CD11b+ monocytes are purified from murine bone marrow cells, obtained by flushing the bone marrow from femora and tibiae of 4-week-old C57BL6 male mice. Mice are anesthetized using isoflurane and euthanized by cervical dislocation. CD11b+ cells are magnetically labeled with CD11b

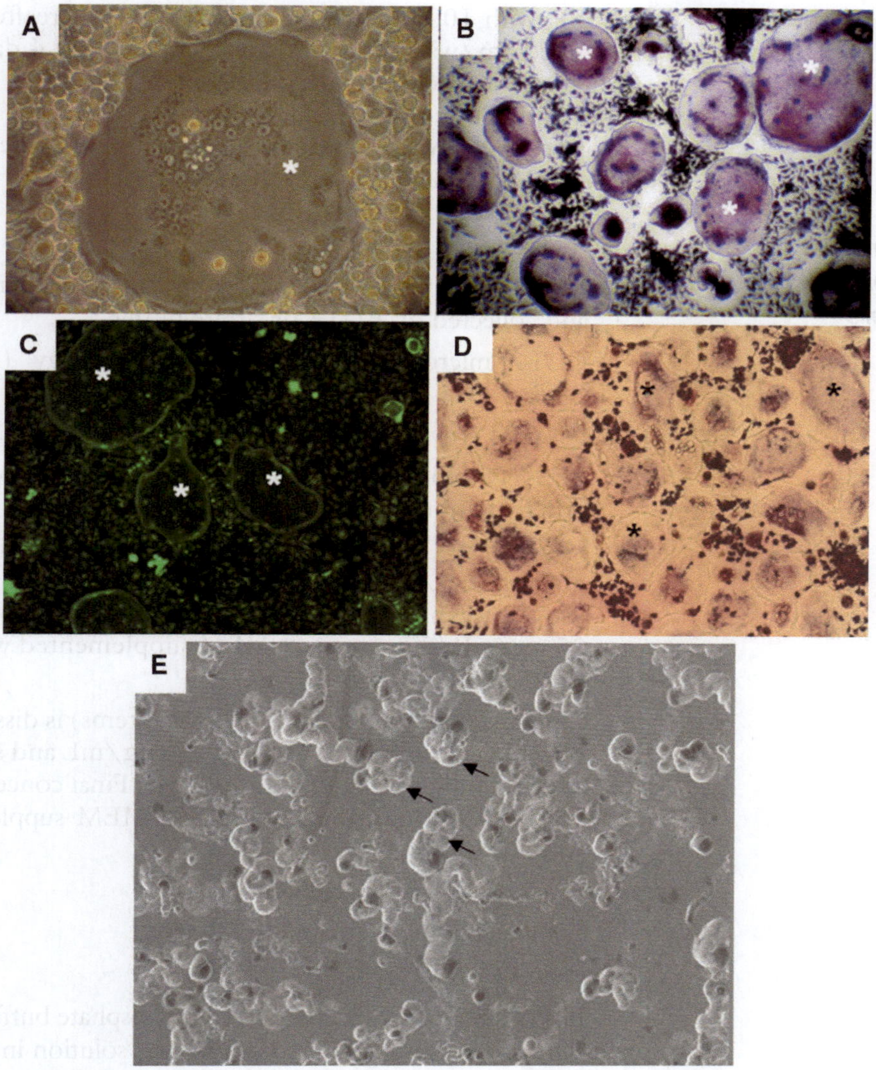

Fig. 1. Study of osteoclastogenesis and osteoclast function from murine and human models. (**a**) Osteoclasts formed from murine RAW264.7 cells after 5 days of culture in the presence of human RANKL (light microscopy, phase contrast); (**b**) Osteoclasts formed from murine $CD11b^+$ cells cultured in the presence of murine M-CSF and human RANKL for 15 days (after TRAP staining); (**c**) Osteoclasts obtained from bone marrow of GFP-mice in the presence of murine M-CSF and human RANKL for 15 days; (**d**) human $CD14^+$ cells differentiated into osteoclasts in the presence of human M-CSF and human RANKL for 11 days (after TRAP staining); (**e**) typical resorption lacunae formed by osteoclasts derived from human $CD14^+$ cells cultured on dentin slice and observed by scanning electron microscopy. *Asterisk* osteoclast; *arrow* resorption lacuna.

microbeads and positively selected by MACS technology. $CD11b^+$ cells are seeded in 24-well plate (500×10^3 cells/well) in phenol red-free α-MEM, containing 10% FCS and 25 ng/mL mM-CSF. This step is absolutely necessary to improve adhesion of osteoclast precursor to the plastic and to stimulate their proliferation. After 3 days of culture, medium is replaced by fresh medium containing 10% FCS, 25 ng/mL murine

M-CSF, with 100 ng/mL human RANKL. Thereafter, complete medium (with cytokines) are changed every 4 days. The formation of osteoclasts occurs between 15 and 21 days of culture and dectects by TRAP staining (see Fig. 1b) (18, 19). Fluorescent osteoclast can be obtained using similar technique with CD11b$^+$ isolated from GFP-mice (see Fig. 1c).

2.3. Differentiation Assay from Human CD14$^+$ Cells

1. Materials and reagents
   - Human peripheral blood from healthy volunteer donors and collected on EDTA or citrate buffer.
   - CD14 microbeads and MACS technology (Miltenyi Biotec).
   - α-MEM (Invitrogen).
   - FCS (Perbio), batch specifically selected for osteoclast differentiation.
   - Human M-CSF (R&D Systems) is dissolved in phosphate buffer/0.1% BSA at 1 mg/mL and stored in single use aliquots at −80°C until use. Final concentration used is 25 ng/mL (dilution in α-MEM supplemented with 10% FCS).
   - Human or mouse RANKL (R&D Systems) is dissolved in phosphate buffer/0.1% BSA at 100 μg/mL and stored in single use aliquots at −80°C until use. Final concentration used is 100 ng/mL (dilution in α-MEM supplemented with 10% FCS).
   - Ficoll® solution, $d = 1.077$ (Sigma).
   - Leukocyte (TRAP) staining (Sigma).
2. Selection of CD14$^+$ cells

   Blood samples are first diluted with phosphate buffer at 50% and diluted samples are layered onto Ficoll solution in a centrifuge tube. Human peripheral blood mononuclear cells (PBMCs) were then isolated by centrifugation over Ficoll gradient for 25 min at 500 × *g*. Whether osteoclasts can be differentiated directly from PBMC or from purified monocytes obtained after 45 min adhesion followed by a differentiation step in the presence of M-CSF and RANKL, enrichment, and purification of osteoclast precursors (CD14$^+$) allow the differentiation of high number of osteoclasts. CD14$^+$ cells are magnetically labeled with CD14 microbeads and positively selected by MACS technology.
3. Osteoclast differentiation

   To induce osteoclast formation, CD14$^+$ cells are seeded at 250×10^3 cells/well in 24-well plates or 45×10^3 cells/well in 96-well plates in α-MEM supplemented with 10% FCS and

25 ng/mL M-CSF. From day 3 of the culture, medium is changed twice a week with fresh medium containing 10% FCS, 25 ng/mL human M-CSF, and 100 ng/mL human RANKL. Then medium is changed every 4 days with cytokines at the same concentration. The formation of osteoclasts occurs after around 11 days and is confirmed by TRAP staining (see Fig. 1d) (20, 21). The two main factors involved in osteoclast differentiation process and survival are: (1) M-CSF which modulates cell adhesion, differentiation, fusion, and resorbing activity and (2) RANKL which is dedicated to the osteoclast fusion, activation, and survival. Osteoclastogenesis can be observed from 30 ng/mL and mouse RANKL can replace human RANKL with twofold higher concentration. RANKL and M-CSF then represent the canonical pathway of osteoclastogenesis and they can be substituted by other protagonists (23). It has been shown that several cytokines can be substituted for RANKL to promote osteoclastogenesis in vitro (TNF-α, IL-11, IL-8) (22). However, osteoclast differentiation is absolutely dependent on RANKL in vivo as confirmed by RANKL knock-out mice which completely lack TRAP-positive immature and mature multinucleated osteoclasts (23). In contrast, M-CSF can be replaced in vitro and in vivo by VEGF, HGF, FLt-3 ligand, or IL-34 for instance (22, 24).

4. Resorption assay

The best validation of the osteoclastic phenotype is to assess the ability of differentiated cells to resorb a mineralized matrix in vitro. For this, $CD14^+$ cells are cultured on dentine or cortical bone slices (for bovine bone, for instance, animal dentine: horse, bovine, etc.) in the conditions previously described. At the end of the culture period, osteoclasts are removed by bleach; dentin/bone slices are fixed with 4% glutaraldehyde in 0.2 M sodium cacodylate solution for 30 min, followed by staining with 1% toluidine blue in 0.5% sodium tetraborate solution for 3 min (25). Resorption lacunae are identified by light stereomicroscopy (STEMI 2000-C, ZEISS) and area of resorbed surfaces is measured using QWin software (Leica). To study the resorption ability of fully mature osteoclasts, the technique established by Fuller et al. (26) can be used. Briefly, after formation of osteoclasts as described above, the medium is removed and the cell layer is washed three times with PBS without calcium and magnesium. Six hundred microliters of 0.02% EDTA are added per well (6-well plate) and cells are incubated for 20 min at room temperature. EDTA is then removed from the well and replaced by 600 μL of calcium/magnesium-free PBS. A cell scraper is used to harvest the cells in PBS, and the resulting cell suspension is mixed using a pipette to ensure uniform cell dispersal. Two hundred and fifty microliters of this cell suspension are then added to each well (24-well plate) on a dentin slice in 250 μL αMEM, 10% FCS. Cells are allowed to sediment for 20 min at 37°C before dentin/bone slices are washed. Cells are incubated in 500 μL αMEM, 10% FCS in the presence or the absence of tested compounds/drugs. After incubation, resorption surfaces are assessed as described above (see Fig. 1e) (27). Resorption lacunae and resorbed surface area can be also revealed and measured by scanning electron microscopy (28, 29).

In all models described, the main markers used to determine the presence of osteoclasts are TRAP, calcitonin receptor, vitronectin receptor, cathepsin K, and the capacity for resorbing mineralized matrix. In all models, RANKL-induced osteoclastogenesis is specifically inhibited by addition of recombinant OPG or RANK-Fc (4).

3. Functional Activities of PGs and GAGs on Osteoclasts

Numerous growth factors/cytokines/receptors carry a heparin-binding domain and consequently can bind to isolated GAGs or GAGs from PGs. Thus, GAGs have many biological activities by holding various extracellular molecules which in bone play key roles in bone metabolism and bone remodeling. Indirect evidence of the role of GAGs in bone remodeling has been published by Kram et al. (30). These authors have shown that heparanase, a heparin sulfate-degrading endoglycosidase, is weakly expressed throughout the bone marrow with a substantial increase in osteoblasts and osteocytes, and in contrast, heparanase is absent from osteoclasts. Interestingly, heparanase transgenic mice exhibit a marked increase of trabecular bone mass, cortical thickness, and bone formation rate, but no difference in osteoclast number. Their data suggest that proteoglycans in bone reduce osteoblast function heparanase limits this reduction by degrading heparin sulfate (30).

The effect of GAGs on osteoclastogenesis in vitro is controversial. For example, Ariyoshi et al. (31) and Shinmyouzu et al. (32) showed an inhibition of osteoclastogenesis after a direct interaction of GAGs with RANKL. In contrast, Irie et al. (33) showed a stimulation of osteoclastic bone resorption by inhibiting OPG activity. However, more recently, using unfractionated osteoblast-derived GAGs that reflect the complex tissue microenvironment within which osteoclasts reside, Ling et al. (34) demonstrate that GAGs block the osteoclastogenic activity of

RANKL. Similarly, Baud'huin et al. (27) demonstrated using three various models of osteoclastogenesis (RAW264.7, murine CD11b$^+$ cells, and human CD14$^+$ cells) that GAGs downregulate RANKL-induced osteoclastogenesis. The mechanism by which GAGs control osteoclastogenesis remains unclear. Baud'huin et al. (27) gave some arguments indicating that GAGs inhibit consecutively osteoclast precursor-adhesion and the fusion of these precursors. Size and sulfation of GAGs are key parameters for the inhibition of RANKL-induced osteoclastogenesis (27), but GAGs (heparin, chondrotin sulfate, dermatan sulfate, heparin sulfate, or oligosaccharides) do not bind to RANKL as studied by surface plasmon resonance experiments (27, 35, 36). Shinmyouzu et al. (32) published that dermatan sulfate inhibits osteoclast formation by binding to RANKL. However, these authors used nonrelevant physiological concentrations of dermatan sulfate (300 μg/mL) and nonpurified osteoclast precursors to study osteoclastogenesis. To continue in the controversial data, Ariyoshi et al. (31) observed that hyaluronic acid increases osteoclastogenesis through activation of CD44 signaling pathway, whereas Chang et al. (37) demonstrated opposite activities and showed an activation of TLR4 signaling pathway without any involvement of CD44. Finally, using fully differentiated osteoclasts derived from human peripheral blood monocytes, Pivetta et al. (38) revealed that hyaluronan inhibits their migration on collagen as well as their ability to resorb bone matrix. These effects are mainly due to a decrease of TRAP, MMP-9, and cathepsin K activities and to the increased levels of TIMP-1. The role of CD44 was confirmed by using blocking anti-CD44 antibodies which fully abrogated hyaluronan effects. Hyaluronan then hampers osteoclast migration through its activity on CD44 (39). Overall, the data published show that GAGs inhibit osteoclastogenesis and their resorption activity by inhibiting the adhesion and fusion of osteoclast precursors. These activities appear independent of RANKL signaling pathway, but may involve CD44 and TLR4 depending on the GAGs used.

In contrast to RANKL, OPG has a heparin-binding domain. OPG belongs to the family of the TNF receptor family and contains three structural domains specifically influencing its biological activities. The first one is a cysteine-rich domain in the N-terminal position which is essential for the inhibition of osteoclastogenesis as well as for the dimerization of OPG via the Cys400. The second domain corresponds to two death domain homologous regions. The third domain is a heparin-binding domain which is able to interact with numerous proteoglycans (40). Full-length OPG bind to GAGs with a high affinity (Kd: 0.28 nM for heparin) in contrast to OPG-Fc in which the heparin-binding domain is lacking (35, 36).

Therefore, the first role of the OPG heparin-binding domain has been revealed by Standal et al. (41) who demonstrated that myeloma cells internalize and degrade OPG through its binding to syndecan-1 and consequently induce osteolysis in patients. Thus, PGs control the bioavailability of OPG, one of the main inhibitors of osteoclastogenesis and bone resorption. PGs are involved in OPG-induced chemotaxis of monocytes (42). Indeed, OPG can interact with syndecan-1 expressed by monocytes (potential osteoclast precursors) and can stimulate the cell migration. In this context, OPG is a chemotactic factor for monocytes which can be recruited in inflammatory context or during osteolysis process. In light of these studies, PGs and GAGs exert a very complex pattern of activities which are not arguable if the biological context is taken into account (inhibition of osteoclastogenesis in vitro, bioavailability of OPG, and monocyte chemotaxis in favor of pro-osteolytic activity).

4. Functional Activities of PGs and GAGs on Osteoblasts

Bone remodeling is a balance between osteoblast and osteoclast activation and the functional activity of the first is influenced by the other one. In this context, similarly to their activities on osteoclasts, PGs and GAGs strongly modulate osteoblast metabolism (9, 43). In bone microenvironment, membrane or soluble forms of RANKL are mainly expressed by stromal cells and osteoblasts which control osteoclastogenesis by this pathway (2, 4). OPG regulates the half-life of membrane RANKL and GAGs inhibit the OPG-induced shortened half-life of RANKL (37). In this specific context, GAGs may increase the half-life of RANKL and by inhibiting OPG activity and thus act as a pro-osteoclastic factor. Furthermore, RANKL significantly reduces ERK activity, a putative suppressor of osteoclastogenesis, but unfractionated osteoblast-derived GAGs abolish the inhibitory effects of RANKL on ERK activity (34) underlining the fact that osteoblast microenvironment is a potent source of GAGs that promote bone anabolic activities. Although the exact mechanism by which GAGs regulate RANKL activity remains unclear, Cao et al. (44) showed that hyaluronan increases RANKL expression in bone marrow stromal cells through CD44 which in turn could stimulate osteoclast activity.

GAGs can be considered as polysaccharides containing protein-binding domains that coordinate mesenchymal stem cell commitment and growth, and ultimately, osteoblast phenotype (43). Among the heparan sulfate-binding factors known to be essential for this process, FGFs, their receptors and members of the

TGF superfamily, are the most important molecule families. Fibroblast growth factor-2 (FGF-2) is a crucial growth factor family driving the proliferation of osteoblasts as many other cell types. Robinson et al. demonstrated that heparin and heparan sulfate are essential for the activity of the fibroblast growth factor (FGF) family (45). GAGs promote FGF oligomerization that, in turn, triggers FGFRs dimerization and signal transduction (46). Like OPG, heparan sulfate PGs (HSPGs) mediate cell internalization of FGF and possibly its nuclear delivery (47). When FGF binds to free heparin/HSPGs, FGF is sequestered in the extracellular environment. Similar observations have been made for BPM2 and the depletion of cell surface HSPGs by enzymatic treatment enhances BMP2 bioavailability and osteogenic activity (48). FGFs also bind to transmembrane HSPGs and then enhance osteoblast proliferation and mineralization, effect partly abolished by an anti-syndecan 4 antibody (48). Another example is given by the paper of Haupt et al. (49). These authors demonstrated that MC3T3-E1 cells under osteogenic conditions decrease their chondroitin and dermatan sulfate PGs (biglycan, decorin, and versican), but increase glypican-3. This shift in expressed HSPGs is concomitant to the switch of FGR1 to FGR3 expression related to the commitment to osteoblast differentiation (50). Similarly to FGF, TGFs stored into the bone matrix could be released during bone resorption and modulate in turn osteoblast and ostoclast metabolism (51). Bi et al. (52) revealed that the absence of the critical TGFβ-binding proteoglycans, biglycan and decorin, prevents TGF-β from proper sequestration within the extracellular matrix. Thus, proteoglycans appear essential for maintaining an appropriate number of osteoblasts and osteoclasts by modulating their proliferation and/or differentiation. More recently, Bi et al. revealed that biglycan deficiency upmodulates osteoclast differentiation and activation due to defective osteoblasts, but independently of RANKL and OPG production (52). The effects of GAGs on osteoblast lineage are dependent on their sulfation. Indeed, sulfation strongly enhances the biological activity of BMPs (TGF member family) by continuously binding the ligands to their receptors and by enhancing osteoblast differentiation (53). In agreement with these data, desulfation of GAGs expressed by MG63 cells delayed in vitro mineralization process (54). Overall, these data point out the key role of GAGs in bone formation and their ability to modulate osteoblast differentiation by indirect mechanism and more specifically by controlling bone cytokines/growth factors. It is also important to keep in mind that osteolysis is the result of both osteoblast and osteoclast activities, and even if osteoblasts are bone cells specialized in bone formation, they contribute in part to the degradation of osseous organic matrix.

5. PGs–GAGs and Bone Remodeling: A Complex Dysregulation of the Anabolic–Catabolic Balance

It is well known that long-term administration of heparin was shown to lead to the development of osteoporosis (55–57). Thus, rats treated once daily by subcutaneous injections of heparin exhibited decreased trabecular bone volume both by decreasing the rate of bone formation and increasing the rate of bone resorption. Barbour et al. (58) also showed that 36% of pregnant women undergoing long-term heparin treatment had a 10% reduction in femoral bone mineral density. However, the mechanism sustaining this osteoporosis was unclear and it was difficult to determine if the effects on bone resorption were due to the direct effect of heparin on osteoclasts or indirectly via its osteoblast activity. Furthermore, these controversial findings on GAG effects on osteoclastogenesis are intensified by the study of Folwarczna et al. (59) who showed that in a rat model, low concentrations of heparin increased the formation of osteoclasts, whereas it decreased with the highest concentrations. In mouse bone marrow cell cultures, heparin suppressed the formation of osteoclasts, with the exception of low concentrations of standard heparin which intensified this process (59). In fact, heparin activity on bone remodeling probably results from a more complex mechanisms altering simultaneously osteoclast and osteoblast metabolisms. Heparin may increase the resorption process through the release of proresorptive factor by osteoblast/stromal cells (61) explaining in part the discrepancy between the in vivo and in vitro results and heparin may also exert an inhibitory activity on bone formation by decreasing osteoblast number and by inhibiting the mineralization process (55–57, 60).

6. Conclusions

GAGs and PGs exert a broad panel of action targeting simultaneously osteoblasts and osteoclasts. Unfortunately, long-term administration of heparin leads to the development of osteoporosis. In this context, although the mechanisms of action of low-molecular-weight heparins are not yet totally elucidated, their use is preferred to unfractionated heparin (60). More specifically, the effect of low-molecular-weight heparins on osteoblasts and on osteoblast–osteoclast communications needs to be investigated and complementary studies to determine whether the effects of heparin on bone are reversible are needed.

References

1. Bruzzaniti, A. and Baron, R. (2006) Molecular regulation of osteoclast activity. *Rev Endocr Metab Disord.* 7, 123–139.
2. Theoleyre, S., Wittrant, Y., Kwan Tat, S., Fortun, Y., Rédini, F., and Heymann, D. (2004) The molecular triad OPG/RANK/RANKL: involvement in the orchestration of pathophysiological bone remodeling. *Cytokine Growth Factor Rev.* **15**, 457–475.
3. Heymann, D., Guicheux, J., Gouin, F., Passuti, N., and Daculsi, G. (1998) Cytokines, growth factors and osteoclasts. *Cytokine* **10**, 155–168.
4. Baud'huin, M., Lamoureux, F., Duplomb, L., Rédini, F., and Heymann, D. (2007) RANKL, RANK, osteoprotegerin: key partners of osteoimmunology and vascular diseases. *Cell Mol Life Sci* **64**, 2334–2350.
5. Rousselle, A.-V. and Heymann, D. (2002) Osteoclastic acidification pathways during bone resorption. *Bone* **30**, 533–540.
6. Georges, S., Ruiz Velasco, C., Trichet, V., Fortun, Y., Heymann, D., and Padrines, M. (2009) Proteases and bone remodelling. *Cytokine Growth Factor Rev* **20**, 29–41.
7. Marie, P. J. and Fromigué, O. (2006). Osteogenic differentiation of human marrow-derived mesenchymal stem cells. *Regen Med* **1**, 539–548.
8. Damiens, C., Fortun, Y., Charrier, C., Heymann, D., and Padrines, M. (2000) Modulation by soluble factors of gelatinase activities released by osteoblastic cells. *Cytokine* **12**, 1727–1731.
9. Lamoureux, F., Baud'huin, M., Duplomb, L., Heymann, D., and Rédini, F. (2007) Proteoglycans: key partners in bone cell biology. *Bioessays* **29**, 758–771.
10. Boyle, W. J., Simonet, W. S., and Lacey, D. L. (2003) Osteoclast differentiation and activation. *Nature* **423**, 337–342.
11. Wittrant, Y., Theoleyre, S., Couillaud, S., Dunstan, C., Heymann, D., and Rédini, F. (2004) Relevance of an in vitro osteoclastogenesis system to study receptor activator of NF-kB ligand and osteoprotegerin biological activities. *Exp Cell Res.* **293**, 292–301.
12. Simonet, W. S., Lacey, D. L., Dunstan, C. R., Kelley, M., Chang, M. S., Luthy, R., et al (1997). Osteoprotegerin: a novel secreted protein involved in the regulation of bone density. *Cell* **89**, 309–319.
13. Yasuda, H., Shima, N., Nakagawa, N., Mochizuki, S. I., Yano, K., Fujise, N., et al. (2004) Identity of osteoclastogenesis inhibitory factor (OCIF) and osteoprotegerin (OPG): a mechanism by which OPG/OCIF inhibits osteoclastogenesis in vitro. *Endocrinology* **139**, 1329–1337.
14. Corsi, A., Xu, T., Chen, X. D., Boyde, A., Liang, J., Mankani, M., et al (2002) Phenotypic effects of biglycan deficiency are linked to collagen fibril abnormalities, are synergized by decorin deficiency, and mimic Ehlers-Danlos-like changes in bone and other connective tissues. *J Bone Miner Res.* **17**, 1180–1189.
15. Duplomb, L., Dagouassat, M., Jourdon, P., and Heymann, D. (2007) Concise review: embryonic stem cells: a new tool to study osteoblast and osteoclast differentiation. *Stem Cells* **25**, 544–552.
16. Raschke, W. C., Baird, S., Ralph, P., and Nakoinz, I. (1978) Functional macrophage cell lines transformed by Abelson leukemia virus. *Cell* 15, 261–267.
17. Theoleyre, S., Wittrant, Y., Couillaud, S., Vusio, P., Berreur, M., Dunstan, C., et al (2004) Cellular activity and signaling induced by osteoprotegerin in osteoclasts: involvement of receptor activator of nuclear factor kappaB ligand and MAPK. *Biochim Biophys Acta* **1644**, 1–7.
18. Blin-Wakkach, C., Breuil, V., Quincey, D., Bagnis, C., and Carle, G. F. (2006) Establishment and characterization of new osteoclast progenitor cell lines derived from osteopetrotic and wild type mice. *Bone* **39**, 53–60.
19. Lézot, F, Thomas, B. L., Blin-Wakkach, C., Castaneda, B., Bolanos, A., Hotton, D., et al. (2010) Dlx homeobox gene family expression in osteoclasts. *J Cell Physiol.* **223**, 779–787.
20. Baud'huin, M., Duplomb, L., Téletchéa, S., Charrier, C., Maillasson, M., Fouassier, M., and Heymann, D. (2009) Factor VIII-von Willebrand factor complex inhibits osteoclastogenesis and controls cell survival. *J Biol Chem.* **284**, 31704–31713.
21. Duplomb, L., Baud'huin, M., Charrier, C., Berreur, M., Trichet, V., Blanchard, F., and Heymann, D. (2008) Interleukin-6 inhibits receptor activator of nuclear factor kappaB ligand-induced osteoclastogenesis by diverting cells into the macrophage lineage: key role of Serine727 phosphorylation of signal transducer and activator of transcription 3. *Endocrinology* **149**, 3688–3697.
22. Knowles, H. J. and Athanasou, N. A. (2009) Canonical and non-canonical pathways of osteoclast formation. *Histol Histopathol* **24**, 337–346.

23. Kong, Y. Y., Yoshida, H., Sarosi, I., Tan, H. L., Timms, E., Capparelli, C., et al. (1999) OPGL is a key regulator of osteoclastogenesis, lymphocyte development and lymph-node organogenesis. *Nature* **397**, 315–323.
24. Baud'Huin, M., Renault, R., Charrier, C., Riet, A., Moreau, A, Brion R, et al. (2010) Interleukin-34 is expressed by giant cell tumours of bone and plays a key role in RANKL-induced osteoclastogenesis. *J Pathol.* **221**, 77–86.
25. Chu, K., Snyder, R., Econs, M. J., and Facp, M. D. (2006) Disease status in autosomal dominant osteopetrosis type 2 is determined by osteoclastic properties. *J Bone Miner Res.* **21**, 1089–1097.
26. Fuller, K., Kirstein, B., and Chambers, T. J. (2006) Murine osteoclast formation and function: differential regulation by humoral agents. *Endocrinology* **147**, 1979–1985.
27. Baud'huin, M., Ruiz-Velasco, C., Jego, G., Charrier, C., Gasiunas, N., Gallagher J et al (2011) Glycosaminoglycans inhibit the adherence and the spreading of osteoclasts and their precursors: Role in osteoclastogenesis and bone resorption. *Eur J Cell Biol.* **60**, 49–57.
28. Gouin, F., Couillaud, S., Cottrel, M., Godard, A., Passuti, N., and Heymann, D. (1999) Presence of leukaemia inhibitory factor (LIF) and LIF-receptor chain (gp190) in osteoclast-like cells cultured from human giant cell tumour of bone. Ultrastructural distribution. *Cytokine* 11, 282–289.
29. Guicheux, J., Heymann, D., Rousselle, A.-V., Gouin, F., Pilet, P., Yamada ,S., et al (1998) Growth hormone stimulatory effects on osteoclastic resorption are partly mediated by insulin-like growth factor I: an in vitro study. *Bone* **22**, 25–31.
30. Kram, V., Zcharia, E., Yacoby-Zeevi, O., Metzger, S., Chajek-Shaul, T., Gabet, Y., et al (2006) Heparanase is expressed in osteoblastic cells and stimulates bone formation and bone mass. *J Cell Physiol.* **207**, 784–792.
31. Ariyoshi, W., Takahashi, T., Kanno, T., Ichimiya, H., Shinmyouzu, K., Takano, H., et al (2008) Heparin inhibits osteoclastic differentiation and function. *J Cell Biochem.* **103**, 1707–1717.
32. Shinmyouzu, K., Takahashi, T., Ariyoshi, W., Ichimiya, H., Kanzaki, S., and Nishihara, T. (2007) Dermatan sulfate inhibits osteoclast formation by binding to receptor activator of NF-kappa B ligand. *Biochem Biophys Res Commun.* **354**, 447–452.
33. Irie, A., Takami, M., Kubo, H., Sekino-Suzuki, N., Kasahara, K., and Sanai, Y. (2007) Heparin enhances osteoclastic bone resorption by inhibiting osteoprotegerin activity. *Bone* **41**, 165–174.
34. Ling, L., Murali, S., Stein, G. S., van Wijnen, A.J.,and Cool,S.M.(2010)Glycosaminoglycans modulate RANKL-induced osteoclastogenesis. *J Cell Biochem.* **109**, 1222–1231.
35. Théoleyre, S., Kwan Tat, S., Vusio, P., Blanchard, F., Gallagher, J., Ricard-Blum, S., et al (2006) Characterization of osteoprotegerin binding to glycosaminoglycans by surface plasmon resonance: role in the interactions with receptor activator of nuclear factor kappaB ligand (RANKL) and RANK. *Biochem Biophys Res Commun.* **347**, 460–467.
36. Lamoureux, F., Picarda, G., Garrigue-Antar, L., Baud'huin, M., Trichet, V., Vidal, A., et al (2009) Glycosaminoglycans as potential regulators of osteoprotegerin therapeutic activity in osteosarcoma. *Cancer Res.* **69**, 526–536.
37. Chang, E. J., Kim, H. J., Ha, J., Kim, H. J., Ryu, J., Park, K. H., et al. (2007) Hyaluronan inhibits osteoclast differentiation via Toll-like receptor 4. *J Cell Sci.* **120**, 166–176.
38. Pivetta, E., Scapolan, M., Wassermann, B., Steffan, A., Colombatti, A., and Spessotto, P. (2011) Blood-derived human osteoclast resorption activity is impaired by hyaluronan-CD44 engagement via a p38-dependent mechanism. *J Cell Physiol.* **226**, 769–779.
39. Spessotto, P., Rossi, F. M., Degan, M., Di Francia, R., Perris, R., Colombatti, A., and Gattei, V. (2002) Hyaluronan-CD44 interaction hampers migration of osteoclast-like cells by down-regulating MMP-9. *J Cell Biol.* **158**, 1133–1144.
40. Yamaguchi, K., Kinosaki, M., Goto, M., Kobayashi, F., Tsuda, E., Morinaga, T., et al (1998) Characterization of structural domains of human osteoclastogenesis inhibitory factor. *J Biol Chem.* **273**, 5117–5123.
41. Standal, T., Seidel, C., Hjertner, O., Plesner, T., Sanderson, R. D., Waage, A., et al (2002) Osteoprotegerin is bound, internalized, and degraded by multiple myeloma cells. *Blood* **100**, 3002– 3007.
42. Mosheimer, B. A., Kaneider, N. C., Feistritzer, C., Djanani, A. M., Sturn, D. H., Patsch, J. R., et al (2005) Syndecan-1 is involved in osteoprotegerin-induced chemotaxis in human peripheral blood monocytes, *J Clin Endocrinol Metab.* **90**, 2964–2971.
43. Cool, S. M. and Nurcombe, V. (2005) The osteoblast-heparan sulfate axis: control of the bone cell lineage. *Int J Biochem Cell Biol.* 37, 1739–1745.

44. Cao, J. J., Singleton, P. A., Majumdar, S., Boudignon, B., Burghardt, A., Kurimoto, P., et al. (2005) Hyaluronan increases RANKL expression in bone marrow stromal cells through CD44. *J Bone Miner Res.* **20**, 30–40.
45. Robinson, C. J., Harmer, N. J., Goodger, S. J., Blundell, T. L., and Gallagher, J. T. (2005) Cooperative dimerization of fibroblast growth factor 1 (FGF1) upon a single heparin saccharide may drive the formation of 2:2:1 FGF1. FGFR2c.heparin ternary complexes. *J Biol Chem.* **280**, 42274–42282.
46. Presta, M., Dell'Era, P., Mitola, S., Moroni, E., Ronca, R., and Rusnati, M. (2005) Fibroblast growth factor/fibroblast growth factor receptor system in angiogenesis. *Cytokine Growth Factor Rev* 16, 159–178.
47. Jiao, X., Billings, P. C., O'Connell, M. P., Kaplan, F. S., Shore, E. M., and Glaser, D. L. (2007) Heparan sulfate proteoglycans (HSPGs) modulate BMP2 osteogenic bioactivity in C2C12 cells. *J Biol Chem.* **282**, 1080–1086.
48. Song, S. J., Cool, S. M., and Nurcombe, V. (2007) Regulated expression of syndecan-4 in rat calvaria osteoblasts induced by fibroblast growth factor-2. *J Cell Biochem.* **100**, 402–411.
49. Haupt, L. M., Murali, S., Mun, F. K., Teplyuk, N., Mei, L. F., Stein, G. S., et al (2009) The heparan sulfate proteoglycan (HSPG) glypican-3 mediates commitment of MC3T3-E1 cells toward osteogenesis. *J Cell Physiol.* **220**, 780–791.
50. Fox, S. W. and Lovibond, A. C. (2005) Current insights into the role of transforming growth factor-beta in bone resorption. *Mol Cell Endocrinol.* **243**, 19–26.
51. Bi, Y., Stuelten, C. H., Kilts, T., Wadhwa, S., Iozzo, R. V., Robey, P. G., et al (2005) Extracellular matrix proteoglycans control the fate of bone marrow stromal cells. *J Biol Chem.* **280**, 30481–30489.
52. Bi, Y., Nielsen, K. L., Kilts, T. M., Yoon, A., Karsdal, M. A., Wimer, H. F., et al (2006) Biglycan deficiency increases osteoclast differentiation and activity due to defective osteoblasts, *Bone* **38**, 30481–30489.
53. Miyazaki, T., Miyauchi, S., Tawada, A., Anada, T., Matsuzaka, S., and Suzuki, O. (2008) Oversulfated chondroitin sulfate-E binds to BMP-4 and enhances osteoblast differentiation. *J Cell Physiol.* **217**, 769–777.
54. Kumarasuriyar, A., Lee, I., Nurcombe, V., and Cool, S. M. (2009) De-sulfation of MG-63 cell glycosaminoglycans delays in vitro osteogenesis, up-regulates cholesterol synthesis and disrupts cell cycle and the actin cytoskeleton. *J Cell Physiol.* **219**, 572–583.
55. Muir, J. M., Andrew, M., Hirsh, J., Weitz, J. I., Young, E., Deschamps, P., and Shaughnessy, S. G. (1996) Histomorphometric analysis of the effects of standard heparin on trabecular bone in vivo. *Blood* **88**, 1314–1320.
56. Muir, J. M., Hirsh, J., Weitz, J. I., Andrew, M., Young, E., and Shaughnessy, S. G. (1997) A histomorphometric comparison of the effects of heparin and low-molecular-weight heparin on cancellous bone in rats. *Blood* **89**, 3236–3242.
57. Ruiz-Velasco, C., Baud'huin, M., Sinquin, C., Maillasson, M., Heymann, D., Colliec-Jouault, S., and Padrines, M. (2011) Effects of a sulphated 'heparin-like' exopolysaccharide produced by *Altermonas infernus* on bone biology. *Glycobiology*, in press.
58. Barbour, L. A., Kick, S. D., Steiner, J. F., LoVerde, M. E., Heddleston, L. N., Lear, J. L., et al. (1994) A prospective study of heparin-induced osteoporosis in pregnancy using bone densitometry. *Am J Obstet Gynecol.* **170**, 862–869.
59. Folwarczna, J., Sliwiński, L., Janiec, W., and Pikul, M. (2005) Effects of standard heparin and low-molecular-weight heparins on the formation of murine osteoclasts *in vitro*. *Pharmacol Rep.* **57**, 635–645.
60. Rajgopal, R., Bear, M., Butcher, M. K., and Shaughnessy, S. G. (2008) The effects of heparin and low molecular weight heparins on bone. *Thromb Res.* **122**, 293–298.
61. Nakano, K., Okada, Y., Saito, K., and Tanaka, Y. (2004) Induction of RANKL expression and osteoclast maturation by the binding of fibroblast growth factor 2 to heparan sulfate proteoglycan on rheumatoid synovial fibroblasts. *Arthritis Rheum.* **50**, 2450–2458.

Chapter 22

Proteoglycans and Cartilage Repair

Mohamed Ouzzine, Narayanan Venkatesan, and Sylvie Fournel-Gigleux

Abstract

Repair of damaged articular cartilage in osteoarthritis (OA) is a clinical challenge. Because cartilage is an avascular and aneural tissue, normal mechanisms of tissue repair through recruitment of cells to the site of tissue destruction are not feasible. Proteoglycan (PG) depletion induced by the proinflammatory cytokine interleukin-1β, a principal mediator in OA, is a major factor in the onset and progression of joint destruction. Current symptomatic treatments of OA by anti-inflammatory drugs do not alter the progression of the disease. Various therapeutic strategies have been developed to antagonize the effect of proinflammatory cytokines. However, relatively few studies were conducted to stimulate anabolic activity, in an attempt to enhance cartilage repair. To this aim, a nonviral gene transfer strategy of glycosyltransferases responsible for PG synthesis has been developed and tested for its capacity to promote cartilage PG synthesis and deposition. Transfection of chondrocytes or cartilage explants by the expression vector for the glycosyltransferase β-1,3-glucuronosyltransferase-I (GlcAT-I) enhanced PG synthesis and deposition in the ECM by promoting the synthesis of chondroitin sulfate GAG chains of the cartilage matrix. This indicates that therapy mediated through GT gene delivery may constitute a new strategy for the treatment of OA.

Key words: Cartilage repair, Chondrocytes, Proteoglycans, Osteoarthritis, Glycosaminoglycan, Gene transfer, Glycosyltransferases

1. Introduction

In degenerative joint diseases such as osteoarthritis (OA), the loss of proteoglycan (PG) anabolism is an early crucial event of cartilage damage (1). Indeed, PGs play key roles in both the structural organization of the extracellular matrix (ECM) and in cell regulation mechanisms, due to their potency to interact with growth factors and cytokines, mainly via their glycosaminoglycan (GAG) chains. The loss of PGs arises from an altered anabolic response of the chondrocytes, which is not only imperative for normal cartilage homeostasis, but also of potential significance during metabolic changes in arthritis, especially when the chondrocytes attempt to

Françoise Rédini (ed.), *Proteoglycans: Methods and Protocols*, Methods in Molecular Biology, vol. 836,
DOI 10.1007/978-1-61779-498-8_22, © Springer Science+Business Media, LLC 2012

repair the degraded ECM. In this regard, disruption of anabolic capacity of chondrocytes, in particular synthesis and assembly of GAG chains of PGs, is critical for the progression of cartilage degeneration.

PGs are a family of complex macromolecules present in the ECM and on the cell surface (2). They are characterized by the presence of one or multiple GAG side chains covalently linked to a core protein. These GAG chains are important regulators in a wide range of biological events, such as matrix deposition, intracellular signaling, morphogenesis, cell migration, normal, and tumor cell growth (3). The heparan-sulfate (HS) GAG chains of PGs located in the plasma membrane are increasingly implicated in the regulation of signal transduction. The key role played by HS in the control of the signaling of morphogens such as Hedgehog proteins as well as growth factors during development has been established in *Drosophila melanogaster*, mouse and human (4). In cartilage matrix, aggrecan is a major PG with many attached chondroitin sulfate (CS) GAG chains. These chains provide a high fixed charge density to the ECM that maintains tissue hydration and gives cartilage its characteristic ability to withstand compressive loads (5). Noteworthy, modification in the composition and/or the structure of these chains is associated with alteration of ECM biomechanical properties that lead to loss of cartilage function during the development of OA.

Several GT enzymes involved in CS and HS biosynthesis have been identified and characterized. The biosynthesis of PG-GAG chains is initiated by the formation of the linkage tetrasaccharide sequence, Galβ1,3Galβ1,4Xyl-*O*- attached to specific serine residues of the core protein. The synthesis of this linkage structure is achieved by the stepwise addition of each sugar residue from the corresponding UDP-sugar catalyzed by O-xylosyltransferase I/II (XylTI/II), galactosyltransferase I (β4GalT7), galactosyltransferase II (β3GalT6), and β1,3-glucuronosyltransferase I (GlcAT-I). The transfer of either an α*N*-acetylglucosamine (αGlcNAc) or β*N*-acetylgalactosamine (βGalNAc) residue on the terminal glucuronic acid (GlcA) residue of the linkage region initiates the polymerization of heparin/HS chains or CS/DS chains, respectively. Polymerization of HS and CS chains is then achieved by the stepwise addition of individual monosaccharides from UDP-sugars by a series of glycosyltransferase (GT)-catalyzed reactions (6).

Several studies have highlighted the important role of GlcAT-I, which catalyzes the transfer of a glucuronosyl moiety from UDP-glucuronic acid onto the nonreducing end of the second galactose of the PG trisaccharide primer Galβ1,3Galβ1,4Xyl-*O*-Serine in PG synthesis (7, 8). This enzyme has received much attention since it plays a central role at a branching point common to CS and HS GAG chains and it has been suggested to be a regulatory factor in the biosynthesis of GAGs in Chinese hamster ovary cells (9).

Therefore, gene transfer of GTs into cartilage could safeguard against progression of OA by promoting PG production and deposition. For these purposes, gene transfer of the glycosyltransferase has been conducted and the effect on PG synthesis has been determined in both chondrocytes and in cartilage explants (10).

Transfection of recombinant vectors expressing the glycosyltransferase GlcAT-I in primary chondrocytes and cartilage explants is carried out using a lipid-based transfection method. Western blot and immunohistochemistry allow detection of the expressed recombinant protein (GlcAT-I) in cultured chondrocytes and in cartilage explants, respectively. Measurement of GlcAT-I enzyme activity permits to assess the functionality of the recombinant enzyme. Radiolabeled sulfate ([^{35}S]SO_4) incorporation into GAG chains allows to evaluate the rate of PG synthesis in recombinant and nonrecombinant chondrocytes and cartilage explants, and fluorophore-assisted carbohydrate electrophoresis (FACE) of GAGs (11) allows to determine the abundance and the composition of the disaccharides units in order to assess the quality of neosynthesized GAG chains. Toluidine blue staining allows monitoring PG deposition in the cartilage matrix.

2. Materials

2.1. Chondrocytes Isolation and Culture

1. Cartilage explants.
2. Dulbecco's modified Eagle's F12 medium (DMEM-F12) supplemented with 10% v/v fetal bovine serum (FBS) (Gibco-Invitrogen).
3. Six-well culture plates.
4. 2 mg/mL pronase solution in 0.9% (w/v) NaCl.
5. 1.5 mg/mL collagenase solution in DMEM-F12 supplemented with 10% FBS.
6. 0.4% (w/v) trypan blue stain solution.

2.2. Chondrocytes and Explants Transfection

1. Exgen 500 (Euromedex).
2. 0.9% NaCl solution.
3. DMEM-F12 medium.
4. Opti-MEM medium (Gibco-Invitrogen).
5. Six-well and 24-well culture plates.

2.3. In Vitro Measurement of GlcAT-I Activity

1. UDP-GlcA (UDP-glucuronic acid), sodium salt (Roche-Boehringer).
2. UDP-[^{14}C]-GlcA (>180 mCi (6.66 GBq)/mmol) (NEN Perkin-Elmer) in 7/3 ethanol-water (v/v).
3. Incubation buffer: 100 mM acetate buffer (pH 6.5), 10 mM $MgCl_2$.

4. Acceptor substrate: Galβ1,3Galβ1-*O*-methoxyphenyl (Galβ1,3-Galβ1-*O*-MP) (12); stock solution 40 mM in DMSO.
5. Thin layer chromatographty (TLC) plates Partisil® LK6DF silica gel with fluorescent indicator, thickness 250 μm, (Whatman).
6. Mobile phase composed of *n*-butanol, acetone, acetic acid, 28% aqueous ammoniac, water (70/50/18/1.5/60, v/v).
7. X-Omat Kodak films for autoradiography.
8. Fluoran-Safe Ultima Gold scintillation cocktail (Packard).

2.4. Immunohistochemical Localization of GlcAT-I and Chondroitin Sulfate Glycosaminoglycans

1. Xylene.
2. 100 and 95% ethanol, anhydrous denatured, histological grade.
3. Double-distilled water (ddH_2O).
4. Mayer hematoxylin (Sigma-Aldrich).
5. Wash buffer: 1× 0.1% TBS/Tween-20 (1× TBST): To prepare 1 L, add 100 mL 10× TBS to 900 mL ddH_2O. Add 1 mL Tween-20 and mix well.
6. 10× Tris-buffered saline (TBS): Dissolve 24.2 g Trizma base and 80 g sodium chloride in 900 mL of distilled water. Adjust pH to 7.6 with concentrated HCl and bring final volume to 1 L with ddH_2O.
7. Antibody diluent: TBST/1% bovine serum albumin (BSA). To 5 mL 1× TBST, add 50 mg BSA (fraction V, Sigma-Aldrich).
8. Antigen retrieval buffer: Sodium citrate buffer (0.01 M sodium citrate buffer, 0.05% Tween 20, pH 6.0). Dissolve 2.94 g tri-sodium citrate dehydrate salt in 900 mL of ddH_2O water. Adjust pH to 6.0 with 1 N HCl. Add 0.5 mL of Tween-20 and mix well. Bring volume to 1 L with ddH_2O. Store at room temperature for 2–3 months or at 4°C for longer duration.
9. 5% Hydrogen peroxide in TBS: Add 10 mL of 30% H_2O_2 to 50 mL TBS.
10. Chondroitinase avidin–biotin complex (ABC) buffer (0.1 U/mL of chondroitinase ABC lyase in 50 mM Tris–HCl buffer, pH 7.4, containing 50 mM sodium acetate).
11. Normal serum.
12. Blocking solution: TBST/10% normal serum + 2% (w/v) BSA. To 5 mL 1 × TBST, add 0.5 mL normal serum and 0.1 g BSA and mix well. Filter the solution using 0.45-μm syringe filter.
13. Primary antibodies: 3B3 (Anti-CS stubs) (Sigma) and GlcAT-I (13).
14. Biotinylated secondary antibody.
15. ABC reagent: (Novostain Super ABC kit, Novo-Castra, Newcastle, UK). Prepare according to manufacturer's instructions 30 min before use and store at 4°C.

16. DAB (3,3′-diaminobenzidine tetrahydrochloride) reagent: Prepare according to the manufacturer's recommendations.
17. Stainless steel pressure cooker.
18. Hot plate or electrical stove.
19. A metal rack for slides.

2.5. Metabolic Labeling of Glycosaminoglycans on Proteoglycans with Radioactive Sulfate [^{35}S]-SO_4

1. Carrier-free sodium [^{35}S]-SO_4 (Amersham, Les Ulis, France).
2. Cell culture labwares: tissue culture flasks, tissue culture plates, pipettes, pipette tips.
3. Cetylpyridinium chloride (Sigma-Aldrich).
4. Dulbecco's modified Eagle's Medium-F12 mix (DMEM-F12; Invitrogen).
5. Fetal bovine serum.
6. L-Glutamine.
7. 4 M Guanidine hydrochloride.
8. Penicillin (10,000 U/mL) stock solution.
9. 10% Buffered formalin.
10. Formic acid.
11. Liquid scintillation fluid (Packard).
12. Phosphate-buffered saline (PBS).
13. Radioactive protective wears, including lab coats, glasses, gloves, etc.
14. Soluene-350 (Packard).
15. Streptomycin (10,000 μg/mL) stock solution.
16. Trypsin solution containing ethylenediamine tetraacetic acid (EDTA).
17. Whatman chromatography paper, 3 MM.

2.6. Toluidine Blue Staining of Cartilage Proteoglycans

1. Xylene.
2. 100 and 95% ethanol, anhydrous denatured, histological grade.
3. 70% ethanol.
4. Deionized water (dH_2O).
5. A metal rack for slides.
6. Staining jars.
7. Toluidine blue O powder *caution*: Avoid contact and inhalation.
8. Whatman No. 1 filter paper.

2.6.1. Preparation of Toluidine Blue Stain

1. Stock toluidine blue stain.
   (a) Add 1.0 g toluidine blue O to 100 mL of 70% alcohol and mix thoroughly.

(b) Filter the stain using Whatman filter paper, No. 1.

(c) Leave the stain to mature for 1–2 weeks.

(d) Store the solution in a dark place, preferably at 4°C. If properly stored, stain is stable for 6 months (see Note 1).

2. Working toluidine blue stain.

(a) Mix one part of toluidine blue stock with nine parts of dH_2O.

(b) Always make fresh before staining.

(c) Discard after use.

2.7. Fluorophore-Assisted Carbohydrate Electrophoresis of Glycosaminoglycan Disaccharides

1. Xylene.
2. 2-Aminoacridone hydrochloride (Molecular Probes).
3. Ammonium acetate.
4. Ammonium persulfate.
5. Mini gel system for protein electrophoresis.
6. Boric acid.
7. Chondroitinase ABC (Sigma-Aldrich).
8. Chondroitin sulfate C (Sigma-Aldrich).
9. Dimethyl sulfoxide.
10. Ethanol.
11. Glacial acetic acid.
12. D-glucuronolactone (Sigma-Aldrich).
13. Glycerol.
14. Hyaluronidase SD (Seikagaku).
15. *N,N'*-methylenebisacrylamide (Sigma-Aldrich).
16. Phenol red (0.5% w/v; Gibco-Invitrogen).
17. Proteinase K (Gibco-Invitrogen).
18. Sodium cyanoborohydride (95%; Sigma-Aldrich).
19. *N,N,N',N'*-tetramethylethylenediamine (Sigma-Aldrich).
20. Unsaturated hyaluronan (ΔDiHA), nonsulfated (ΔDi0S), and monosulfated chondroitin (ΔDi6S and ΔDi4S) disaccharide standards (Dextra Laboratories).

3. Methods

In order to assess the effect of gene transfer of the glycosyltransferase GlcAT-I on PG synthesis and deposition in cartilage explants, it is important to determine whether the plasmid construct is able to produce the recombinant GlcAT-I protein in chondrocytes and to determine the effect of expression of this glycosyltransferase on PG synthesis. For these purposes, chondrocytes in culture should be transfected with the expression vector and then analyzed for expression of the recombinant human GlcAT-I by western blotting before analyzing the functionality of the enzyme by activity test. The effect of expression of GlcAT-I on PG synthesis is then determined by measuring the content of radioactivity associated to GAG chains produced by chondrocytes. The next step is to determine whether GlcAT-I transfection in cartilage explants stimulates PG synthesis and deposition, a prerequisite to cartilage reparation strategies in vivo.

3.1. Isolation and Culture of Chondrocytes from Cartilage Explants

1. Aseptically excise the articular cartilage from femoral head caps of male Wistar rats (150–175 g).
2. Add 2 mL pronase solution per gram of explants and incubate at 37°C for 2 h in a 15-mL falcon tube.
3. Transfer the explants into the same volume of collagenase solution in DMEM-F12 supplemented with 10% FBS and 1% (v/v) antibiotics.
4. Incubate the mixture overnight at 37°C.
5. Pull down the cells by centrifugation at 1,000 × *g* for 5 min.
6. Resuspend the cell pellet in complete DMEM-F12 medium and plate the chondrocytes at 1×10^5 cells/cm² in a 6-well plate.
7. Incubate at 37°C in a humidified atmosphere supplemented with 5% CO_2.

3.2. Cartilage Explants Culture

1. Aseptically excise the articular cartilage from femoral head caps of male Wistar rats (150–175 g).
2. Wash three times with DMEM-F12 medium.
3. Transfer explants into a 24-well plate (one explant per well).
4. Add 1 mL of DMEM-F12 supplemented with 10% FBS and 1% (v/v) antibiotics.
5. Incubate the plate at 37°C in a humidified atmosphere supplemented with 5% CO_2.

3.3. Transfection of Chondrocytes and Cartilage Explants

1. Grow chondrocytes to 80% confluency in 6-well plates.
2. Add in an Eppendorf tube 2 μg DNA of GlcAT-I expression vector or empty plasmid (control) to 0.9% NaCl in a volume of 50 μL.
3. Add in an Eppendorf 10 μL of Exgen 500 to 40 μL of 0.9% NaCl.
4. Transfer solution from step 3 to the solution step 2 and mix by vortexing.
5. Leave at room temperature for 10 min.

6. Remove the cell culture medium and replace it with the transfection mixture from step 4 and add 900 μL of OPTI-MEM medium.
7. Two hours later, replace the medium with complete DMEM-F12.
8. Culture the cells for 48 h before analysis of the expression of recombinant GlcAT-I.

3.4. Transfection Cartilage Explants

1. Culture cartilage explants (rat femoral head cap) in 6-well plate (one explant per well) for 24 h in DMEM-F12 medium.
2. Add in an Eppendorf tube 5 μg of GlcAT-I expression vector or empty plasmid (control) to 0.9% NaCl in a volume of 50 μL.
3. Add in an Eppendorf 25 μL of Exgen 500 to 40 μL of 0.9% NaCl.
4. Proceed as above for steps 4–8.

3.5. Analysis of the Expression and Activity of Recombinant GlcAT-I in Chondrocytes

The method used is based on the measurement of the incorporation of the [^{14}C]-radiolabeled donor substrate, GlcA, into the acceptor substrate such as Galβ1,3Galβ1-*O*-MP separated by HPLC or thin layer chromatography (TLC).

1. Perform incubation in Eppendorf tubes in a final volume of 40 μL, containing 30 μg of homogenate chondrocyte proteins, 4 μL 1.0 mM UDP-GlcA (0.1 mM final), and UDP-[^{14}C]-GlcA (0.1 μCi) in incubation buffer.
2. Start the reaction by addition of 5 μL stock solution of acceptor substrate.
3. Incubate for 60 min at 37°C. Control assays in which either the donor substrate UDP-GlcA or the acceptor substrate was omitted are simultaneously run under the same conditions.
4. Stop the reaction by addition of 40 μL ethanol on ice.
5. Remove the precipitated proteins by centrifugation for 10 min at 4,000 × *g* in a table-top centrifuge (4°C).
6. Separate the radiolabeled reaction product by TLC as follows: 60 μL of supernatant is loaded onto TLC plates and developed with mobile phase for 3–4 h.
7. Dry the plates and spray with 1% (v/v) 2-(4-*t*-butylphenyl)-5(-4-biphenyl)-1,3,4-oxadiazole in toluene. Radiolabeled product is detected by autoradiography for 3 days at −20°C, or by a Phosphor Imager Typhoon 9410.
8. Scrap the silicagel area corresponding to the radioactive spots out from the TLC plate, and quantify the associated radioactivity in a scintillation counting spectrometer in vials containing 5 mL scintillation cocktail (see Note 2).

9. Calculate the activity as following:

$$A = \frac{(\text{dpm assay}) - (\text{dpm control assay})}{\text{total dpm}} \times \frac{\text{nanomoles of UDP - GlcA}}{\text{Time (min)} \times \text{mg protein}} \times \text{dilution factor}$$

A: activity in nmol/min/mg protein

3.6. Immunohistochemical Localization of GlcAT-I and Chondroitin Sulfate GAG

3.6.1. Deparaffinization/Rehydration

1. Deparaffinize sections in xylene for 5 min. Repeat twice in new xylene for 5 min each.
2. Hydrate sections twice in 100% ethanol for 5 min each, followed by 95 and 70% ethanol for 5 min each.
3. Rinse slides for 5 min in dH_2O water followed by two changes of TBS buffer 5 min each.

3.6.2. Antigen Unmasking

1. Place the pressure cooker on a hot plate and transfer the antigen retrieval buffer into the pressure cooker. Turn on the hotplate and heat the buffer to boiling. Do not lock lid of the pressure cooker at this moment.
2. When the buffer boils, transfer the metal rack with slides into the pressure cooker. Ensure slides are well immersed in buffer for 5 min. Lock the pressure cooker lid according to the manufacturer's instructions (see Note 3).
3. Once the pressure cooker has attained full pressure (varies according to the manufacturer), i.e., 10 min, turn off the hotplate. Remove the cooker, place it in an empty sink, and cool it down under cold running water.
4. Once the pressure valve sinks (which means it is depressurized), open the lid and remove the slides from the cooker. Allow slides to cool and then continue with immunohistochemistry protocol (see Note 4).

3.6.3. Staining Protocol

1. Incubate sections in 5% hydrogen peroxide for 30 min.
2. Wash sections in TBS three times for 5 min each.
3. Incubate sections in chondroitinase ABC buffer for 1 h at 37°C (see Notes 5–7).
4. Wash sections in TBS three times for 5 min each.
5. Block sections in 10% normal serum (from the species in which the secondary antibody was raised) in TBST containing 2% BSA for 1 h at room temperature.
6. Discard blocking solution and add primary antibodies (monoclonal 3B3 antibody at a dilution of 5 μg/mL or a polyclonal anti-GlcAT-I antibody at a dilution of 1:200) diluted in TBST/1% BSA to each section and incubate overnight at 4°C in a humid chamber (the 3B3 monoclonal IgM antibody reacts

with a nonreducing Δ-unsaturated disaccharides in chondroitin 6-sulfate following chondroitinase ABC digestion, which exposes the specific epitopes in native PGs).

7. After overnight incubation with the primary antibodies, wash sections in TBS thrice for 5 min each.
8. Incubate sections with secondary antibodies (biotinylated goat anti-mouse IgM or biotinylated goat anti-rabbit IgG diluted (1:100)) in TBST for 1 h at room temperature.
9. Prepare (see Note 8) HRP/ABComplex (according to the manufacturer's instructions) (2 mL/10 slides). For example, 10 μL of reagent A+10 μL of reagent B in 1 mL of TBS.
10. Wash sections three times with TBS for 5 min each.
11. Apply 200 μL HRP-ABC complexes to each section and incubate for 1 h at room temperature.
12. Remove ABC reagent and wash sections three times in TBS for 5 min each.
13. Develop with DAB stain (see Note 9). Add 20 μL of DAB stain to 1 mL of buffer (provided with the kit by the manufacturer). Add 200 μL DAB substrate to each section.
14. Monitor color development under the microscope. Positive cells stain brown, which takes 20 s to 2 min. After color development, wash slides in TBS for 5 min followed by ddH_2O for 5 min.
15. If desired, counterstain sections in Mayer hematoxylin according to manufacturer's instructions.
16. Wash sections in ddH_2O three times for 5 min each.

3.6.4. Dehydration and Mounting

1. Incubate sections in 95% ethanol for 2 min.
2. Repeat twice in 100% ethanol, incubating sections for 2 min each.
3. Clear sections in xylene twice for 2 min each. Note: slides can be left in the last xylene bath until ready to coverslip.
4. Mount coverslips with Permount in a fume hood and leave slides to air-dry overnight.

3.7. Toluidine Blue Staining of Cartilage Proteoglycans

3.7.1. Staining Procedure

1. Deparaffinize paraffin-embedded sections of femoral head cartilage and knee joints of rats in xylene for 10 min.
2. Repeat twice in new xylene for 5 min each.
3. Rehydrate tissue specimens twice in 100% ethanol for 5 min each.
4. Rehydrate sections in 95% for 5 min.
5. Rehydrate sections in 70% ethanol for 5 min.
6. Rinse slides for 5 min in deionized water.
7. Immerse slides in working toluidine blue for 1–2 min.

8. Rinse in tap water (no running water) for 5 min.
9. Immerse sections in 95% ethanol for 1 min.
10. Immerse sections twice in 100% ethanol for 1 min each (see Note 10).
11. Clear sections in xylene twice for 2 min each.
12. Mount coverslips with Permount in a fume hood and leave slides to air-dry overnight.

3.8. Metabolic Labeling of Glycosaminoglycans on Proteoglycans with Radioactive Sulfate [^{35}S]-SO_4

3.8.1. General Practice

1. Prepare chondrocytes or cartilage explants using standard cell culture techniques and suitable cell culture media and growth supplements.
2. On the day of radioactive sulfate incorporation, replace old spent culture medium with fresh medium and necessary supplements and allow cultures for equilibration prior to metabolic labeling.
3. Dilute the required amount of radiolabel (see Note 11) in fresh medium and add to cultures. In general, final [^{35}S]-sulfate concentration ranges from 5 to 500 μCi/mL. [^{35}S]-sulfate concentration may also depend on the use of sulfate-depleted or sulfate-repleted medium. For sulfate-depleted medium, a final concentration of not less than 100 μM is recommended. Labeling time ranges from 3 to 24 h, depending on the endpoint to be analyzed. Normally, [^{35}S]-sulfate incorporation into sulfated GAGs on PGs will be linear for 10–24 h.
4. At the end of metabolic labeling, separate the medium from the cell layer or cartilage explants. Always wash the cell layer or cartilage explants with PBS free of radioisotope.

3.8.2. Proteoglycan Synthesis in Femoral Head Cartilage Explants

1. Remove articular cartilage from femoral head caps under sterile conditions from rat hip joint.
2. Clean aseptically with sterile PBS to remove blood and other contaminating tissues.
3. Culture cartilage explants in DMEM-F12 medium supplemented with 2 mM L-glutamine, 100 μg/mL streptomycin, and 100 IU/mL penicillin at 37°C in a humidified atmosphere and 5% CO_2.
4. Pulse cartilage explants with 5 μCi/mL of $Na_2[^{35}SO_4]$ in DMEM-F12 at 37°C for 6 h.
5. At the end of pulse, separate culture supernatants from the cartilage explants. Wash the explants with PBS to remove any unbound $Na_2[^{35}SO_4]$.
6. Cut the cartilage explants into thin slices (20-μm sections) with a cryostat. Extract PGs for 48 h at 4°C with 10 volumes of 4 M guanidine hydrochloride and 100 mM sodium acetate, pH 6.0, containing protease inhibitors.

7. Centrifuge at 10,000 × *g* for 15 min, dialyze the supernatants exhaustively against 10 mM sodium acetate/10 mM Tris–HCl, pH 7.3, and distilled water.
8. Measure PG synthesis rate following cetylpyridinium chloride precipitation, as described below.

3.8.3. Proteoglycan Synthesis in Rat Articular Chondrocytes

1. Isolate primary chondrocytes from rat articular cartilage (femoral head caps) using sequential digestions with pronase and collagenase in order to release cells.
2. Determine cell viability by trypan blue dye exclusion method and count cells in a hemocytometer.
3. Plate chondrocytes at high density (1×10^5 cells/cm^2) in culture flasks with DMEM-F12 supplemented with 10% FBS, 2 mM L-glutamine, and antibiotics.
4. Feed cultures with fresh medium every 2 days and maintain in DMEM-F12 mix medium at 37°C in a humidified atmosphere supplemented with 5% CO_2. Continue culture until 80% confluence and then subculture by trypsinization, using 0.05% trypsin containing 0.02% EDTA.
5. Once chondrocytes are ready for metabolic labeling, after experimental treatment, pulse cells with 5 μCi/mL [^{35}S]-sulfate for the last 6 h of the experimental period.
6. At the end of metabolic labeling, separate pulsed medium from cell monolayers. Wash cell layers with isotope-free PBS to remove unincorporated radioactive sulfate.
7. Extract PGs from cell layer with 4 M guanidine hydrochloride containing proteinase inhibitors for 48 h at 4°C in a rocking platform. Use safe-lock tubes to extract PGs.
8. After 48 h extraction of PGs, centrifuge the extract for 15 min at 10,000 × *g* and take clear supernatant (this contains cell-associated PGs).
9. Precipitate the isolated PGs using cetylpyridinium chloride, and measure the PG synthesis rate by liquid scintillation counting as described below.

3.8.4. Cetylpyridinium Chloride Precipitation and Proteoglycan Synthesis Rate

1. To quantify total newly synthesized sulfated GAGs, spot aliquots of the media (containing secreted PGs) and PG extracts from cell layer or femoral cartilage explants in triplicate on 30 × 15 mm rectangles of Whatman® chromatography paper.
2. Leave the spot to dry at room temperature.
3. Wash the filter five times in a solution containing 1% cetylpyridinium chloride and 50 mM NaCl.

4. Leave the filter paper to total dryness. After drying, place the paper from each sample in scintillation cocktail and count on a liquid scintillation analyzer.
5. Express the radioactive counts as newly synthesized total sulfated GAGs.

3.9. Fluorophore-Assisted Carbohydrate Electrophoresis

3.9.1. Day 1: Proteinase K Digestion of Cartilage Explants

1. Place cartilage explants (20 mg) in a preweighed Eppendorf tubes on ice.
2. Add 180 μL of 0.1 M ammonium acetate, pH 7.0, containing 0.0005% phenol red.
3. Digest cartilage explants with 10 μL of proteinase K (2.5 mg/mL stock solution in 0.1 M ammonium acetate, pH 7.0, containing 0.0005% phenol red) at 60°C for 2 h. Mix the digests every 30 min.
4. Add a second aliquot of a fresh solution of 10 μL of proteinase K (2.5 mg/mL stock solution) and continue digestion at 60°C for another 2 h. Mix the digests every 30 min.
5. Inactivate proteinase K by boiling cartilage digests for 10 min in a water bath.
6. Centrifuge at 10,000 × *g* for 15 min at room temperature and transfer clear supernatant without disturbing the pellet to a fresh 1.5-mL tube.
7. Run a control sample using buffer and enzyme only (without cartilage explants) as described above.
8. Precipitate proteinase K supernatants with −20°C absolute ethanol to a 77% final ethanol concentration. Mix samples well and incubate overnight at −20°C.

3.9.2. Day 2: Hyaluron idase + Chondroitinase ABC Lyase Digestion

1. Centrifuge samples at 10, 000 × *g* for 15 min at 4 C to pellet precipitated GAGs. Discard the supernatant as this fraction contains minor amount of the total GAGs.
2. Wash pellet with 1 mL of −20°C absolute ethanol, mix them thoroughly, and incubate for an additional 4 h at −20°C. Centrifuge samples at 10,000 × *g* for 15 min at 4°C to pellet precipitated GAGs. Clear away supernatants.
3. Resuspend the samples in (100 μL) 0.1 M ammonium acetate, pH 7.0, containing 0.0005% phenol red. Take an aliquot of each digest to estimate the amount of uronic acid to determine total GAG content, in order to determine the volume required for further analysis.
4. Transfer aliquots containing equal amounts of uronic acid to a new microfuge tube. Dilute to 0.1 mL with 0.1 M ammonium acetate, pH 7.0, containing 0.0005% phenol red.

5. Digest one aliquot with 0.1 U/mL of hyaluronidase SD for 1 h at 37°C followed by 0.1 U/mL chondroitinase ABC for an additional 3 h at 37°C. Both enzymes are prepared in 0.0005% phenol red, 0.1 M ammonium acetate, pH 7.0. The pH of enzyme digestion buffer should be maintained at pH 7.0.
6. Use another aliquot as a control sample without enzyme digestion.
7. After digestion, the samples are freeze-dried overnight prior to fluorescent labeling as described below.

3.9.3. Day 3: Preparation of Cartilage Digests and ΔDisaccharide Standards for Fluores cent Derivatization with 2-Aminoacridone

1. Resuspend the dried standard disaccharides from hyaluronan (ΔDiHA), and chondroitin/dermatan sulfate (ΔDi0S, ΔDi6S, ΔDi4S) in ultrapure water.
2. Freeze equal aliquots of each Δdisaccharide (25 nmol of disaccharide reducing equivalents based on uronic acid content) on dry ice, and lyophilize until dry on a freeze dryer. Samples must be always kept frozen on dry ice and then lyophilized in order to prevent drastic change in pH.
3. Derivatize the dried samples (both cartilage digests and standard Δdisaccharides) by adding 40 μL of 12.5 mM aminoacridone (AMAC) in 3:17 acetic acid:dimethyl sulfoxide for 15 min at room temperature. Next, add 40 μL of 1.25 M sodium cyanoborohydride (freshly prepared in ultrapure water) and incubate for 16 h at 37°C.

3.9.4. Day 4: Fluorophore-Assisted Carbohydrate Electrophoresis, Image Acquisition, and Data Analysis

After derivatization, quench sodium cyanoborohydride by adding 20 μL glycerol (final concentration 20%) to each sample and mix well. All derivatized products must be stored in the dark at −70°C.

3.9.5. Preparation of Monocomposition Gel Buffer and Stock Solutions

1. Separating gel stock solution: 38% (w/v) acrylamide and 2% (w/v) N,N′-methylenebisacrylamide.
2. Separating gel buffer (stock solution): 4×, 0.75 M Tris–0.5 M boric acid, adjust pH to 7.0 with concentrated HCl.
3. Stacking gel stock solution: 10% (w/v) acrylamide and 2.5% (w/v) N,N′-methylenebisacrylamide.
4. Stacking gel buffer (stock solution): 4×, 0.5 M Tris–0.5 M boric acid, adjust pH to 6.8 with concentrated HCl.
5. Electrophoresis buffer (stock solution): 5×, 0.5 M glycine, 0.6 M Tris base, and 0.5 M boric acid, final pH 8.3.
6. 10% (w/v) ammonium persulfate.

3.9.6. Gel Casting and Electrophoresis

1. Separate the derivatized digestion products on monosaccharide composition mini gel.
2. Clean thoroughly the glass plates and combs with ultrapure water and dry. Assemble the gel electrophoresis apparatus.
3. Prepare separating gel solution by adding 4 mL separating gel stock solution to 2 mL stock separating gel buffer, 2 mL ultrapure water, and 30 μL 10% ammonium persulfate.
4. Add 10 μL TEMED, mix gently, and use immediately (since gel polymerization will start soon after TEMED is added).
5. Carefully pour the gel solution into the casting apparatus. Clear any air bubbles. Pour the gel solution to a height of 0.75 cm below where the base of the teeth of the comb will come into contact with the separating gel when it is in position.
6. Carefully overlay the separating gel solution with water-saturated *n*-butanol to form a smooth surface of the gel. Leave the separating gel to polymerize for 20 min.
7. Prepare stacking gel solution by adding 2 mL stacking gel stock solution to 1 mL stock stacking gel buffer, 1 mL ultrapure water, and 15 μL 10% ammonium persulfate. Add 5 μL TEMED, mix gently and use immediately.
8. Pour off the water-butanol overlay from the polymerized separating gel. Wash the gel top with ultrapure water, and fill the remaining space in the casting plate with the stacking gel mixture. Insert the comb to form sample wells. Leave the stacking gel to polymerize for 20 min.
9. When the stacking gel polymerization is complete, carefully remove the comb without destroying the formed wells. Wash the wells of gel with running buffer stored at 4°C. Transfer the gel plates into the gel apparatus.
10. Dilute the stock electrophoresis buffer fivefold and cool at 4°C. Fill apparatus with chilled electrophoresis buffer. Insert the gel sandwich into the electrophoresis apparatus and equilibrate the gel.
11. Load the lanes of a gel with 1–2 μL aliquots of sample. The concentration of sample loaded in the gel should be such as to give adequate separation of fluorescent bands.
12. Connect to a power supply and start electrophoresis on ice for 90 min at a constant 500 V with a starting current of ~25 mA/gel, and a final current of ~10 mA/gel.
13. Check the gel regularly with a long-wave hand-held UV lamp to determine whether the position of the fastest-migrating band reached the bottom of the gel.

3.9.7. Gel Image Acquisition and Data Analysis

1. After electrophoresis, turn off the current. Do not remove the gel from the glass plates. Remove one gel at a time for imaging while leaving the other gel (if you are running two gels) in the chilled electrophoresis tank buffer to avoid diffusion of fluorescent bands.
2. Carefully clean the glass plates of each gel with ultrapure water. Cover the area around the wells with dark tape to cover the fluorescence from the unreacted AMAC that is retained in the wells during electrophoresis.
3. Place gels on the gel imager (with both glass plates intact). Expose gels to UV light and acquire images of the fluorescent bands without delay using a Bio-Rad Gel Doc imaging system. Save electronic images and quantify and analyze images with Bio-Rad QUANTITY ONE image analyzer software.
4. Fluorescent bands generated by derivatized digestion products are compared by coelectrophoresis with hyaluronan and CS disaccharide standards. Disaccharides present in small amount are visualized by exposure over saturated pixel intensity.

3.10. Conclusion

In this chapter, we describe various techniques that have been successfully applied to the analysis of chondrocyte and cartilage-associated GT and PG-GAG expression. These techniques will have broad application in many areas of chondroitin sulfate-related research, including cartilage destruction and repair during osteoarthritis.

4. Notes

1. Given that dehydrating alcohol series and toluidine blue O stain are fairly stable during storage, these reagents can be prepared in large volumes and used as desired.
2. The method has to be performed in strict conditions and in a controlled area. The laboratory should have all authorizations to handle radiolabeled chemicals.
3. Caution with boiling buffer solution—wear safety gloves and use long forceps.
4. Optimum antigen retrieval time for each antibody should be determined by the user. Less than optimal time may lead to underretrieved antigens, while longer time in general yields background staining due to overretrieved antigens.
5. Do not allow slides to dry at any time during enzyme digestion.
6. Use humid chamber (either the incubator itself or a container with a wet paper towel).
7. Do not digest tissue sections with chondroitinase ABC to stain GlcAT-I protein. After antigen retrieval, proceed directly to

hydrogen peroxide treatment step followed by TBS washing and blocking step.

8. Prepare 30 min in advance HRP/ABComplex and leave it in the fridge until ready to use.
9. DAB is carcinogenic. Wear appropriate gloves and face mask.
10. As toluidine blue is soluble in alcohol, dehydration for longer duration is not recommended.
11. Persons working with radioactive substances must undergo proper training and must be certified to handle radioactive substances. Conform to radioactive regulatory requirements.

Acknowledgments

This work was supported by the Programme National de Recherches sur les Maladies Ostéo-Articulaires from the Institut National de la Santé et de la Recherche Médicale, The Agence Nationale de la Recherche (ANR BLAN08-3_313970), the Ligue Régionale contre le Cancer, and the Contrat de Programme de Recherche Clinique.

References

1. Mankin, H.J., and Lippiello, L. (1970). Biochemical and metabolic abnormalities in articular cartilage from osteo-arthritic human hips. *J. Bone Joint Surg. Am.* **52**, 424–434.
2. Iozzo, R.V. (1998). Matrix proteoglycans: from molecular design to cellular function. *Annu. Rev. Biochem.* **67**, 609–652.
3. Sasisekharan, R. and Venkataraman, G. (2000). Heparin and heparan sulfate: biosynthesis, structure and function. *Curr. Opin. Chem. Biol.* **4**, 626–631.
4. Selleck, S.B. (2000). Proteoglycans and pattern formation: sugar biochemistry meets developmental genetics. *Trends Genet.* **16,** 206–212.
5. Roughley, P.J. and Lee, E.R. (1994). Cartilage proteoglycans: structure and potential functions. *Microsc. Res. Tec.* **28**, 385–397.
6. Prydz, K. and Dalen KT. (2000). Synthesis and sorting of proteoglycans. *J. Cell Sci.* **113**, 193–205.
7. Kitagawa, H., Tone, Y., Tamura, J., Neumann, K.W., Agawa, T., Oka, S., et al. (1998). Molecular cloning and expression of glucuronyltransferase I involved in the biosynthesis of the glycosaminoglycan-protein linkage region of proteoglycans. *J. Biol. Chem.* **273**, 6615–6618.
8. Kitagawa, H., Ujikawa, M., and Sugahara, K. (1996). Developmental changes in serum UDP-GlcA:chondroitin glucuronyltransferase activity. *J. Biol. Chem.* **271**, 6583–6585.
9. Bai, X., Wei, G., Sinha, A., and Esko, J.D. (1999). Chinese hamster ovary cell mutants defective in glycosaminoglycan assembly and glucuronosyltransferase I. *J. Biol. Chem.* **274**, 13017–13024.
10. Venkatesan, N., Barré, L., Benani, A., Netter, P., Magdalou, J., Fournel-Gigleux, S., and Ouzzine, M. (2004). Stimulation of proteoglycan synthesis by glucuronosyltransferase-I gene delivery: a strategy to promote cartilage repair. *Proc. Natl. Acad. Sci. USA.* **101**, 18087–18092.
11. Calabro, A., Midura, R., Wang, A., West, L., Plaas, A., and Hascall, V.C. (2001). Fluorophore-assisted carbohydrate electrophoresis (FACE) of glycosaminoglycans. *Osteoarthritis Cartilage.* **9 Suppl A**, S16–S22.
12. Jacquinet, J.C. (2004). An expeditious preparation of various sulfoforms of the disaccharide beta-D-Galp-(1->3)-D-Galp, a partial structure of the linkage region of proteoglycans, as their 4-methoxyphenyl beta-D-glycosides. *Carbohydr. Res.* **339**, 349–359.
13. Ouzzine, M., Gulberti, S., Netter, P., Magdalou, J., and Fournel-Gigleux, S. (2000). Structure/function of the human Galbeta1,3-glucuronosyltransferase. Dimerization and functional activity are mediated by two crucial cysteine residues. *J. Biol. Chem.* **275**, 28254–28260.

INDEX

Françoise Rédini (ed.), *Proteoglycans: Methods and Protocols*, Methods in Molecular Biology, vol. 836, DOI 10.1007/978-1-61779-498-8, © Springer Science+Business Media, LLC 2012

F

G

H

I

K

L

M

N

O

MIX
Papier aus verantwortungsvollen Quellen
Paper from responsible sources
FSC® C105338

FSC
www.fsc.org

If you have any concerns about our products,
you can contact us on
ProductSafety@springernature.com

In case Publisher is established outside the EU,
the EU authorized representative is:
Springer Nature Customer Service Center GmbH
Europaplatz 3, 69115 Heidelberg, Germany

Printed by Libri Plureos GmbH
in Hamburg, Germany